Materials Technology for the Energy and Environmental Nexus, Volume 1

Online at: https://doi.org/10.1088/978-0-7503-5724-1

Materials Technology for the Energy and Environmental Nexus, Volume 1

Edited by
R V Mangalaraja
Faculty of Engineering and Sciences, Universidad Adolfo Ibáñez, Diagonal las Torres 2640, Peñalolén, Santiago, Chile

Arunachalam Arulraj
Departamento de Electricidad, Facultad de Ingeniería, Universidad Tecnológica Metropolitana, Santiago, Chile

Nalandhiran Pugazhenthiran
Laboratorio de Fotoquímica y Fotofísica, Departamento de Química, Universidad Técnica Federico Santa María, Av. España 1680, Valparaíso, Chile

Chan Siew Hwa
School of Mechanical & Aerospace Engineering, Nanyang Technological University, Singapore

IOP Publishing, Bristol, UK

ISBN 978-0-7503-5724-1 (ebook)
ISBN 978-0-7503-5722-7 (print)
ISBN 978-0-7503-5725-8 (myPrint)
ISBN 978-0-7503-5723-4 (mobi)

DOI 10.1088/978-0-7503-5724-1

Version: 20230901

IOP ebooks

British Library Cataloguing-in-Publication Data: A catalogue record for this book is available from the British Library.

Published by IOP Publishing, wholly owned by The Institute of Physics, London

IOP Publishing, No.2 The Distillery, Glassfields, Avon Street, Bristol, BS2 0GR, UK

US Office: IOP Publishing, Inc., 190 North Independence Mall West, Suite 601, Philadelphia, PA 19106, USA

The editor Prof. R.V. Mangalaraja would like to dedicate this book on the remembrance and honor of (Late) Prof. Manohar Paramasivam, Head (Retired) of the Department of Ceramic Technology, Anna University, Chennai, India.

Contents

Acknowledgements

The editors gratefully acknowledge all the contributors for their sincere efforts and kind support for the successful completion of the book. The editor Prof. R.V. Mangalaraja sincerely acknowledges the Adolfo Ibáñez University, Santiago, Chile for the kind support and management.

Editor biographies

R V Mangalaraja

Prof. R V Mangalaraja is a Full Professor in the Faculty of Engineering and Sciences, Universidad Adolfo Ibáñez, Peñalolén, Santiago, Chile. He worked as the Director of International Relations at the University of Concepción for the years 2018 and 2019 and as the Director for the Department of Materials Engineering at University of Concepcion for six years from 2010 to 2016. Further, he also served as the President of the Chilean Metallurgy and Materials Society (SOCHIMM) for the period from 2015 to 2017. Prof. Mangalaraja is well recognized in the scientific community, and he also serves in many professional bodies like Life-Member of Indian Ceramic Society (ICS), Fellow of Indian Institute of Ceramics (IIC), Member of The Mineral, Metals and Materials Society (TMS), Singapore Materials Research Society (MRS-S) during 2017-2018; member of American (ACerS) and PAN-American Ceramic Societies.

Prof. Mangalaraja earned his B.Sc. degree in Physics from Madurai Kamaraj University and M.Sc. degree in Materials Science, M.Tech. in Ceramic Technology. Later, he continued his Ph.D. in Materials Technology (Advanced Ceramics) from Anna University-Chennai, India. Prior to joining Universidad Adolfo Ibáñez, he served as Professor for 17 years at the University of Concepcion, he was a Postdoctoral researcher at AIST-Nagoya, Japan, Lulea University of Technology-Lulea, Sweden and Anna University-Chennai, India. His primary research focuses on the areas of nanotechnology- driven advanced ceramics (functional ceramics for energy and environmental applications) and micro-nano-structure-property relationship of functional and structural ceramics and metallic composite materials.

Prof. Mangalaraja has published over 280 refereed international journal publications, 40 book chapters, three patents, edited two Books, and more than 200 research presentations/invited talks in several International Conferences. Further, he has participated as Responsible and Co-Investigator of 40 research projects funded by ANID-National Agency for Research Development (formerly CONICYT - National Commission of Scientific and Technological Research), Government of Chile and two projects funded by Universidad Adolfo Ibáñez, Peñalolén, Santiago, Chile. For his credits, he has guided 30 Postdoctoral researchers and 13 doctoral candidates in the area of materials science and technology. Also, he had been acted as a member in the expert committee of FONDECYT-ANID for two times (four years) the area of Materials Science and Engineering, and now extending his expertise as a Reviewer of various projects in the area of Materials Science and Engineering.

Arunachalam Arulraj

Dr. Arulraj Arunachalam is an Assistant Professor in the Department of Electricidad, Universidad Tecnológica Metropolitana, Santiago, Chile. Priorly, he worked as a FONDECYT Postdoctoral Researcher in the Faculty of Engineering and Sciences at the Universidad Adolfo Ibáñez, Peñalolén, Santiago, Chile. Also, he worked as a Research Fellow, in the Graphene and Advanced 2D Materials Research Group, at Sunway University, Malaysia, and Institute of Physics, National Taiwan University, Taiwan. He graduated his B. E (Electrical and Electronics Engineering), M. Tech (Nanoscience and Nanotechnology), and Ph.D. (Technology) degrees from Anna University, India. For his academic achievement, he was awarded the meritorious TEQIP-II fellowship, India for the period of 2013 and 2016.

Dr. Arulraj also serves in many professional bodies like the American Ceramics Society (Associate Member), Royal Society of Chemistry (Affiliated Member), American Chemical Society (Member), International Society for Development & Sustainability (Associate Member). To his credit, he has published 30 research articles including 7 book chapters in various peer-reviewed international journals. His research interests are focused on designing and studying various two-dimensional materials which include graphene, chalcogenides, and MXenes for energy conversion and storage.

Nalandhiran Pugazhenthiran

Dr. N. Pugazhenthiran currently working as an Assistant Professor of Chemistry in the Universidad Técnica Federico Santa Maria, Valparaiso, Chile. Dr. N. Pugazhenthiran obtained his doctoral degree (Ph. D., in Chemistry) at School of Chemistry, Madurai Kamaraj University, Madurai, India. He has worked on a Ph. D., the project in the area of nanomaterials for photocatalytic degradation of environmental pollutants and successfully completed in November 2014. In addition, during the period of his Ph.D. program, he was awarded the CSIR-Senior Research Fellowship from Council of Scientific and Industrial Research (CSIR), New Delhi, India and International travel support for Young Scientist award from Department of Science & Technology (DST), New Delhi, India. He has continued to maintain a high academic standing in the field of nanomaterials for energy and environmental applications. He got four research grants from ANID- FONDECYT and ANID-FONDEF, the Government of Chile from the period of November 2015 to March 2023. Moreover, he has published 32 scientific articles in WoS indexed Journals, and h-index of 16, with a total citation of 1285 in prestigious international journals. Additionally, he has published four book chapters and granted two patents for his credit.

Chan Siew Hwa

Prof. CHAN Siew Hwa is a Fellow of Academy of Engineering, Singapore (SAEng), a Fellow of ASEAN Academy of Engineering and Technology (AAET), President's Chair in Energy and a Professor in Nanyang Technological University (NTU). He joined NTU in 1991 after obtaining his PhD and subsequently working as a post-doctoral researcher at Imperial College London. He leads the hydrogen and fuel cell research at Energy Research Institute @ NTU. Since 2017, he has been appointed as Senior Vice President of China-Singapore International Joint Research Institute in China- Singapore Guangzhou Knowledge City, focusing on incubation and commercialization of technologies. Prof. Chan has been active in serving academic & research communities and industries such as Singapore's covering focal point for the Sub- committee on Sustainable Energy Research (SCSER) under the ASEAN Committee on Science, Technology, and Innovation (COSTI), Ministry of Trade and Industry (MTI) Singapore's Future Energy Technology (Hydrogen) Watch Group, NUS's Governing Board of Centre for Hydrogen Innovation, Advisory Board of Total SA and Sydrogen Energy, etc. He is also active in commercialization of technologies including 12 technology licensing. Dr Chan was a Director of Maz Energy Pte Ltd (2004 – 2022) with core business in nitroparaffin-based fuel additives and founded Xin Xiang (Guangzhou) Hydrogen Technologies Co., Ltd., which is a tech-company manufacturing key components of PEMFC. He has published ~300 refereed journal papers with total citations of >15K and h-index of 64. He was a recipient of NTU's Teacher-of-the-year, Nanyang Award (Research Excellence), Nanyang Award (Innovation Entrepreneurship), George-Stephenson Medal from IMechE, UK, Outstanding Scientific Achievement from International Association of Hydrogen Energy (IAHE), USA, Star of Innovation Talent award from Guangzhou government, World's Most Influential Scientific Minds from Thomson- Reuters, etc.

List of contributors

Mehdi Abedi
Pharmaceutical Sciences Research Center, Shiraz University of Medical Sciences, Shiraz, Iran
Research and Innovation Division, Danesh Salamat Kowsar Co., Shiraz, Iran

Vahid Alimardani
Student Research Committee, Shiraz University of Medical Sciences, Shiraz, Iran

Sambandam Anandan
Department of Chemistry, National Institute of Technology, Tiruchirappalli 620015, India

Padmanaban Annamalai
Faculty of Engineering, Universidad Catolica de la Santisima Concepcion, Concepcion, Chile

Mohammad Ashfaq
Department of Biotechnology and University Centre for Research and Development (UCRD), Chandigarh University, Gharaun, Mohali, 140413, Punjab, India

Karthikeyan Balasubramanian
Nanophotonics Laboratory, Department of Physics, National Institute of Technology, Tiruchirappalli 620015, India

Divya Chauhan
Department of Drinking Water and Sanitation, Ministry of Jal Shakti, 1208-A, Pandit Deendayal Antyodaya Bhawan, CGO Complex, Lodhi Road, New Delhi 110003, India

Tatianne Ferreira de Oliveira
Universidade Federal de Goias, Escola de Agronomia, Campus Samambaia, Rodovia Goiania-Nova Veneza Km-0, Caixa Postal 131, 74690-900, Goiania, Brazil

Younes Ghasemi
Pharmaceutical Sciences Research Center, Shiraz University of Medical Sciences, Shiraz, Iran
Department of Pharmaceutical Biotechnology, School of Pharmacy, Shiraz University of Medical Sciences, Shiraz, Iran

Anusha Gowri
SRM Institute of Science and Technology, Kattankulathur 603203, Chennai, India

A Gowrisankar
Department of Chemistry, Bharathiar University, Coimbatore 641046, India

Athithyan Gunasekaran
College of Engineering Guindy (CEG), Anna University, Guindy 600025, Chennai, India

Kumaramangalam Jeyalakshmi
Department of Chemistry, M. Kumarasamy College of Engineering, Karur, Tamilnadu, India

Smagul Karazhanov
Department of Solar Energy, Institute for Energy Technology, PO Box 40, NO 2027-Kjeller, Norway

N B Karthik Babu
Department of Mechanical Engineering, Assam Energy Institute, Rajiv Gandhi Institute of Petroleum Technology, Sivasagar 785697, India

Natesan Kumaresan
Department of Physics, SSN Research Centre, Sri Sivasubramaniya Nadar College of Engineering, Chennai 603110, Tamil Nadu, India

Subrata Kundu
Electrochemical Process Engineering Division, CSIR-Central Electrochemical Research Institute (CECRI), Karaikudi 630003, Tamil Nadu, India

Marzieh Mehrzadeh
Department of Pharmacognosy, School of Pharmacy, Shiraz University of Medical Sciences, Shiraz, Iran

Rubina Michael
Department of Electronics and Communication Engineering, Narasu's Sarathy Institute of Technology, Salem, India

J A Moses
National Institute of Food Technology, Entrepreneurship and Management—Thanjavur, Thanjavur 613005, Tamil Nadu, India

Suhail Mubarak
Department of Chemical and Biomolecular Engineering, Chonnam National University, Yeosu, Jeonnam 59626, Republic of Korea

G Murugadoss
Centre for Nanoscience and Nanotechnology, Sathyabama Institute of Science and Technology, Chennai, 600119, India

Pramila Murugesan
National Institute of Food Technology, Entrepreneurship and Management—Thanjavur, Thanjavur 613005, Tamil Nadu, India

Rishabh Anand Omar
Centre for Environmental Science and Engineering, Indian Institute of Technology Kanpur, Kanpur 208016, India

Sivasubramanian Palanisamy
Department of Mechanical Engineering, Kalasalingam Academy of Research and Education, Srivilliputhur, Tamil Nadu 626126, India

M Praveen Kumar
Faculty of Engineering and Sciences, Adolfo Ibáñez University, Diagonal las Torres 2640, Peñalolén, Santiago, Chile

Zahra Rahiminezhad
Research and Innovation Division, Kani Godakht Tahmin Co., Marvdasht, Iran

C Rajkumar
Institute of Scientific and Industrial Research, Osaka University, Osaka, Japan

Kiruthiga Ramakrishnan
Departamento de Ingeniería Mecánica, Facultad de Ingeniería, Universidad Tecnológica Metropolitana, Santiago, Chile

Venkadeshkumar Ramar
Nanophotonics Laboratory, Department of Physics, National Institute of Technology, Tiruchirappalli 620015, India

Srinivasan Ravichandran
Anabond Limited, Thiruvanmiyur, Chennai 600041, India

Udayabhaskar Rednam
Departamento de Ingeniería Mecánica, Facultad de Ingeniería, Universidad Tecnológica Metropolitana, Santiago, Chile

Prabhu Saravanan
Department of Physical Chemistry, Faculty of Chemical Sciences, University of Concepcion, Edmundo Larenas 129, Concepcion, Chile

M Sasikumar
Dapartment of Physics, Bishop Heber College, Thiruchirapalli, Tamilnadu 620023, India

Preethi Selvamuthu
University College of Engineering, Anna University, BIT Campus, Trichy 620024, India

T Selvaraju
Department of Chemistry, Bharathiar University, Coimbatore 641046, India

Sam Sankar Selvasundarasekar
Academy of Scientific and Innovative Research (AcSIR), Ghaziabad 201002, India
Electrochemical Process Engineering Division, CSIR-Central Electrochemical Research Institute (CECRI), Karaikudi 630003, Tamil Nadu, India

Nidhi Sharma
Department of Physics, Dhanauri PG College, Dhanauri, Roorkee, Haridwar, Uttrakhand, 247667, India

K Sureka
Department of Chemistry, Bharathiar University, Coimbatore 641046, India

Neetu Talreja
Department of Science, Faculty of Science and Technology, Alliance University, Anekal, Bengaluru 562 106, Karnataka, India

Victor Vinoth
Laboratorio de Fotoquímica y Fotofísica, Departamento de Química, Universidad Técnica Federico Santa María, Av. España 1680, Valparaíso, Chile
Faculty of Engineering and Sciences, Universidad Adolfo Ibáñez, Diagonal Las Torres 2640, Peñalolén, Santiago, Chile

Arunachalam Arulraj
Departamento de Electricidad, Facultad de Ingeniería, Universidad Tecnológica Metropolitana (UTEM), Santiago, Chile Faculty of Engineering and Sciences, Universidad Adolfo Ibáñez, Peñalolén, Santiago, Chile

R V Mangalaraja
Faculty of Engineering and Sciences, Universidad Adolfo Ibáñez, Diagonal las Torres 2640, Peñalolén, Santiago, Chile

Chan Siew Hwa
School of Mechanical and Aerospace Engineering, Nanyang Technological University, Singapore

Nalandhiran Pugazhenthiran
Laboratorio de Fotoquímica y Fotofísica, Departamento de Química,
Universidad Técnica Federico Santa María, Av. España 1680, Valparaíso, Chile

IOP Publishing

Materials Technology for the Energy and Environmental Nexus, Volume 1

R V Mangalaraja, Arunachalam Arulraj, Nalandhiran Pugazhenthiran and Chan Siew Hwa

Chapter 1

A sustainable energy transition to build a safer future: an introduction

A Arulraj, C Rajkumar, Chan Siew Hwa and R V Mangalaraja

The sustainability challenge of recent years is mainly associated with greenhouse gas emissions due to the use of fossil fuels, primarily coal, natural gas, and oil, which account for 61.3% of global electricity generation. The use of such existing sources is becoming unfavorable due to their scarce supplies and environmental impact, requiring a transition to sustainable energy. Although there have been historical shifts between different primary energy sources, most of these changes took more than a century or even longer and were brought about by high labor costs, resource scarcity, and technological advancements. The transition of the energy system, including conversion and storage, needs to be more rapid in the twenty-first century. Unfortunately, the knowledge on hastening the energy transition is limited. Three significant technological advancements are typically included in sustainable transition strategies: reduced energy demand, production efficiency, and replacement of fossil fuels with varied renewable energy sources. To make the global energy transition technically and economically feasible, policy initiatives are needed to direct the transition toward a sustainable energy system. Measures to increase the effectiveness of currently used non-renewable sources, which continue to play a significant cost-cutting and cost-stabilization role, should be included in the large-scale adoption of renewable energy. In this chapter the significance of sustainable energy systems, the recent generation of energy conversion (thin-film solar cells), hydrogen production with advanced functional materials, CO_2 capture, energy storage systems (supercapacitors and batteries), and environmental remediation are outlined.

doi:10.1088/978-0-7503-5724-1ch1

1.1 Introduction

Humans have always strived to find and use energy resources to satisfy their basic needs for food, warmth, shelter, and other necessities. The Sun, wind, water, human and animal muscle, wood, grasses, dung, agricultural crops, animal fats and tallow, and, eventually, fossil fuels have been among these sources. Humans continue to seek a better lifestyle, causing the movement in the energy transition toward good energy quality, technical innovation, reduced cost, a flexible system, energy storage, and the other associated factors which come into play. However, biomass energy remains dominant in the history of energy transitions owing to its ease of use and abundance. The impetus to change from one major energy resource to another may depend on a number of connected factors. First, there might be a shortage due to a depletion of the local or regional supply. Both non-renewable fossil fuels and renewable resources are susceptible to this. The second factor is the fluctuation of cost in the global market before the depletion effect of existing energy resources. The next factor is that the use of most non-renewable resources causes adverse effects which make a rapid transition towards renewable energy more appealing. In order to benefit from innovative modern technology (such as electrification, flexible electronic devices, fuel cells, and nanotechnology) to improve the energy efficiency or energy storage, etc, especially on a pilot scale, a change in the energy sources may be necessary. Finally, by leveraging the energy efficiency, improvements in the existing economic activity may speed-up the switch to the alternative energy sources and result in further cost savings [1–3].

Moreover, the switching of energy towards renewable systems can create further transformational opportunities for those who are in need, in particular clean, sustainable energy systems may provide several vital services, thereby increasing the economic value and reducing poverty. The different historical energy transitions in various time frames are presented schematically in figure 1.1. Sustainable energy transitions are not only needed to hasten economic growth and development, but also urgently needed to reduce the emissions that are warming our planet drastically. To date, 73% of the greenhouse gas emissions have been caused by the humans and are dominated by fossil fuels. Global CO_2 emissions must be cut in half by 2030 in order to prevent an increase in the frequency and severity of hazardous and unprecedented weather extremes, such as heatwaves, disastrous floods and droughts, threats to food and water security, population displacement, and loss of livelihoods.

1.2 Energy conversion

In energy conversion from renewable resources, solar energy plays a major role in fulfilling the aims of modern energy transition systems. There is an enormous amount of solar radiation in the form of heat and light that falls over the surface of the Earth, which can be utilized to meet the increased energy demands of the world. With the aid of solar energy, a rapid movement in the energy transition systems towards sustainability can be achieved owing to its inexhaustibility and abundance. Moreover, the utilization of solar energy does not adversely impact the environment. One of the most common approaches to convert the solar radiation into electrical

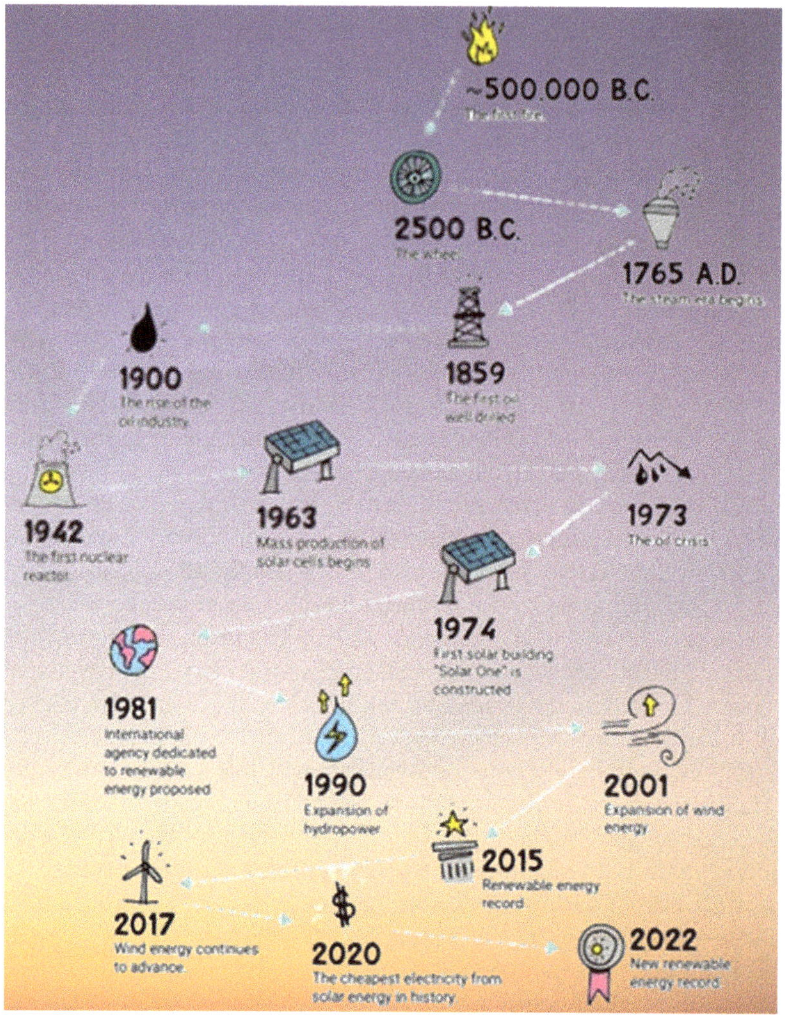

Figure 1.1. Schematic representation of the various energy transitions in different time periods.

energy takes place by means of utilizing photovoltaic (PV) technology. PV technology has the benefits of low-cost, a non-toxic nature, low maintenance requirements, and minimal adverse effects on the environment to produce clean energy. In PV technology, silicon has been used extensively in the fabrication of panels and is considered to be a dominant material in the energy and semiconducting industries. However, the scarcity of silicon, its relatively high processing costs, and the requirement for high temperature and sophisticated instruments in its fabrication remain major concerns. Thus research communities continue to develop novel materials/technologies for achieving high energy conversion efficiency with minimal cost. In this regard, metal and non-metal-based semiconductor materials, polymers, and organic materials have the potential for use in high-performance modern PV devices.

Over the decades it has become apparent that a transition to a second generation of thin-film cell technology is imminent. Thin-films present the opportunity for a significant decrease in the material costs by negating the need for silicon wafers, including semiconductor. This second-generation technology could be anticipated to eventually close the current gap between itself and first-generation products in terms of the energy conversion efficiency. As thin-film second-generation technology develops, the costs of the constituent materials—in this case the top cover sheet and other encapsulants needed to maintain a 30 year operating life—which will be gradually rise-up after certain times. Assuming that the module costs are half of the system costs and a variety of insolation and cost-of-money scenarios exist, there will be a lower limit on such costs (say US\$30/m^2) which, when combined with likely cell efficiency (15%), determine the lower limit on the photovoltaic module costs (US \$0.20/Wp) and electricity generation costs (US\$0.02). However, the conversion efficiency must be increased significantly in order to advance. The theoretical upper limit for a typical solar cell is 33%, whereas the Carnot limit is 95% for the photon to electron conversion rate. This suggests that by combining different concepts to create a third generation of high-performance, low-cost photovoltaic products, solar cell performance could be improved by two to three times. Fortunately, given the likely evolution of the new materials and technologies over the next few decades, which include dye-sensitized solar cells, organic solar cells, and perovskite solar cells, discovering thin-film cells based on new ideas appears quite promising [4–15]. The detailed characteristics of different materials for thin-film solar cells, their architectures, and fabrication processes are discussed in the next chapter (chapter 2) with an emphasis on the next-generation applicability of 2D nanosheet based solar cells.

1.3 Sustainable technologies (H_2 generation, CO_2 capture)

The promotion of the shift to clean and sustainable energy systems depends heavily on renewable energy sources, which are variable and intermittent in terms of their supply, so there is huge need to explore the benefits of hydrogen [16–19]. Hydrogen (H_2) fuel addresses many of the problems of energy production in the modern world while being an energy-efficient and economically promising response to the rising global energy demands [20, 21]. As a clean fuel, H_2 offers several benefits, including the absence of combustion by-products, high energy density, and ease of storage and transporting over longer ranges of distance, enabling it to be a viable renewable energy system [22–24]. The accelerated energy transition away from conventional energy systems depends on the inclusion of H_2 in energy policies thereby exploring its full market penetration [25–28]. The provision of H_2 as a reliable energy source across all regions could meet about 18% of the final energy demand, creating an integrated approach associated with resilience and a reduction in carbon dioxide emission, allowing the extensive use of sustainable energy technologies [29]. By 2030 H_2 is anticipated to satisfy 15%–18% of the world's energy demand, and further advancements should lower production costs to \$1.80/kg [30]. As a result of widespread usage, H_2 energy is advancing among the sustainable energy technologies. Chapters 4 and 7 discuss the state-of-the-art of sustainable H_2 (hydrogen evolution reaction) technology and its by-products of oxygen evolution reaction (OER) research by comparing various innovative experiments, development, and emerging technologies.

In the near future, it is essentially impossible to completely replace conventional power plants with the alternative clean energy generation technologies that emit zero carbon dioxide (CO_2). In this regard, carbon capture and storage (CCS) technology has become a complex energy system taking into account the pertinent political, social, economic, and environmental factors. These uncertainties are linked to the variety of system behaviors, components, and parameters to effectively manage such a complex system, and trade-offs between these influential factors must be balanced under numerous uncertainties. Thus, the implementation of CCS may be complicated by these technical and economic uncertainties, safety concerns, as well as risks associated with transport and storage. Although CCS has been discussed since the 1980s, its high cost as well as regulatory and technological uncertainties have slowed down the technology's rapid global adoption. Several techniques for extracting CO_2 from gaseous mixtures have been developed and put to use. The most recent methods for capturing CO_2 are presented in figure 1.2. Depending on the gaseous states and purity in relation with the release of CO_2 a different refining technology will be applied. During the process of refining natural gas, CO_2 capturing systems aid in the removal of pollutants as well as the production of H_2, NH_3, and other chemicals for use in industry [31–35]. Chapter 5 describes the state-of-the-art of solid sorbents for carbon capture applications and their opportunities and future perspectives.

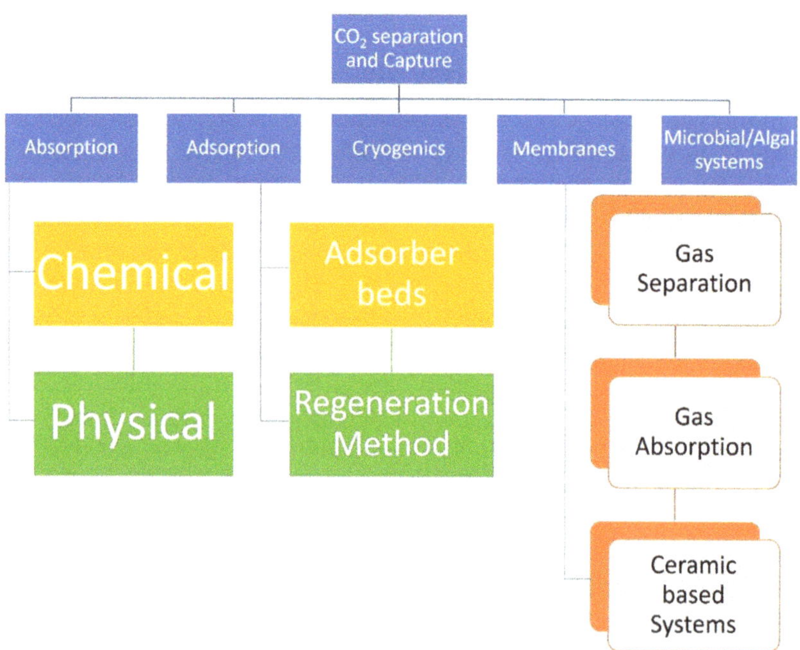

Figure 1.2. Schematic representation of various CO_2 capture technologies.

1.4 Energy storage (supercapacitors)

Devices for energy conversion and storage have become leading technologies in the explosive growth of electric vehicles, portable electronics, grids, and other systems. In order to predict and analyze the electrochemical performance of various energy storage and conversion systems, such as batteries, supercapacitors, and fuel cells, a fundamental understanding of thermodynamics is essential. When there is no access to environmental energy, energy storage devices such as supercapacitors can store energy for its later use. The primary function of a storage device is to hold a designated amount of energy and should go without a call's ability to store more energy making the device (supercapacitors) as the better piece of technology. Including all the end-user applications, whether portable or stationary, the device solely depends on the stored energy on both the specific and density scales by favoring a fabricated cell with higher capacity. Any amount of energy can be obtained from a power pack by simply connecting numerous cells (supercapacitors) in series or parallel and, as a result, the maximum energy can be acquired depending on the desired usage. Therefore, the amount of energy itself is not significant, whereas the energy and energy density are the matters of concern. For the proper integration of renewable energy into existing energy facilities, there must be large-scale sophisticated energy storage systems that can solve these issues by capturing and releasing energy at various times.

One of the most advanced capacitor technologies in recent years, the supercapacitor (or ultracapacitor), offers extremely high DC capacitance and high energy densities. The supercapacitor has demonstrated its dependability and flexibility in terms of design by offering a variety of energy storage solutions for small wearables, commercial applications, automobiles, and large energy backup systems for the power networks. There are two basic mechanisms in electrochemical charge storage: electrostatic charge storage and pseudocapacitance.

The electrochemical charge storage via accumulation of charge in charge traps within so-called Helmholtz layers is the basis for electrostatic storage (inner and outer). Pseudocapacitance storage is an electrochemical process in which charged ions are adsorbed. Because there is no fixed dielectric material, the charge accumulates at the interface between the active electrodes and electrolyte. In detail, an electrical double-layer capacitor (EDLC) stores electrical energy by intercalating charges at the electrode–electrolyte interface, resulting in a double layer of charges. An electrostatic attraction physically deposits the charges, resulting in rapid charge–discharge kinetics, high power density, and a long-life cycle (as no chemical reaction is involved). Graphene, carbon nanotubes, activated carbon, graphite, and other carbonaceous materials are frequently used as electrode materials to store charges. The second kind of supercapacitor is called a pseudocapacitor because it stores electric energy through faradaic reactions. The specific capacitance of the well-known pseudocapacitive materials RuO and MnO is close to that of their theoretical limit. It is referred to as a pseudocapacitor because its electrochemical signature resembles that of an EDLC. Although this supercapacitor has a higher energy density than an EDLC, the involvement of chemical reactions shortens both its life cycle and power density. In

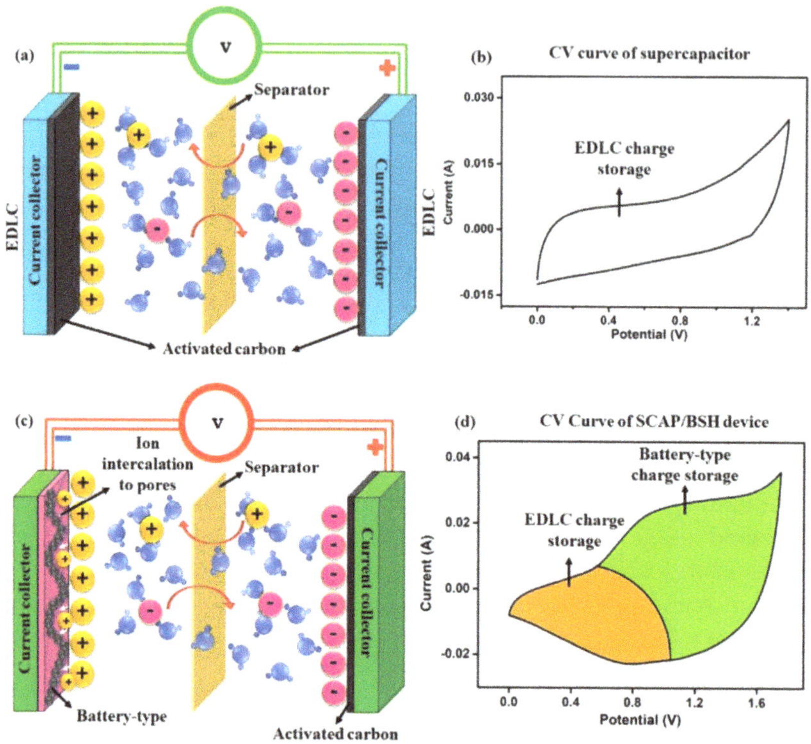

Figure 1.3. Schematic representations of the components and working mechanisms of energy storage devices. (Reproduced with permission from [40]. Copyright 2021 Elsevier.)

comparison with EDLCs and pseudocapacitive materials alone, a hybrid energy storage device can deliver high energy and power densities (Figure 1.3). Supercapacitors can deliver higher power densities than batteries overall, but they still lag behind them in terms of energy density [36–45]. In this regard, various innovative functional materials which are in development to fabricate highly stable active electrodes for energy storage devices are discussed in detail in chapter 8. Similarly, chapter 9 describes the significance of the transition metal nitrides for the fabrication of highly active electrode materials for supercapacitors.

1.5 Sustainable environmental technologies

Like solar energy, water is an invaluable natural renewable energy resource. Indeed, safe and unpolluted drinking water is one of the necessities for all living beings including humans. However, in recent decades, the increasing contamination of wastewater has created severe problems for the environment. The existing conventional industrial methods fall short of producing wastewater of a quality that can be used again after treatment, which is the 'zero effluent' discharge of industrial effluent. Therefore, the development of innovative technologies and materials for use in wastewater treatment needs to be improved while considering socioeconomic principles and contributing to a reduced footprint of greenhouse gas emissions towards achieving the target of sustainability [46]. One of the sustainable emerging technologies is the advanced oxidation technology (AOT) or process (AOP), which mineralizes toxic pollutants into non-toxic intermediates, CO_2, and water, instead of converting them from one phase to another as in adsorption. The AOP typically results in the production of the hydroxyl radical (•OH), the second-most potent oxidant ($E_0 = 2.8$ V) after fluorine. The hydroxyl radicals (•OH) are highly reactive oxidants, and are non-toxic and non-corrosive, and therefore will be easy to control within any system [47–49]. Depending on the type of compound, the •OH radical can degrade by any of three different mechanisms: dehydrogenation, hydroxylation, and charge transfer [49]. An individual bacteria can degrade or decolorize specific recalcitrant organic compounds such as reactive triphenylmethane, azo, and anthraquinone, however, their effectiveness varies with other associated factors such as heavy metal contaminants, effluents, etc [50, 51]. The main disadvantage of the biological process is excess sludge generation, which can be decreased by using an inexpensive co-substrate made of agricultural waste, a substance that encourages microbe growth, and nutritional supplements. The other significant approach is combination of conventional wastewater treatment with the AOP, which tends to boost the degradation efficiency of the contaminants. For instance, by using AOP (specifically electro-Fenton), bio-refractory compounds in the dairy-based wastewater are transformed into straightforward biodegradable intermediates which can be removed using a sequential batch reactor [52]. Chapters 6, 10, 11, and 12 of this book discuss recent innovative materials and the different techniques used in the treatment of wastewater effluents and microplastics in contaminated water. In particular, the sustainable technologies that address remediation for environmental protection are described in detail.

1.6 Conclusion

Sustainable generation/development cannot be accomplished without sustainable renewable energy systems. Generally, sustainable development can be evaluated based on three main factors—social, economic and environment—however, it is best evaluated using five strategies, namely social, economic, technical, policy (institutional or political), and environment. Moreover, the three main technological changes that make-up sustainable transition strategies are energy savings on the demand side, increased production efficiency, and the replacement of fossil fuels with low-carbon nuclear and other renewable energy sources. In the transition to low-carbon sustainable energy systems and economies, the role of technology is a crucial one. The rapid digitalization of the energy sector, which will increase its adaptability and resilience to absorb variable renewable energy sources, particularly solar and hydrogen (clean energy fuel), will be the best technologies to support in the energy transition. Innovative research pushing forward to the pilot-scaling of technologies is recommended to be the actual measure in guiding the resources to facilitate a sustainable energy future.

Acknowledgments

The authors AA and RVM acknowledge FONDECYT Project No. 3200076, Government of Chile and Universidad Tecnológica Metropolitana, and Universidad Adolfo Ibáñez, Santiago, Chile for the financial and administrative support.

References

[1] Kabeyi M J B and Olanrewaju O A 2022 Sustainable energy transition for renewable and low carbon grid electricity generation and supply *Front. Energy Res.* **9** 743114

[2] Neofytou H, Nikas A and Doukas H 2020 Sustainable energy transition readiness: a multicriteria assessment index *Renew. Sustain. Energy Rev.* **131** 109988

[3] Solomon B D and Krishna K 2011 The coming sustainable energy transition: history, strategies, and outlook *Energy Policy* **39** 7422–31

[4] Senthilkumar N, Arulraj A, Nandhakumar E, Ganapathy M, Vimalan M and Vetha I 2018 Potheher, Green mediated synthesis of plasmonic nanoparticle (Ag) for antireflection coating in bare mono silicon solar cell *J. Mater. Sci., Mater. Electron.* **29** 12744–53

[5] Nandhakumar E, Selvakumar P, Arulraj A, Vivek E, Venkatraman M R, Sasikumar A, Prem Kumar M, Mangalaraja R V, Kamatchi R and Senthilkumar N 2023 Investigation on rod like SnO_2@$CdCO_3$ nanocomposite-based electron transport layer for $CsPbBr_3$ hetero-junction perovskite solar cell applications *Mater. Lett.* **330** 133396

[6] Eswaramoorthy N, Arulraj A, Mangalaraja R V, Pitchaiya S and Rajaram K 2022 Nanoscale interfacial engineering of 1D g-C_3 N_4 enables effective and thermally stable HTL-free carbon-based perovskite solar cells with aging for 100 hours *Int. J. Energy Res.* **46** 20194–205

[7] Arulraj A, Thachnatharen N, Rajkumar C and Mangalaraja R V 2022 Engineering 2D chalcogenides for energy and environmental remediation *Nanostructured Materials for Sustainable Energy and Environmental Remediation* (Bristol: IOP Publishing) pp 2-1–2-28

[8] Arulraj A, Govindan S, Vadivel S and Subramanian B 2017 Photovoltaic performance of TiO_2 using natural sensitizer extracted from *Phyllanthus reticulatus J. Mater. Sci., Mater. Electron.* **28** 18455–62

[9] Kesavan M, Arulraj A, Rajendran K, Anbarasu P, Ganesh P A, Jeyakumar D and Ramesh M 2018 Performance of dye-sensitized solar cells employing polymer gel as an electrolyte and the influence of nano-porous materials as fillers *Mater. Res. Express* **5** 115305

[10] Arulraj A and Ramesh M 2018 Rational strategies for large-area perovskite solar cells *Rational Design of Solar Cells for Efficient Solar Energy Conversion* (Hoboken, NJ: Wiley) pp 307–37

[11] Karthick R, Arulraj A, Ramesh M and Selvaraj M 2018 Free-standing graphene/NiMoS paper as cathode for quasi-solid state dye-sensitized solar cells *J. Colloid Interface Sci.* **530** 179–88

[12] Arulraj A, Subramanian B, Ramesh M and Senguttuvan G 2019 Effect of active sites in pulsed laser deposited bimetallic $NiMoS_2$ thin films for solar energy conversion *Mater. Lett.* **241** 132–5

[13] Arulraj A, Ramesh M, Subramanian B and Senguttuvan G 2018 *In-situ* temperature and thickness control grown 2D-MoS_2 via pulsed laser ablation for photovoltaic devices *Sol. Energy* **174** 286–95

[14] Rajkumar C and Arulraj A 2018 Seed mediated synthesis of nanosized zinc oxide and its electron transporting activity in dye-sensitized solar cells *Mater. Res. Express* **5** 015029

[15] Arulraj A, Senguttuvan G, Veeramani S, Sivakumar V and Subramanian B 2019 Photovoltaic performance of natural metal free photo-sensitizer for TiO_2 based dye-sensitized solar cells *Optik (Stuttg)* **181** 619–26

[16] Gielen D, Boshell F, Saygin D, Bazilian M D, Wagner N and Gorini R 2019 The role of renewable energy in the global energy transformation *Energy Strateg. Rev.* **24** 38–50

[17] Parra D, Valverde L, Pino F J and Patel M K 2019 A review on the role, cost and value of hydrogen energy systems for deep decarbonisation *Renew. Sustain. Energy Rev.* **101** 279–94

[18] Pleßmann G, Erdmann M, Hlusiak M and Breyer C 2014 Global energy storage demand for a 100% renewable electricity supply *Energy Procedia* **46** 22–31

[19] Pugazhendhi A and Chen W H 2022 Hydrogen energy technology for future *Int. J. Hydrogen Energy* **47** 37153

[20] Arulraj A, Murugesan P K, Zamorano R C A T and Mangalaraja R V 2023 Nanoarchitectonics of layered metal chalcogenides-based ternary electrocatalyst for water splitting *Energies* **16** 1669

[21] Praveen Kumar M, Sasikumar M, Arulraj A, Rajasudha V, Murugadoss G, Kumar M R, Gouse Peera S and Mangalaraja R V 2022 NiFe layered double hydroxide electrocatalyst prepared via an electrochemical deposition method for the oxygen evolution reaction *Catalysts* **12** 1470

[22] Sharma S and Ghoshal S K 2015 Hydrogen the future transportation fuel: from production to applications *Renew. Sustain. Energy Rev.* **43** 1151–8

[23] Abe J O, Popoola A P I, Ajenifuja E and Popoola O M 2019 Hydrogen energy, economy and storage: review and recommendation *Int. J. Hydrogen Energy* **44** 15072–86

[24] Dutta S 2014 A review on production, storage of hydrogen and its utilization as an energy resource *J. Ind. Eng. Chem.* **20** 1148–56

[25] Staffell I, Scamman D, Velazquez Abad A, Balcombe P, Dodds P E, Ekins P, Shah N and Ward K R 2019 The role of hydrogen and fuel cells in the global energy system *Energy Environ. Sci.* **12** 463–91

[26] Loisel R, Baranger L, Chemouri N, Spinu S and Pardo S 2015 Economic evaluation of hybrid off-shore wind power and hydrogen storage system *Int. J. Hydrogen Energy* **40** 6727–39

[27] Uyar T S and Beşikci D 2017 Integration of hydrogen energy systems into renewable energy systems for better design of 100% renewable energy communities *Int. J. Hydrogen Energy* **42** 2453–6

[28] Abdalla A M, Hossain S, Nisfindy O B, Azad A T, Dawood M and Azad A K 2018 Hydrogen production, storage, transportation and key challenges with applications: a review *Energy Convers. Manag.* **165** 602–27

[29] Dincer I and Acar C 2015 A review on clean energy solutions for better sustainability *Int. J. Energy Res.* **39** 585–606

[30] Ahmed S F, Mofijur M, Nuzhat S, Rafa N, Musharrat A, Lam S S and Boretti A 2022 Sustainable hydrogen production: technological advancements and economic analysis *Int. J. Hydrogen Energy* **47** 37227–55

[31] Wilberforce T, Baroutaji A, Soudan B, Al-Alami A H and Olabi A G 2019 Outlook of carbon capture technology and challenges *Sci. Total Environ.* **657** 56–72

[32] Jin S W, Li Y P, Nie S and Sun J 2017 The potential role of carbon capture and storage technology in sustainable electric-power systems under multiple uncertainties *Renew. Sustain. Energy Rev.* **80** 467–80

[33] López R, Díaz M J and González-Pérez J A 2018 Extra CO_2 sequestration following reutilization of biomass ash *Sci. Total Environ.* **625** 1013–20

[34] Porter R T J, Fairweather M, Kolster C, Mac Dowell N, Shah N and Woolley R M 2017 Cost and performance of some carbon capture technology options for producing different quality CO_2 product streams *Int. J. Greenh. Gas Control.* **57** 185–95

[35] Ioannou I, Galán-Martín Á, Pérez-Ramírez J and Guillén-Gosálbez G 2023 Trade-offs between sustainable development goals in carbon capture and utilisation *Energy Environ. Sci.* **16** 113–24

[36] Vivek E, Arulraj A, Khalid M and Potheher I V 2022 Samarium hydroxide nanorolls anchored graphitic carbon nitride nanosheets: an active electrode material for supercapacitors *J. Alloys Compd.* **908** 164541

[37] Krishnan S G, Arulraj A, Jagadish P, Khalid M, Nasrollahzadeh M, Fen R, Yang C C and Hegde G 2023 Pore size matters!—a critical review on the supercapacitive charge storage enhancement of biocarbonaceous materials *Crit. Rev. Solid State Mater. Sci.* **48** 1–56

[38] Kumaresan N *et al* 2023 Nitrogen implanted carbon nanosheets derived from *Acorus calamus* as an efficient electrode for the supercapacitor application *Mol. Catal.* **538** 112978

[39] Vivek E, Arulraj A, Khalid M and Vetha Potheher I 2021 Facile synthesis of 2D $Ni(OH)_2$ anchored $g\text{-}C_3N_4$ as electrode material for high-performance supercapacitor *Inorg. Chem. Commun.* **130** 108704

[40] Krishnan S G, Arulraj A, Khalid M, Reddy M V and Jose R 2021 Energy storage in metal cobaltite electrodes: opportunities and challenges in magnesium cobalt oxide *Renew. Sustain. Energy Rev.* **141** 110798

[41] Arulraj A and Nishanthi S T 2020 2D materials–based flexible supercapacitors for high energy storage devices *Nanostructured, Functional, and Flexible Materials for Energy Conversion and Storage Systems* (Amsterdam: Elsevier) pp 417–36

[42] Vivek E, Arulraj A, Krishnan S G, Khalid M and Potheher I V 2021 Novel nanostructured $Nd(OH)_3/g\text{-}C_3N_4$ nanocomposites (nanorolls anchored on nanosheets) as reliable electrode material for supercapacitors *Energy Fuels* **35** 15205–12

[43] Krishnan S G, Arunachalam A and Jagadish P 2021 Applications of supercapattery *Advances in Supercapacitor and Supercapattery* (Amsterdam: Elsevier) pp 311–48

[44] Krishnan S G, Arunachalam A, Jagadish P and Khalid M 2020 2D materials for supercapacitor and supercapattery applications, Adapting 2D Nanomaterials for Advanced Applications, Chapter 2, ACS Publications pp 33–47

[45] Arulraj A, Ilayaraja N, Rajeshkumar V and Ramesh M 2019 Direct synthesis of cubic shaped Ag_2S on Ni mesh as binder-free electrodes for energy storage applications *Sci. Rep.* **9** 10108

[46] Manna M and Sen S 2022 Advanced oxidation process: a sustainable technology for treating refractory organic compounds present in industrial wastewater *Environ. Sci. Pollut. Res.* **30** 25477–505

[47] Miklos D B, Remy C, Jekel M, Linden K G, Drewes J E and Hübner U 2018 Evaluation of advanced oxidation processes for water and wastewater treatment—a critical review *Water Res.* **139** 118–31

[48] Ma D *et al* 2021 Critical review of advanced oxidation processes in organic wastewater treatment *Chemosphere* **275** 130104

[49] Cuerda-Correa E M, Alexandre-Franco M F and Fernández-González C 2019 Advanced oxidation processes for the removal of antibiotics from water: an overview *Water* **12** 102

[50] Mishra S and Maiti A 2018 The efficacy of bacterial species to decolourise reactive azo, anthroquinone and triphenylmethane dyes from wastewater: a review *Environ. Sci. Pollut. Res.* **25** 8286–314

[51] Zuo S, Yang H, Jiang X and Ma Y 2021 Magnetic Fe_3O_4 nanoparticles enhance cyanobactericidal effect of allelopathic p-hydroxybenzoic acid on *Microcystis aeruginosa* by enhancing hydroxyl radical production *Sci. Total Environ.* **770** 145201

[52] Heidari M R, Malakootian M, Boczkaj G, Sun X, Tao Y, Sonawane S H and Mehdizadeh H 2021 Evaluation and start-up of an electro-Fenton-sequencing batch reactor for dairy wastewater treatment *Water Resour. Ind.* **25** 100149

IOP Publishing

Materials Technology for the Energy and Environmental Nexus, Volume 1

R V Mangalaraja, Arunachalam Arulraj, Nalandhiran Pugazhenthiran and Chan Siew Hwa

Chapter 2

Thin-film-based functional materials for solar cell applications

Nidhi Sharma, Rishabh Anand Omar, Neetu Talreja, Divya Chauhan, R V Mangalaraja and Mohammad Ashfaq

The demand for efficient energy sources has continuously encouraged researchers to develop cost-effective, lightweight, and high-performance solar cells. In this respect, the thin-film-based solar cell offers an innovative solution for high-performance solar cells. Thin film-based functional materials provide an effective way to develop newer-generation solar cell devices. Due to their exceptional characteristics, numerous emerging materials based on metals, carbon, polymers, and hybrids are being used effectively for the development of high-performance solar cells. By utilizing these materials as thin films, we can easily design newer solar cells with outstanding performance. The solar cells' efficiency depends on the materials, device design, and fabrication process. This chapter highlights thin films based on different materials, in particular the emerging materials such as two-dimensional (2D) nanosheets and polymeric composites, their device design, and the fabrication process of solar cells. We will also discuss the next-generation applicability of 2D nanosheets for developing thin-film-based solar cells.

2.1 Introduction

Energy demands have been increasing with the growing human population, as energy drives almost everything. The continued economic and social development of the modern era depends mainly on the sustainable production of energy and its storage. The non-renewable resources such as oil, gas, and coal that are used to produce energy can become depleted. Moreover, the use of fossil fuels as energy sources causes adverse effects through environmental pollution, in particular global warming, which is one of the most significant challenges facing the world

doi:10.1088/978-0-7503-5724-1ch2

(Gielen *et al* 2019, Arulraj *et al* 2022, Saidi and Omri 2020). Researchers are focused on reducing environmental pollution, including greenhouse effects, by means of reducing its emission. Thus, there is a need to focus on environmental friendly clean energy through relying on renewable sources that mitigate environmental pollution and the emission of greenhouse gases. In this respect, solar energy is one of the promising renewable sources of clean energy that can effectively reduce greenhouse gases emissions and fulfill energy requirements.

Solar energy is the primary energy source and reaches the earth in numerous forms, but mainly light and heat. With the help of solar energy, we can quickly fulfill global energy demands as it is inexhaustible and freely available. Moreover, the utilization of solar energy does not adversely impact the environment. Furthermore, incorporating solar energy based technologies in our daily life might be one of the best solutions that easily avoids harmful consequences for future generations (Kannan and Vakeesan 2016, Mekhilef *et al* 2011). In this respect, photovoltaic (PV) cells/devices can produce energy without any adverse effects on the environment.

Recently, PV cells/devices have been used extensively to produce energy using solar irradiation due to their low cost, non-toxic nature, low maintenance requirements, and having no adverse effect on the environment. Thus, the use of PV devices is one of the most promising methods for the conversion of solar energy into electrical energy. Silicon (Si) has been used extensively to develop PV cells/devices that are reported to have high conversion efficiency. However, the relatively high cost, lower stability, limited availability, and the requirement for high temperature fabrication of Si remain concerns (Sharaf *et al* 2022, Schwartz 1982, Senthilkumar *et al* 2018, Okil *et al* 2022, Battaglia *et al* 2016). Researchers continue to focus on the development of new materials with high conversion efficiency and minimal cost. In this respect, metal and non-metal-based semiconducting materials have the potential for use in developing high-performance PV devices.

Several materials, such as carbon-based materials (carbon nanotubes (CNTs), carbon nanofibers (CNFs), and graphene), metals and their oxides (TiO_2, Cd, Pb, Fe, Bi, and Zn), nanosheets (WS_2, MoS_2, MXene, hBN, and black phosphorus) have been used for the development of PV devices demonstrating higher conversion performance. However, the PV devices' conversion efficiency and their stability remain concerns (Arulraj *et al* 2018, Singh and Nalwa 2015, Gao *et al* 2019, Jayaweera *et al* 2018, Pan *et al* 2020, Xu *et al* 2018, Ahmad *et al* 2021, Zhang *et al* 2022, Yin *et al* 2021, Xie *et al* 2020, Batmunkh *et al* 2018). Researchers are looking toward a new class of PV devices with high conversion efficiency, durability, cost-effectiveness, and flexibility, that can be used easily in real-time applications. In this respect, incorporating polymers within carbon-based materials/metals and their oxides/nanosheets might be beneficial for resolving issues associated with existing PV devices.

Several conducting polymers, such as polypyrroles, polythiophenes, and polyanilines, have been used to develop high-performance, flexible, and cost-effective organic PV devices, which are easily replaced with inorganic PV devices. These polymeric PV devices are considered as the fourth-generation PV devices/cells with the characteristics of low cost and high mechanical flexibility (Mayer *et al* 2007, You *et al* 2013, Fu *et al* 2021). For example, the polymer solar cells include planar heterojunction polymer solar cells, bulk heterojunction (BHJ) polymer solar cells, hybrid planar heterojunction solar cells, and BHJ tandem solar cells. Furthermore, various semiconducting devices such as light-emitting diodes (LEDs), solar cells, and thin-film transistors (TFTs) were developed using polymers (Tamilselvan *et al* 2019, Arabpour Roghabadi *et al* 2018, Zhang *et al* 2014a, Wang *et al* 2015, Chiang and Wu 2016). However, the commercialization of fourth-generation PV devices/cells has not been possible to date. In this chapter we focus on thin-film-based functional materials and their application in the fabrication process of solar cells. The performance of solar cells depends on the materials, device design, and fabrication process. Initially, this chapter discusses the newer material synthesis processes based on carbon, metal, polymer, and hybrid materials. Further, it also discusses the different types of device designs with their performance. The applicability of nanosheets in improving thin-film-based solar cells is also explained in this chapter. Furthermore, this chapter offers newer insights into nanosheets and their uses in solar cells that might encourage researchers to develop this new class of materials and flexible solar cells for the development of society.

2.2 Thin-layered materials for solar cells

To attain performance efficiency, in thin-film-based organic solar cells 2D nanosheets have been incorporated into the active layer of the devices as a third and important component, which has generated extensive attention. Graphene, a well-know carbon-based 2D layer which has excellent carrier mobility, has received considerable attention since it was first synthesized in 2004 (Novoselov *et al* 2004) and later its derivatives by liquid exfoliation. Graphene has been applied in the field of solar panel design due to its high carrier movement, higher conductivity, and dispersion ability in solvents (Eda *et al* 2008, Eda and Chhowalla 2010). Among all the 2D-nanostructured materials, graphene and its derivatives, such as graphene oxide (GO) and reduced graphene oxide (rGO), and transition metal dichalcogenides (TMDs), which can be synthesized using transition elements or d-block elements (W, Mo, Pd, and Pt) that occupy its position between chalcogenide atoms (Se, S, and Te) such as WSe_2, are the most comprehensively used materials. The band gap of WS_2 is more than 2 eV, whereas MoS_2 possesses a band gap ranging from 1.2 to 1.9 eV. Generally, the TMDs with variable thicknesses are associated with the nanoscale domain. For example, by varying the thickness of WSe_2, the band gap can be tuned from 0.8 eV in bulk to 1.3 eV in a single layer, providing similar energy levels to enhance the movement of charge carriers, exciton dissociation, and ejection of charge from the surface (Wei *et al* 2021, Bin Rafiq *et al* 2020). It is recognized that the integration of WSe_2 flakes with the same size as the domain of the active layer results in the best power conversion efficiency (PCE) for the organic solar cells due to the optimized permeation pathways for charge transport (Baig 2023, Zhou *et al* 2020, Derakhshi and Daemi 2022, He *et al* 2022, Sobayel *et al*

2020). Usually, these thin layered materials are used effectively in the fabrication of different kinds of solar cells, including the dye-sensitized solar cells (DSSCs), organic solar cells, and perovskite solar cells.

2.2.1 DSSCs

DSSCs are a technical and economical replacement for p–n junction based PV devices. In a discovery of the 1960s, it was found that the generation of electricity can be achieved by the illumination of organic dyes on electrochemical cells (Tributsch and Calvin 1971). A chlorophyll-synthesized zinc oxide (ZnO) electrode was developed for the first time in 1972 and it was the first time that photons were utilized to generate electricity through electron injection of excited dye molecules into a semiconductor with a wide band gap. Much research was done on single-crystal ZnO (Tsubomura *et al* 1976). However, the efficiency was less due to the lower absorption (<1%) of incident light (photons) into the DSSCs. In further research, the efficiency of the DSSCs was improved by enhancing the electrodes' porosity using fine oxide powder, which enhances the incident radiation absorption, which further results in strengthening the light-harvesting efficiency (Sharma *et al* 2018, Dissanayake *et al* 2022, Navarro-Pardo *et al* 2017, Shi *et al* 2017b). In 1991 nanoporous titanium oxide (TiO_2) electrodes with a roughness factor of ca. 100 were invented and used in the fabrication of DSSCs, which showed approximately 7% of energy-conversion efficiency (O'Regan and Grätzel 1991). Brian O'Regan and Michael Grätzel at UC Berkeley developed these cells, also known as Grätzel cells, in 1988. Further development was carried out by scientists at Ecole Polytechnique Fédèrale de Lausanne (EPFL) between 1991 and 1997. In other research, a device was fabricated by Brian O'Regan and Michael Grätzel, a thin (~10 μm) transparent film of TiO_2 nanoparticles. The film was coated with a charge transfer dye with ideal spectral characteristics to make it sensitive to light harvest action. The developed device absorbed a high portion of solar flux energy (~46%) and showed a very high efficiency, approximately 80% of photon conversion to electrical current. The overall incident photon to current conversion efficiency (IPCE) yielded in simulated solar light was 7.0%–7.9%, while the IPCE yield in diffuse daylight was 12%. It was reported in 1993 that Grätzel *et al* achieved 9.6% power conversion efficiency using DSSCs and at the National Renewable Energy Laboratory (NREL) researchers achieved 10% efficiency in 1997 (O'Regan and Grätzel 1991). Similarly, Zarick *et al* fabricated Au@SiO_2-incorporated TiO_2-based DSSCs. Figure 2.1 shows transmission electron microscopy (TEM) images and a schematic illustration of the Au@SiO_2-incorporated TiO_2-based DSSCs. Incorporating the Au@SiO_2 based nanocubes significantly improved the power conversion efficiency (Zarick *et al* 2014). Usually, the sensitizers are enriched with numerous functional groups such as -COOH, -$B(OH)_2$, and PO_3H_2 (Nazeeruddin *et al* 2011, Altobello *et al* 2004, Mutashar and Al-bahrani 2021). These functional groups provide the stable adsorption of incident light to the semiconductor substrates (Rafique *et al* 2021, Sun *et al* 2020b). In a recent study in 2018 by Kunzmann *et al*, approximately 8.67% efficiency was reported using hybrid dye-titania nanoparticles in the

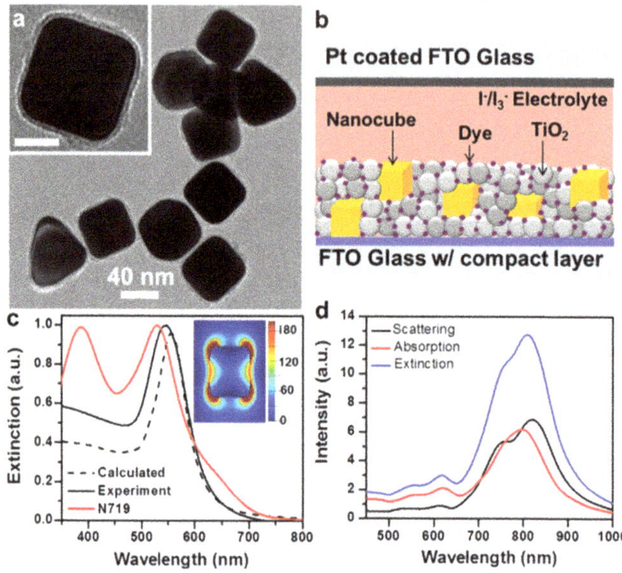

Figure 2.1. (a) TEM images of $Au@SiO_2$, (b) schematic representation TiO_2-based DSSCs, (c) electromagnetic intensity profile, and (d) calculated absorption, scattering, and extinction spectra. (Reproduced with permission from (Zarick *et al* 2014). Copyright 2014 American Chemical Society.)

fabrication of DSSCs in a low-temperature environment (Kunzmann *et al* 2018). As mentioned previously, in traditional solar cells silica (Si) is used and it plays two important roles: (i) it acts as a photo electronic source and (ii) it separates the charge and creates current by providing an electric field (Praveen *et al* 2020, Fadhilah *et al* 2019, Tsai *et al* 2018). However, in DSSCs the bulk semiconductor is capable of both charge transportation and being a source for photoelectrons provided by the photosensitive dyes (Aneesiya and Louis 2020, Wan *et al* 2019). Theoretically, the power consumption efficiency of DSSCs was determined to be approximately 20% (Snaith 2010, Frank *et al* 2004, Kilic *et al* 2016). Therefore, extensive research has been carried out the years to improve the performance of DSSC and employ it at the commercial level. Some of the past research on DSSCs and their efficiency is provided in table 2.1.

2.2.2 Organic solar cells

Organic solar cells (OSCs) consist of conductive and transparent substrates (generally indium tin oxide (ITO)), an interfacial layer, an active layer, and electrodes. In recent times, OSCs have become a focus of interest among researchers because of their several advantages, including their flexibility, semitransparent nature, and cost efficiency (Hou *et al* 2018, Zhang *et al* 2014b). The high efficiency of OSCs is due to their active layer adopting a bulk heterojunction structure where the acceptor and donor jointly form a bi-continuous network. The controlled phase separation in OSCs confirms their efficient charge transport and exciton separation (Zhang *et al* 2011, Roncali 2009, Qin *et al* 2016). In recent decades, the development of PV

Table 2.1. Different materials used for the development of DSSCs.

S. No.	Material	Structure	Efficiency (η %)	Reference
1	Au/SnO_2/MgO	Nanoparticles	4.69	Dissanayake *et al* (2022)
2	TiO_2/(MWCNT)	Mesoporous networks	8.96	Mutashar and Al-bahrani (2021)
3	Cu polypyrrole multiwall carbon nanotubes	Nanotubes	7.1	Rafique *et al* (2021)
4	TiO_2	Nanofibers	6.36	Sun *et al* (2020b)
5	ZnO:Li	Scale	5.58	Praveen *et al* (2020)
6	Cu-doped ZnO	Hexagonal wurtzite structure	1.34	Aneesiya and Louis (2020)
7	TiO_2	Nanorods	11.14	Wan *et al* (2019)
8	TiO_2	Nanoparticles	0.016	Fadhilah *et al* (2019)
9	Graphene	Nano leaves	9.26	Tsai *et al* (2018)
10	TiO_2	Flower-like	7.69	Shi *et al* (2017b)
11	TiO_2	Nanofibers	6.00	Navarro-Pardo *et al* (2017)
12	TiO_2/(MWCNT)	Mesoporous networks	7.27	Kilic *et al* (2016)
13	TiO_2	Mesoporous	5.30	Apriani *et al* (2016)
14	TiO_2	Hydrangea-like	7.50	Chang *et al* (2016)
15	TiO_2	Nanoparticles	7.65	Mazloum-Ardakani and Khoshroo (2015)
16	TiO_2/Fe_2O_3	Nanoparticles	7.27	Kılıç *et al* (2015)
17	TiO_2+Au/SiO_2 nanocubes	Nanoparticles + nanocubes	7.80	Zarick *et al* (2014)

materials and devices has shown rapid progress in OSCs, and their power conversion efficiency has been determined to be 16%–18% (Sun *et al* 2020a, Liu *et al* 2020). In the past few decades, developing novel photovoltaic materials has been considered an important research direction. A high-performing photovoltaic material should have some unique properties, such as being highly soluble in solvent processing, showing strong and complementary adsorption in the visible–near infrared regions, having high charge mobility (approximately 10^{-5}–10^{-4} cm^2 V^{-1} s^{-1}), having an energy level that should match to maintain the balance between the energy offset for excitation separation and loss of voltage, and having efficient aggregation miscibility and performance between the acceptor and donor to form a nanoscale phase (Guo *et al* 2021b, Yao *et al* 2018, Wang *et al* 2019b, Liu *et al* 2017). In a few reports, the acceptor layers were mainly synthesized from fluorescent materials as they have high electron affinity and charge mobility, making them efficient materials (Liu *et al* 2017, Li *et al* 2018). However, the fluorescent materials have some drawbacks, such as inefficient adsorption, poor modification capability, and fast dimerization. These drawbacks limit the power conversion efficiency in OSCs by up to 13% (Li *et al*

2018). Thus researchers have made many notable efforts to increase the efficiency and stability of OSCs significantly, for instance, the addition of solvent-based additives, using several energetic layered thermal or chemical (solvent annealing) treatments by applying a tannery approach, and using an inverted fabrication scheme (Hu *et al* 2018, Hu *et al* 2020). Lin *et al* developed an efficient organic solar cell using a non-fullerene material ITIC composite as an acceptor with a narrow band gap. The developed organic solar cells showed approximately 6.28% power conversion efficiency (Lin *et al* 2015). Another OSC was designed and reported by Yuan *et al* in 2019, using another Y6-type non-fullerene structure and its various derivatives acting as acceptors. The developed cells showed approximately 15% power conversion efficiency (Yuan *et al* 2019). Figure 2.2 shows the thin-film morphology of single junction organic solar cells. In a comparison between the inorganic and organic semiconductor materials, the band gap modification is easy in organic semiconductor materials near the visible–infrared region allowing a significant improvement in the performance of organic solar cells (Wang *et al* 2019a, Yao *et al* 2019, Ren *et al* 2021). In this context, organic semiconducting materials exhibit a larger adsorption coefficient in comparison to the inorganic semiconducting materials.

The organic material can optimize the photon gathering near the infrared region and the transmittance in the visible region by selecting an appropriate organic material (Fan *et al* 2019, Price *et al* 2011, Liu *et al* 2017). However, the main challenge is the development of semitransparent organic solar cells with efficient visible light transmission and sufficient harvesting of photons in the infrared region for applications in windows and greenhouses with high power conversion efficiency. Some materials with narrow band gaps are used to prepare the active layer to obtain the semitransparent organic solar cells. Semitransparent electrodes replace opaque electrodes to maintain the balance between power conversion efficiency and average visible light transmittance of the organic solar cells (Sun *et al* 2020a, Wu *et al* 2020, Cui *et al* 2020). Additionally, the average visible light transmittance and conductivity can be optimized by replacing the material and/or changing the thickness of

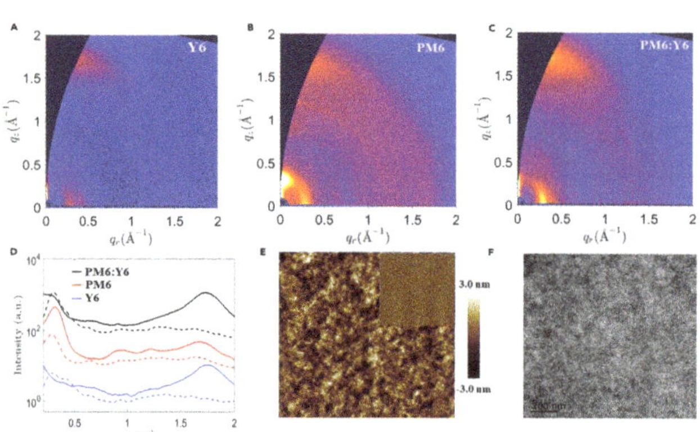

Figure 2.2. Thin film morphology of single junction solar cells. (Reproduced with permission from (Yuan *et al* 2019). Copyright 2019 Elsevier.)

Table 2.2. Different organic materials for solar cells.

S. No.	Acceptor	Donor	Efficiency (*η* %)	Reference
1	PTVT-T	BTP-eC9	16.20	Ren *et al* (2021)
2	PM6	Y6	15.7	Guo *et al* (2021b)
3	D18	Y6	18.22	Liu *et al* (2020)
4	PBDB-TF	BTP-eC9	17.8	Cui *et al* (2020)
5	P1	Y6	17.6	Wu *et al* (2020)
6	PTQ11	TPT10	16.32	Sun *et al* (2020a)
7	PDCBT-Cl	ITIC-Th1	12.38	Wang *et al* (2019a)
8	PTO2	IT-4F	14.7	Yao *et al* (2019)
9	P2F-EHp	Y6	16.02	Fan *et al* (2019)
10	PDCBT-2F	IT-4F	11.9	Yao *et al* (2018)
11	PDTB-EF-T	IT-4F	13.0	Li *et al* (2018)
12	PB3T	IT-M	11.9	Liu *et al* (2017)
13	PDCBT	ITIC	10.16	Qin *et al* (2016)
14	PDCBT	$PC_{71}BM$	7.2	Zhang *et al* (2014b)
15	FTAZ	$PC_{61}BM$	6.81	Price *et al* (2011)
16	PT-C3	$PC_{71}BM$	3.87	Zhang *et al* (2011)

the electrodes. The semitransparent organic solar cells can be applied to buildings, windows, ceilings, and greenhouses for more efficient solar and indoor lights (Guo *et al* 2021a, Liu *et al* 2020). A summary of the recent studies in organic solar cells, photon acceptors, and their power conversion efficiency is provided in table 2.2.

2.2.3 Perovskite solar cells

Perovskite solar cells are a kind of solar cell that includes perovskite-structured material (Manser *et al* 2016, Xue *et al* 2021). These are third-generation advanced solar cells because of their outstanding and effective properties, including high adsorption capacity, adjustable band gap, carrier mobility, anti-polar carrier transport property, and large carrier diffusion length (Jeon *et al* 2014, Wang *et al* 2022, Suresh Kumar and Chandra Babu Naidu 2021). Recently, the halide-based perovskite solar cells were found to be a promising photovoltaic technology owing to the efficient and swift increase in their power conversion efficiency (3%–25%) in past studies and the cost-effective nature of the materials used for the fabrication of perovskite solar cells (Suresh Kumar and Chandra Babu Naidu 2021, Chen *et al* 2021b). The perovskite solar cell market is likely to reach up to 3.2 billion USD by 2026 with a huge growth rate of 32% per year. Organic–inorganic hybrid perovskite materials are used to synthesize efficient and cost-effective perovskite solar cells (Wang *et al* 2021, Shi *et al* 2017a). The perovskite solar cells were first developed for light-emitting diode and transistor applications by Mitazi *et al* using $Bi_2Sr_2CaCu_2O_8$ material with a chemical formula of ABX_3, in which A is the organic citation group such as Cs^+, formamidine (FA^+, $CH_3CH_2NH_3^+$), and methylammonium (MA^+, $CH_3NH_3^+$), and B is the inorganic citation group such as Pb^{2+} and Sn^{2+}, and X is halide ions such as Br, Cl, and I, etc (Mitzi *et al* 1990, Suresh Kumar and Chandra Babu Naidu 2021, Xiao *et al* 2016). Among the available semiconductors, whether it is organic, inorganic, or a hybrid, the combined form of perovskites demonstrates excellent electrical and optical properties such as a high dielectric constant, weak binding energy, high Bohr radius, high diffusion length, and high carrier diffusion velocity (Burschka *et al* 2013, Liu *et al* 2013). Xue *et al* fabricated flexible halide-based perovskite solar cells, and the rough and flexible substrate could improve the photoconversion efficiency by around 19.7% (Xue *et al* 2021). Figure 2.3 shows a schematic illustration of the flexible perovskite cells. All the properties confirm the benefits of the application of organic–inorganic hybrid perovskite materials in synthesizing cost-effective and high-performance solar cells (Kojima *et al* 2009).

Generally, the perovskite solar cells contain an absorber layer (such as $CH_3NH_3PbX_3$) that is inserted between the hole-transporter and electron-transporter layers (Yang *et al* 2015, Liu *et al* 2013). Whenever the absorber is exposed to light, it puts a charge carrier (electron and hole) into p-type and n-type carriers transporting material to produce a free charge carrier. The produced electron moves through the pores of the film and external circuit and reaches the cathode. The condensed section of the hole transporting material reaches the ground state by restoring the oxidized perovskite (Hörantner *et al* 2015). Further, the hole is extorted into a hole transporting material and diffuses into the direction of the counter electrodes. Finally, it combines with electrons to produce the current. The efficiency and current generation in perovskite solar cells also depend on the thickness of the material (Suresh Kumar and Chandra Babu Naidu 2021, Ginting *et al* 2018, Eswaramoorthy *et al* 2022). Kojima *et al* first used dye sensitizer in perovskite-based materials for a dye-sensitized

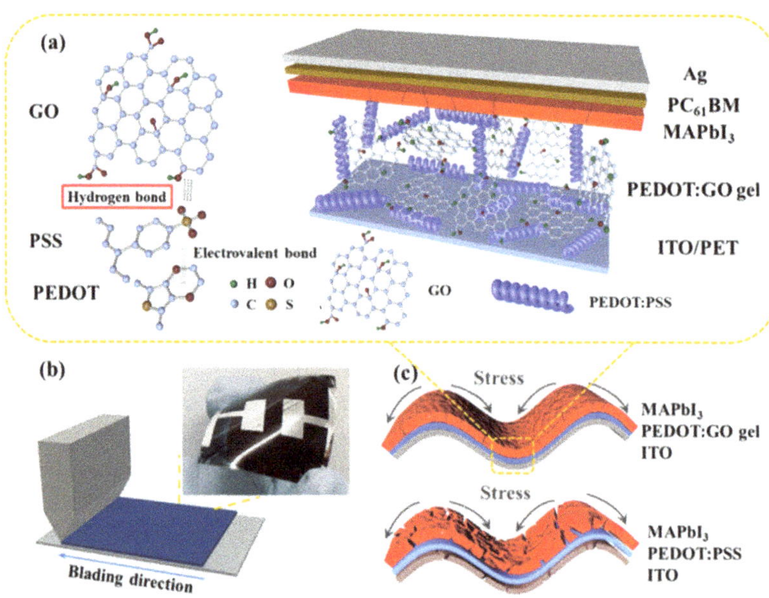

Figure 2.3. Schematic illustration of flexible perovskite cells. (Reproduced with permission from (Xue *et al* 2021). Copyright 2021 American Chemical Society.)

solar cell application in 2009 using $CH_3NH_3PbX_3$, which showed a power conversion efficiency of approximately 3.81% (Kojima *et al* 2009). Further, it is believed that OHIP perovskite materials are potential candidates for solar cell applications because of the richness of the precursor elements and easy synthesis methods, such as the vacuum-deposition method and atmospheric solution processing method (Yang *et al* 2015, Dou *et al* 2014, Wei *et al* 2017). In further studies, the advancement of perovskite solar cells was achieved by improving the properties of the electron-transport layers, which also played a significant role in the power conversion efficiency of the solar cells (Ha *et al* 2017, Jiang *et al* 2019). In previous times the electron-transport layers were made up of thin films of SiO_2, ZnO, and TiO_2, or their mesoporous structures (Lin *et al* 2016, Guo *et al* 2018, Dagar *et al* 2018). These electron-transport layers have weak recombination interference and significant grain boundaries, leading to defects in the semiconductor process due to oxygen defects and trap-assisted recombination (Zheng *et al* 2016, Choi *et al* 2019). The problems were overcome by applying different materials in several studies to develop electron-transport layers. The electron-transport layers of perovskite solar cells were prepared using nanosheets or thick layers of transition metal dichalcogenides, such as MoS_2, TiS_2, and WoS_2 (Singh *et al* 2019, Xu *et al* 2022). These newly synthesized electron-transport layers using different materials showed rapid diffusion of the electrons due to lower thickness of the layers of electron transport (Xiao *et al* 2016). A summary of some of the recent studies on the fabrication and application of perovskite solar cells is provided in table 2.3.

Table 2.3. Different materials used for perovskite solar cells.

S. No.	Material	Efficiency (η %)	Reference
1	$FTO/TiO_2/CsPbI_2Br$/carbon	15.2	Wang *et al* (2022)
2	FTO/TiO_2/Gly-X:$CsPbI_2Br$/spiro-OMeTAD/Au	17.26	Xu *et al* (2022)
3	Flexible PEN/ITO/PTAA/GBA	20.16	Wang *et al* (2021)
4	Flexible PET/ITO/PEDOT:GO/$MAPbI_3$/PC61 BM/Ag	19.70	Xue *et al* (Xue 2021)
5	ITO/PTAA/BHC-modified $MAPbI_3$/C60/BCP/Cu	23.20	Chen *et al* (2021b)
6	ITO/PTAA/CBH-modified $MA_{0.6}FA_{0.4}PbI_3/C_{60}$/ BCP/Cu	23.60	Chen *et al* (2021a)
7	Spiro-OMeTAD	25.5	Green *et al* (2021)
8	poly(3-hexylthiophene)	25.2	Green *et al* (2020)
9	PEAI was applied for surface defect inversion	23.32	Jiang *et al* (2019)
10	2D photonic crystal nanodisk array	18.70	Choi *et al* (2019)
11	EDTA complexed SnO_2/E-SnO_2	21.6	Yang *et al* (2018)
12	Textured inverse micro-cone array	14.83	Ginting *et al* (2018)
13	Bio-inspired nanostructured back electrode	16.31	Wei *et al* (2017)
14	TiO_2 meso-IO scaffold	17.1	Ha *et al* (2017)
15	Textured FTO	13.3	Shi *et al* (2017a)
16	TiO_2 nanobowl array	12.02	Zheng *et al* (2016)
17	Metal oxide honeycomb	9.5	Hörantner *et al* (2015)

2.3 Conclusion

In conclusion, thin-film-based materials have gained superiority in solar cell applications due to their exceptional properties, with high thermal and chemical stability compared to the other available materials used in solar cells. These materials demonstrate excellent properties for solar cells and other applications, including displays, printable electronics, flexible transistors, memories, etc. Due to their excellent characteristic properties, for example their capability to form π–π stacking and weak bonds such as hydrogen bonding, etc, electron/hole transporters, and serve as interfacial layers in different types of solar cells, these materials are gaining attention. Additionally, the band gap tuning and processes for band gap opening have always been a concern for solar cell applications. Hence, it is believed that thin films will play more and more essential roles in solar cells and several other applications, such as energy storage, optoelectronics, electrics, and sensing.

References

Ahmad M S, Rahim N A, Syed S, Mehmood S and Khan A D 2021 Effect of WS_2 nano-sheets on the catalytic activity of polyaniline nano-rods based counter electrode for dye sensitized solar cell *Physica* E **126** 114466

Altobello S, Bignozzi C A, Caramori S, Larramona G, Quici S, Marzanni G and Lakhmiri R 2004 Sensitization of TiO_2 with ruthenium complexes containing boronic acid functions *J. Photochem. Photobiol.* A **166** 91–8

Alturaif H A, Alothman Z A, Shapter J G and Wabaidur S M 2014 Use of carbon nanotubes (CNTs) with polymers in solar cells *Molecules* **19** 17329–44

Aneesiya K R and Louis C 2020 Localized surface plasmon resonance of Cu-doped ZnO nanostructures and the material's integration in dye sensitized solar cells (DSSCs) enabling high open-circuit potentials *J. Alloys Compd.* **829** 154497

Apriani T, Arsyad W S, Wulandari P and Hidayat R 2016 Investigation on the influences of layer structure and nanoporosity of light scattering TiO_2 layer in DSSC *J. Phys. Conf. Ser.* **739** 012134

Arabpour Roghabadi F, Ahmadi N, Ahmadi V, Di Carlo A, Oniy Aghmiuni K, Shokrolahzadeh Tehrani A, Ghoreishi F S, Payandeh M and Mansour Rezaei Fumani N 2018 Bulk heterojunction polymer solar cell and perovskite solar cell: concepts, materials, current status, and opto-electronic properties *Sol. Energy* **173** 407–24

Arulraj A, Thachnatharen N, Rajkumar C and Mangalaraja R V 2022 *Engineering 2D chalcogenides for energy and environmental remediation Nanostructured Materials for Sustainable Energy and Environmental Remediation* (Bristol: IOP Publishing) pp 2-1–28

Arulraj A and Ramesh M 2018 *Rational strategies for large-area perovskite solar cells Rational Design of Solar Cells for Efficient Solar Energy Conversion* (Hoboken, NJ: Wiley) pp 307–37

Baig N 2023 Two-dimensional nanomaterials: a critical review of recent progress, properties, applications, and future directions *Composites* A **165** 107362

Batmunkh M, Bat-Erdene M and Shapter J G 2018 Black phosphorus: synthesis and application for solar cells *Adv. Energy Mater.* **8** 1701832

Battaglia C, Cuevas A and De Wolf S 2016 High-efficiency crystalline silicon solar cells: status and perspectives *Energy Environ. Sci.* **9** 1552–76

Bin Rafiq M K S, Amin N, Alharbi H F, Luqman M, Ayob A, Alharthi Y S, Alharthi N H, Bais B and Akhtaruzzaman M 2020 WS_2: a new window layer material for solar cell application *Sci. Rep.* **10** 771

Burschka J, Pellet N, Moon S J, Humphry-Baker R, Gao P, Nazeeruddin M K and Grätzel M 2013 Sequential deposition as a route to high-performance perovskite-sensitized solar cells *Nature* **499** 316–9

Chang W C, Tang B H, Lu Y W, Yu W C, Lin L Y and Wu R J 2016 Incorporating hydrangea-like titanium dioxide light scatterer with high dye-loading on the photoanode for dye-sensitized solar cells *J. Power Sources* **319** 131–8

Chen S, Dai X, Xu S, Jiao H, Zhao L and Huang J 2021a Stabilizing perovskite-substrate interfaces for high-performance perovskite modules *Science* **373** 902–7

Chen S, Xiao X, Gu H and Huang J 2021b Iodine reduction for reproducible and high-performance perovskite solar cells and modules *Sci. Adv.* **7** eabe8130

Chiang C H and Wu C G 2016 Bulk heterojunction perovskite–PCBM solar cells with high fill factor *Nat. Photonics* **10** 196–200

Choi D H, Nam S K, Jung K and Moon J H 2019 2D photonic crystal nanodisk array as electron transport layer for highly efficient perovskite solar cells *Nano Energy* **56** 365–72

Cui Y *et al* 2020 Single-junction organic photovoltaic cells with approaching 18% efficiency *Adv. Mater.* **32** 1908205

Dagar J, Castro-Hermosa S, Gasbarri M, Palma A L, Cina L, Matteocci F, Calabrò E, Di Carlo A and Brown T M 2018 Efficient fully laser-patterned flexible perovskite modules and solar cells based on low-temperature solution-processed SnO_2/mesoporous-TiO_2 electron transport layers *Nano Res.* **11** 2669–81

Derakhshi M and Daemi S 2022 Two-dimensional nanomaterials beyond graphene for bio-medical applications *J Funct Biomater.* **13**(1) 27

Dissanayake M A K L, Umair K, Senadeera G K R, Jaseetharan T, Weerasinghe A M J S and Wijayasinghe H W M A C 2022 Plasmonic gold nanoparticle incorporated MgO-coated SnO_2 photoanode for efficiency enhancement in dye-sensitized solar cells *Sol. Energy* **233** 363–77

Dou L, Yang Y, You J, Hong Z, Chang W H, Li G and Yang Y 2014 Solution-processed hybrid perovskite photodetectors with high detectivity *Nat. Commun.* **5** 5404

Eda G and Chhowalla M 2010 Chemically derived graphene oxide: towards large-area thin-film electronics and optoelectronics *Adv. Mater.* **22** 2392–415

Eda G, Fanchini G and Chhowalla M 2008 Large-area ultrathin films of reduced graphene oxide as a transparent and flexible electronic material *Nat. Nanotechnol.* **3** 270–4

Eswaramoorthy N, Arulraj A, Mangalaraja R V, Pitchaiya S and Rajaram K 2022 Nanoscale interfacial engineering of 1D g-C3 N4 enables effective and thermally stable HTL-free carbon-based perovskite solar cells with aging for 100 hours *Int. J. Energy Res.* **46** 20194–205

Fadhilah N, Pratama D Y, Sawitri D and Risanti D D 2019 Preparation of Au@TiO_2@SiO_2 core-shell nanostructure and their light harvesting capability on DSSC (dye sensitized solar cells) *AIP Conf. Proc.* **2088** 060007

Fan B, Zhang D, Li M, Zhong W, Zeng Z, Ying L, Huang F and Cao Y 2019 Achieving over 16% efficiency for single-junction organic solar cells *Sci. China Chem.* **62** 746–52

Fu H, Li Y, Yu J, Wu Z, Fan Q, Lin F, Woo H Y, Gao F, Zhu Z and Jen A K Y 2021 High efficiency (15.8%) all-polymer solar cells enabled by a regioregular narrow bandgap polymer acceptor *J. Am. Chem. Soc.* **143** 2665–70

Gao Q, Sun X, Yang X, Wang L, Li X, Zhang X, Duan L and Lü W 2019 Combined photoanodes of TiO_2 nanoparticles and {001}-faceted TiO_2 nanosheets for quantum dot-sensitized solar cells *New J. Chem.* **43** 8551–6

Gielen D, Boshell F, Saygin D, Bazilian M D, Wagner N and Gorini R 2019 The role of renewable energy in the global energy transformation *Energy Strategy Rev.* **24** 38–50

Ginting R T, Jeon E B, Kim J M, Jin W Y and Kang J W 2018 Dual light trapping and water-repellent effects of a flexible-based inverse micro-cone array for organic and perovskite solar cells *ACS Appl. Mater. Interfaces* **10** 31291–9

Green M A, Dunlop E D, Hohl-Ebinger J, Yoshita M, Kopidakis N and Hao X 2020 Solar cell efficiency tables (version 56) *Prog. Photovoltaics Res. Appl.* **28** 629–38

Green M A, Dunlop E D, Hohl-Ebinger J, Yoshita M, Kopidakis N and Hao X 2021 Solar cell efficiency tables (version 58) *Prog. Photovoltaics Res. Appl.* **29** 657–67

Guo Q, Guo Q, Geng Y, Tang A, Zhang M, Du M, Sun X and Zhou E 2021a Recent advances in PM6:Y6-based organic solar cells *Mater. Chem. Front.* **5** 3257–80

Guo X *et al* 2021b Optimized active layer morphologies via ternary copolymerization of polymer donors for 17.6% efficiency organic solar cells with enhanced fill factor *Angew. Chem. Int. Ed.* **60** 2322–9

Guo Z, Gao L, Zhang C, Xu Z and Ma T 2018 Low-temperature processed non-TiO_2 electron selective layers for perovskite solar cells *J. Mater. Chem.* A **6** 4572–89

Ha S J, Heo J H, Im S H and Moon J H 2017 Mesoscopic $CH_3NH_3PbI_3$ perovskite solar cells using TiO_2 inverse opal electron-conducting scaffolds *J. Mater. Chem.* A **5** 1972–7

He X, Iwamoto Y, Kaneko T and Kato T 2022 Fabrication of near-invisible solar cell with monolayer WS_2 *Sci. Rep.* **12** 11315

Hörantner M T, Zhang W, Saliba M, Wojciechowski K and Snaith H J 2015 Templated microstructural growth of perovskite thin films via colloidal monolayer lithography *Energy Environ. Sci.* **8** 2041–7

Hou J, Inganäs O, Friend R H and Gao F 2018 Organic solar cells based on non-fullerene acceptors *Nat. Mater.* **17** 119–28

Hu Z, Wang J, Ma X, Gao J, Xu C, Yang K, Wang Z, Zhang J and Zhang F 2020 A critical review on semitransparent organic solar cells *Nano Energy* **78** 105376

Hu Z, Zhang F, An Q, Zhang M, Ma X, Wang J, Zhang J and Wang J 2018 Ternary nonfullerene polymer solar cells with a power conversion efficiency of 11.6% by inheriting the advantages of binary cells *ACS Energy Lett.* **3** 555–61

Frank J, Kopidakis A, Kopidakis N and Lagemaat J V D 2004 Electrons in nanostructured TiO_2 solar cells: transport, recombination and photovoltaic properties *Coord. Chem. Rev.* **248** 1165–79

Jayaweera E N, Kumara G R A, Kumarage C, Ranasinghe S K, Rajapakse R M G, Bandara H M N, Ileperuma O A and Dassanayake B S 2018 CdS nanosheet-sensitized solar cells based on SnO_2/MgO composite films *J. Photochem. Photobiol.* A **364** 109–15

Jeon N J, Noh J H, Kim Y C, Yang W S, Ryu S and Seok S I 2014 Solvent engineering for high-performance inorganic–organic hybrid perovskite solar cells *Nat. Mater.* **13** 897–903

Jiang Q, Zhao Y, Zhang X, Yang X, Chen Y, Chu Z, Ye Q, Li X, Yin Z and You J 2019 Surface passivation of perovskite film for efficient solar cells *Nat. Photonics* **13** 460–6

Kannan N and Vakeesan D 2016 Solar energy for future world—a review *Renew. Sustain. Energy Rev.* **62** 1092–105

Kılıç B, Gedik N, Mucur S P, Hergul A S and Gür E 2015 Band gap engineering and modifying surface of TiO_2 nanostructures by Fe_2O_3 for enhanced-performance of dye sensitized solar cell *Mater. Sci. Semicond. Process.* **31** 363–71

Kilic B, Turkdogan S, Astam A, Ozer O C, Asgin M, Cebeci H, Urk D and Mucur S P 2016 Preparation of carbon nanotube/TiO_2 mesoporous hybrid photoanode with iron pyrite (FeS_2) thin films counter electrodes for dye-sensitized solar cell *Sci. Rep.* **6** 27052

Kojima A, Teshima K, Shirai Y and Miyasaka T 2009 Organometal halide perovskites as visible-light sensitizers for photovoltaic cells *J. Am. Chem. Soc.* **131** 6050–1

Kunzmann A, Valero S, Sepúlveda Á E, Rico-Santacruz M, Lalinde E, Berenguer J R, García-Martínez J, Guldi D M, Serrano E and Costa R D 2018 Hybrid dye-titania nanoparticles for superior low-temperature dye-sensitized solar cells *Adv. Energy Mater.* **8** 1702583

Li S, Ye L, Zhao W, Yan H, Yang B, Liu D, Li W, Ade H and Hou J 2018 A wide band gap polymer with a deep highest occupied molecular orbital level enables 14.2% efficiency in polymer solar cells *J. Am. Chem. Soc.* **140** 7159–67

Lin Y, Wang J, Zhang Z G, Bai H, Li Y, Zhu D and Zhan X 2015 An electron acceptor challenging fullerenes for efficient polymer solar cells *Adv. Mater.* **27** 1170–4

Lin Z *et al* 2016 2D materials advances: from large scale synthesis and controlled heterostructures to improved characterization techniques, defects and applications *2D Mater.* **3** 042001

Liu D, Yang B, Jang B, Xu B, Zhang S, He C, Woo H Y and Hou J 2017 Molecular design of a wide-band-gap conjugated polymer for efficient fullerene-free polymer solar cells *Energy Environ. Sci.* **10** 546–51

Liu M, Johnston M B and Snaith H J 2013 Efficient planar heterojunction perovskite solar cells by vapour deposition *Nature* **501** 395–8

Liu Q *et al* 2020 18% Efficiency organic solar cells *Sci. Bull.* **65** 272–5

Manser J S, Christians J A and Kamat P V 2016 Intriguing optoelectronic properties of metal halide perovskites *Chem. Rev.* **116** 12956–3008

Mayer A C, Scully S R, Hardin B E, Rowell M W and McGehee M D 2007 Polymer-based solar cells *Mater. Today* **10** 28–33

Mazloum-Ardakani M and Khoshroo A 2015 Enhanced performance of dye-sensitized solar cells with dual-function coadsorbent: reducing the surface concentration of dye–iodine complexes concomitant with attenuated charge recombination *Phys. Chem. Chem. Phys.* **17** 22985–90

Mekhilef S, Saidur R and Safari A 2011 A review on solar energy use in industries *Renew. Sustain. Energy Rev.* **15** 1777–90

Mitzi D B, Lombardo L W, Kapitulnik A, Laderman S S and Jacowitz R D 1990 Growth and properties of oxygen- and ion-doped $Bi_2Sr_2CaCu_2O_8+\delta$ single crystals *Phys. Rev.* B **41** 6564–74

Mutashar I S and Al-bahrani M R 2021 Enhancing of dye-sensitized solar cells efficiency using graphene–MWCNT nanocomposite as photoanode *Mater Today: Proc.* **80** 2660–63

Navarro-Pardo F, Benetti D, Benavides J, Zhao H G, Cloutier S G, Castano V M, Vomiero A and Rosei F 2017 Nanofiber-structured TiO_2 nanocrystals as a scattering layer in dye-sensitized solar cells *ECS J. Solid State Sci. Technol.* **6** N32

Nazeeruddin M K, Baranoff E and Grätzel M 2011 Dye-sensitized solar cells: a brief overview *Sol. Energy* **85** 1172–8

Novoselov K S, Geim A K, Morozov S V, Jiang D, Zhang Y, Dubonos S V, Grigorieva I V and Firsov A A 2004 Electric field effect in atomically thin carbon films *Science* **306** 666–9

O'Regan B and Grätzel M 1991 A low-cost, high-efficiency solar cell based on dye-sensitized colloidal TiO_2 films *Nature* **353** 737–40

Okil M, Salem M S, Abdolkader T M and Shaker A 2022 From crystalline to low-cost silicon-based solar cells: a review *Silicon* **14** 1895–911

Pan J, Zheng R, Wang Y, Ye X, Wan Z, Jia C, Weng X, Xie J and Deng L 2020 A high-performance electrochromic device assembled with hexagonal WO_3 and NiO/PB composite nanosheet electrodes towards energy storage smart window *Sol. Energy Mater. Sol. Cells* **207** 110337

Praveen E, Peter I J, Kumar A M, Ramachandran K and Jayakumar K 2020 Boosting of power conversion efficiency of 2D ZnO nanostructures-based DSSC by the Lorentz force with chitosan polymer electrolyte *J. Inorg. Organomet. Polym. Mater.* **30** 4927–43

Price S C, Stuart A C, Yang L, Zhou H and You W 2011 Fluorine substituted conjugated polymer of medium band gap yields 7% efficiency in polymer–fullerene solar cells *J. Am. Chem. Soc.* **133** 4625–31

Qin Y, Uddin M A, Chen Y, Jang B, Zhao K, Zheng Z, Yu R, Shin T J, Woo H Y and Hou J 2016 Highly efficient fullerene-free polymer solar cells fabricated with polythiophene derivative *Adv. Mater.* **28** 9416–22

Rafique S, Rashid I and Sharif R 2021 Cost effective dye sensitized solar cell based on novel Cu polypyrrole multiwall carbon nanotubes nanocomposites counter electrode *Sci. Rep.* **11** 14830

Ren J, Bi P, Zhang J, Liu J, Wang J, Xu Y, Wei Z, Zhang S and Hou J 2021 Molecular design revitalizes the low-cost PTV-polymer for highly efficient organic solar cells *Natl Sci. Rev.* **8** nwab031

Roncali J 2009 Molecular bulk heterojunctions: an emerging approach to organic solar cells *Acc. Chem. Res.* **42** 1719–30

Saidi K and Omri A 2020 The impact of renewable energy on carbon emissions and economic growth in 15 major renewable energy-consuming countries *Environ. Res.* **186** 109567

Schwartz R J 1982 Review of silicon solar cells for high concentrations *Sol. Cells* **6** 17–38

Senthilkumar N, Arulraj A, Nandhakumar E, Ganapathy M, Vimalan M and Vetha I 2018 Potheher, Green mediated synthesis of plasmonic nanoparticle (Ag) for antireflection coating in bare mono silicon solar cell *J. Mater. Sci., Mater. Electron.* **29** 12744–53

Sharaf M, Yousef M S and Huzayyin A S 2022 Review of cooling techniques used to enhance the efficiency of photovoltaic power systems *Environ. Sci. Pollut. Res.* **29** 26131–59

Sharma K, Sharma V and Sharma S S 2018 Dye-sensitized solar cells: fundamentals and current status *Nanoscale Res. Lett.* **13** 381

Shi B *et al* 2017a Enhanced light absorption of thin perovskite solar cells using textured substrates *Sol. Energy Mater. Solar Cells* **168** 214–20

Shi Y, Yang Y, Dong G, Jiang Y, Wei L, Su T and Fan R 2017b Three-dimensional flower-like rutile TiO_2 microsphere composed of nanorods: a potential material as light scattering layer for DSSCs *Chem. Res. Chin. Univ.* **33** 298–304

Singh E and Nalwa H S 2015 Stability of graphene-based heterojunction solar cells *RSC Adv.* **5** 73575–600

Singh R, Giri A, Pal M, Thiyagarajan K, Kwak J, Lee J J, Jeong U and Cho K 2019 Perovskite solar cells with an MoS_2 electron transport layer *J. Mater. Chem.* A **7** 7151–8

Snaith H J 2010 Estimating the maximum attainable efficiency in dye-sensitized solar cells *Adv. Funct. Mater.* **20** 13–9

Sobayel K, Shahinuzzaman M, Amin N, Karim M R, Dar M A, Gul R, Alghoul M A, Sopian K, Hasan A K M and Akhtaruzzaman M 2020 Efficiency enhancement of CIGS solar cell by WS_2 as window layer through numerical modelling tool *Sol. Energy* **207** 479–85

Sun C *et al* 2020a High efficiency polymer solar cells with efficient hole transfer at zero highest occupied molecular orbital offset between methylated polymer donor and brominated acceptor *J. Am. Chem. Soc.* **142** 1465–74

Sun J, Yang X, Zhao L, Dong B and Wang S 2020b Ag-decorated TiO_2 nanofibers for highly efficient dye sensitized solar cell *Mater. Lett.* **260** 126882

Suresh Kumar N and Chandra Babu Naidu K 2021 A review on perovskite solar cells (PSCs), materials and applications *J. Mater.* **7** 940–56

Tamilselvan M, Byregowda A, Su C Y, Tseng C J and Bhattacharyya A J 2019 Planar heterojunction solar cell employing a single-source precursor solution-processed Sb_2S_3 thin film as the light absorber *ACS Omega* **4** 11380–7

Tributsch H and Calvin M 1971 Electrochemistry of excited molecules: photo-electrochemical reactions of chlorophylls *Photochem. Photobiol.* **14** 95–112

Tsai C H, Chuang P Y and Hsu H L 2018 Adding graphene nanosheets in liquid electrolytes to improve the efficiency of dye-sensitized solar cells *Mater. Chem. Phys.* **207** 154–60

Tsubomura H, Matsumura M, Nomura Y and Amamiya T 1976 Dye sensitised zinc oxide: aqueous electrolyte: platinum photocell *Nature* **261** 402–3

Wan J, Tao L, Wang B, Zhang J, Wang H and Lund P D 2019 A facile method to produce TiO_2 nanorods for high-efficiency dye solar cells *J. Power Sources* **438** 227012

Wang K, Liu C, Du P, Zheng J and Gong X 2015 Bulk heterojunction perovskite hybrid solar cells with large fill factor *Energy Environ. Sci.* **8** 1245–55

Wang K *et al* 2019a Simple near-infrared nonfullerene acceptors enable organic solar cells with >9% efficiency *ACS Appl. Mater. Interfaces* **11** 6717–23

Wang K, Sun W, Liu W, Huo X, Yin R, Liu J, Gao Y, You T and Yin P 2022 Mitigating interfacial and bulk defects via chlorine modulation for HTL-free all-inorganic $CsPbI_2Br$ carbon-based perovskite solar cells with efficiency over 14% *Chem. Eng. J.* **445** 136781

Wang Q, Li M, Zhang X, Qin Y, Wang J, Zhang J, Hou J, Janssen R A J and Geng Y 2019b Carboxylate-substituted polythiophenes for efficient fullerene-free polymer solar cells: the effect of chlorination on their properties *Macromolecules* **52** 4464–74

Wang Z, Lu Y, Xu Z, Hu J, Chen Y, Zhang C, Wang Y, Guo F and Mai Y 2021 An embedding 2D/3D heterostructure enables high-performance FA-alloyed flexible perovskite solar cells with efficiency over 20% *Adv. Sci.* **8** 2101856

Wei J, Xu R P, Li Y Q, Li C, Chen J D, Zhao X D, Xie Z Z, Lee C S, Zhang W J and Tang J X 2017 Enhanced light harvesting in perovskite solar cells by a bioinspired nanostructured back electrode *Adv. Energy Mater.* **7** 1700492

Wei Z, Chen L, Liu K, Liu S, Li X, Zhang Q and Shuai J 2021 Two-dimensional nanomaterials for boosting the performance of organic solar cells *Coatings* **11** 1530

Wu J *et al* 2020 Random terpolymer based on thiophene-thiazolothiazole unit enabling efficient non-fullerene organic solar cells *Nat. Commun.* **11** 4612

Xiao Z *et al* 2016 Photovoltaic properties of two-dimensional $(CH_3NH_3)_2Pb(SCN)_2I_2$ perovskite: a combined experimental and density functional theory study *J. Phys. Chem. Lett.* **7** 1213–8

Xie B, Li C, Chen J and Wang N 2020 Exfoliated 2D hexagonal boron nitride nanosheet stabilized stearic acid as composite phase change materials for thermal energy storage *Sol. Energy* **204** 624–34

Xu J *et al* 2022 Stable high-efficiency $CsPbI_2Br$ solar cells by designed passivation using multifunctional 2D perovskite *Adv. Funct. Mater.* **32** 2202829

Xu Z, Shi J, Haroone M S, Chen W, Zheng S and Lu J 2018 Zinc-aluminum oxide solid solution nanosheets obtained by pyrolysis of layered double hydroxide as the photoanodes for dye-sensitized solar cells *J. Colloid Interface Sci.* **515** 240–7

Xue T, Chen G, Hu X, Su M, Huang Z, Meng X, Jin Z, Ma J, Zhang Y and Song Y 2021 Mechanically robust and flexible perovskite solar cells via a printable and gelatinous interface *ACS Appl. Mater. Interfaces* **13** 19959–69

Yang D, Yang R, Wang K, Wu C, Zhu X, Feng J, Ren X, Fang G, Priya S and Liu S 2018 High efficiency planar-type perovskite solar cells with negligible hysteresis using EDTA-complexed SnO_2 *Nat. Commun.* **9** 3239

Yang D, Yang Z, Qin W, Zhang Y, Liu S and Li C 2015 Alternating precursor layer deposition for highly stable perovskite films towards efficient solar cells using vacuum deposition *J. Mater. Chem.* A **3** 9401–5

Yao H *et al* 2019 14.7% Efficiency organic photovoltaic cells enabled by active materials with a large electrostatic potential difference *J. Am. Chem. Soc.* **141** 7743–50

Yao H, Qian D, Zhang H, Qin Y, Xu B, Cui Y, Yu R, Gao F and Hou J 2018 Critical role of molecular electrostatic potential on charge generation in organic solar cells *Chin. J. Chem.* **36** 491–4

Yin L, Li Y, Yao X, Wang Y, Jia L, Liu Q, Li J, Li Y and He D 2021 MXenes for solar cells *Nano-Micro Lett.* **13** 78

You J *et al* 2013 A polymer tandem solar cell with 10.6% power conversion efficiency *Nat. Commun.* **4** 1446

Yuan J *et al* 2019 Single-junction organic solar cell with over 15% efficiency using fused-ring acceptor with electron-deficient core *Joule* **3** 1140–51

Zarick H F, Hurd O, Webb J A, Hungerford C, Erwin W R and Bardhan R 2014 Enhanced efficiency in dye-sensitized solar cells with shape-controlled plasmonic nanostructures *ACS Photon.* **1** 806–11

Zhang H, Azimi H, Hou Y, Ameri T, Przybilla T, Spiecker E, Kraft M, Scherf U and Brabec C J 2014a Improved high-efficiency perovskite planar heterojunction solar cells via incorporation of a polyelectrolyte interlayer *Chem. Mater.* **26** 5190–3

Zhang M, Guo X, Ma W, Ade H and Hou J 2014b A polythiophene derivative with superior properties for practical application in polymer solar cells *Adv. Mater.* **26** 5880–5

Zhang M, Guo X, Yang Y, Zhang J, Zhang Z G and Li Y 2011 Downwards tuning the HOMO level of polythiophene by carboxylate substitution for high open-circuit-voltage polymer solar cells *Polym. Chem.* **2** 2900–6

Zhang Z *et al* 2022 Size-tunable MoS_2 nanosheets for controlling the crystal morphology and residual stress in sequentially deposited perovskite solar cells with over 22.5% efficiency *J. Mater. Chem.* A **10** 3605–17

Zheng X, Wei Z, Chen H, Zhang Q, He H, Xiao S, Fan Z, Wong K S and Yang S 2016 Designing nanobowl arrays of mesoporous TiO_2 as an alternative electron transporting layer for carbon cathode-based perovskite solar cells *Nanoscale* **8** 6393–402

Zhou X, Sun H and Bai X 2020 Two-dimensional transition metal dichalcogenides: synthesis, biomedical applications and biosafety evaluation *Front. Bioeng. Biotechnol.* **8** 236

IOP Publishing

Materials Technology for the Energy and Environmental Nexus, Volume 1

R V Mangalaraja, Arunachalam Arulraj, Nalandhiran Pugazhenthiran and Chan Siew Hwa

Chapter 3

An overview of noble-metal-free nanostructured electrocatalysts for overall water splitting

M Praveen Kumar, R V Mangalaraja, Smagul Karazhanov, Tatianne Ferreira de Oliveira, M Sasikumar, G Murugadoss, Suhail Mubarak, Sivasubramanian Palanisamy and Natesan Kumaresan

The increasing demand for energy coupled with the effects of greenhouse gases from fossil fuels has triggered extensive research on the development of sustainable and renewable energy sources. Toward this aim, hydrogen has enormous potential to replace fossils fuel as a clean and alternative fuel of the future. Moreover, the commonly used electrochemical water splitting is emerging as a very hopeful solution to generate hydrogen in the production of carbon-free and environmentally friendly green energy. Water is a widely available resource on Earth and is the key source of hydrogen, and thus the main component in the water splitting process. Catalysts play an important role in the oxygen evolution reaction (OER) and the hydrogen evolution reaction (HER) in water oxidation. In this chapter we detail the synthesis and characterization of electrocatalysts, and the water splitting applications thereof.

3.1 Introduction

The focus of concern of human society is the global energy crisis and environmental pollution (CO_2 emission) due to the utilization of fossil fuels for energy [1]. New advanced methods must be utilized to produce green and clean energy which can exhibit the novel properties required under a real-world environment. Triggered by the development of new technologies and the consequent decrease in the cost of devices for applications, the use of renewable energy, including solar and wind power, has evolved rapidly over the last few decades. However, both technologies suffer from the disadvantage of intermittency, which limits their use as a primary

doi:10.1088/978-0-7503-5724-1ch3

feed for the grid, requiring the use of mediating systems to buffer their output. Several methods for storage have been proposed which include hydroelectric, compressed air, thermal energy, and batteries, etc [2–7]. Indeed, the energy that can be produced by the chemical bonds of small molecules represents a very attractive approach. Stimulated by this knowledge, a great deal of effort is being dedicated to the generation of environmentally friendly clean energy using hydrogen [8]. Hydrogen gas (H_2) has great promise and is considered to be an ideal fuel for the future. Hydrogen gas could be generated from clean and renewable energy sources using various methods [9].

Currently, the scientific community is focusing on the use of hydrogen gas as the main fuel in the industrial and transportation sectors. In such way H_2 is being utilized for stationary, mobile, and transportation applications, using fuel cell technology. The problems associated with the usage of fossil fuels are leading the world towards what is known as the hydrogen economy, and if the predictions are true, over the next few decades we will begin to see an amazing shift away from the fossil fuel economy.

The fuels produced and distributed using H_2 fuel cells could significantly support reducing environmental pollution and greenhouse gas emissions [8, 10]. Further, if the H_2 is produced using domestic energy supplies, it could help to reduce the dependence on imported petroleum. There are several environmental, economical, and social implications associated with the hydrogen economy. The burning of H_2 as a fuel would result in a reduction of pollutants and an elimination of carbon emissions, such as carbon monoxide and carbon dioxide, thus making hydrogen a fuel that does not contribute to global warming [9–12]. The worldwide market for H_2 is booming up due to the adundance of its source, causing it to be one of the great competitors in the energy sector. The element hydrogen is as an energy carrier, which could theoretically be produced by commercial nuclear fusion power plants using tritium or deuterium. In a natural context, hydrogen is produced by the Sun during the nuclear fusion process, however, it is very difficult to replicate this reaction in a practical Earth-bound context, since it involves a great input of energy. As an alternative, H_2 can be obtained from natural sources such as biomass and methane which are sustainable and more available sources [13].

The energy density of H_2 gas per unit volume at several practicable pressures is considerably less compared to existing conventional fuel sources, even though the energy density of H_2 gas per unit fuel mass is higher. For example, the practicable pressures for the carbon dioxide (CO_2) capture and storage are relatively high compared with the H_2 gas. Also, there are negligible carbon emissions with only trace NO_x emissions in the use of H_2 gas in the transportation sector. Furthermore, the advantages of using H_2 as a green fuel include its high electrochemical reactivity for applications, higher theoretical energy density, infinite accessibility resources, and eco-friendliness. Also, H_2 gas is considered to be a excellent future energy source to meet energy demands, one of the major advantages being that it is so readily available. Due to its properties, H_2 gas has been used recently in NASA's space-ships; the product of the burned H_2 gas is used in a recycling process to provide pure drinking water for the astronauts. Hydrogen gas is non-toxic, which is makes it a rarity among all the fuel sources. The use of coal, nuclear energy, and gasoline is ultimately toxic to their end-users. Further, they are known to be harmful to the environment. This makes the use of H_2 gas ideal in many applications, such that other fuel sources cannot compete against it. Fossil fuels are non-renewable sources of energy, so they cannot be produced again and again for applications, but hydrogen fuel can be produced on demand.

The use of H_2 is associated primarily with the fertilizer, paint, food, and chemical industries. In the food industry H_2 is used to make hydrogenated vegetable oils (margarine and butter). Vegetable oils are combined with H_2 in this procedure, and the solid fat substances are produced using nickel (Ni) as catalyst. Similarly, the H_2 is essential for crude oil refinement in the petrochemical industry. Welding companies use H_2 in the welding torches which are utilized for melting steel. In the chemical industry H_2 is necessary as a reducing agent for metal extraction [14]. It is also used for producing a number of chemical compounds. It can be utilized to make fertilizers, an assortment of bases, and hydrochloric acids. For instance, H_2 is essential for making the methyl alcohol which is used in inks, varnishes, and paints. Further, it is more affordable than using any other similar product to achieve the same energy per pound of fuel. See figure 3.1 for a summary of the used of hydrogen [15].

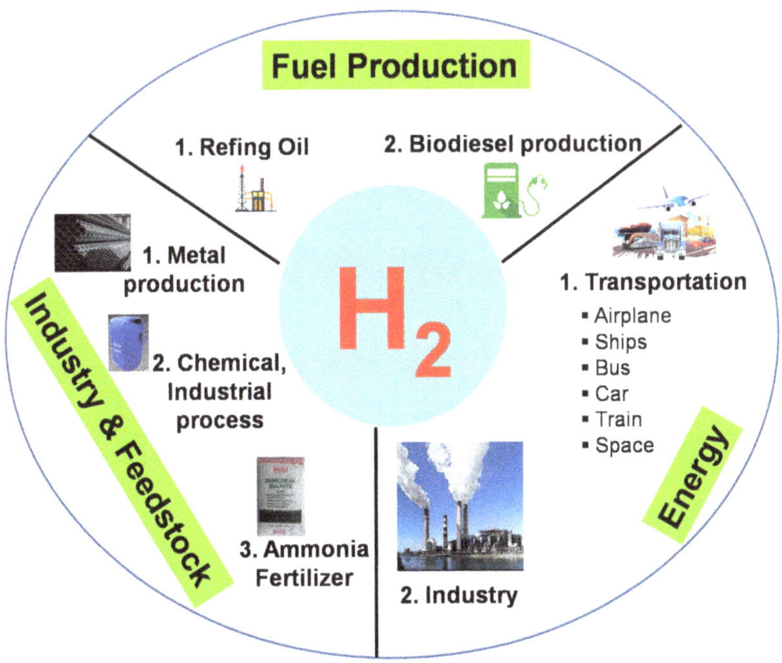

Figure 3.1. The potential uses of hydrogen as clean energy.

Since H_2 gas is light in weight, meteorologists use it in the weather balloons that carry the apparatus to record information to study the climatic changes. Airships during the First World War were filled with hydrogen.

3.2 Methods for the production of hydrogen

There is increasing interest in the generation of H_2 as an energy carrier, and we can produce H_2 using several methods, as shown in figure 3.2.

3.2.1 Pyrolysis of wood

Biomass gasification is a mature technology that applies a controlled method. The process involves heat, steam, and oxygen to convert biomass to H_2 and some other products, without combustion [16]. Because increasing biomass removes hydrocarbon (CO_2, CO etc) from the atmosphere, the net carbon emissions of this process can be low, particularly if coupled with carbon capture, utilization, and storage in the long term. The gasification processes for biofuel plants that are being built and operated can provide the best take-away and lessons learned for hydrogen generation. The example reaction is

$$C_6H_{12}O_6 + O_2 + H_2O \rightarrow CO + CO_2 + H_2 + \text{ other species.}$$

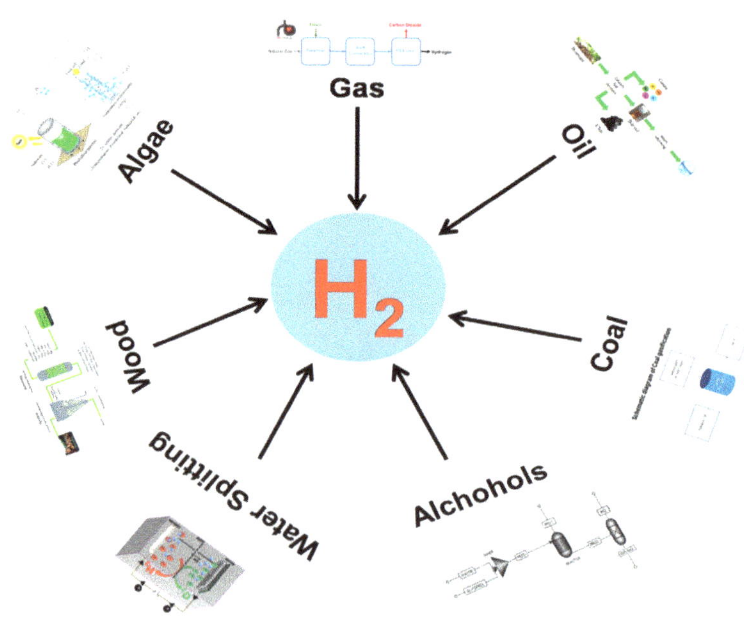

Figure 3.2. The various methods for hydrogen generation.

3.2.1.1 Disadvantages

- The generation of hydrogen is not entirely clean.
- There is a danger of deforestation.
- Requires a great deal of water.

3.2.2 Algae method for hydrogen generation

Biological H_2 generation using algae is a process of photobiological water splitting; it has been carried out in a closed photobioreactor for H_2 generation using solar energy and algae [17] under specific conditions. Algae cultivation can be carried out in the open, in covered ponds, and in closed photobioreactors using solar energy. It is based on tubular, flat plate, and other designs.

3.2.2.1 Disadvantages

- The closed systems are higher in cost than ponds.
- There are currently considerable operating challenges (e.g. overheating, fouling).
- It cannot be scaled-up much beyond about 100 square meters for an individual growth unit due to gas exchange limitations.

3.2.3 Steam reforming method for hydrogen production

The steam reforming of natural gas is a highly developed and mature production method that builds upon the existing natural gas (methane, ethane, etc) pipeline delivery infrastructure. Currently, 95% of the H_2 produced using natural gas reforming by the United States is made in large central plants [18]. This is a significant technology lane for near-term hydrogen generation.

3.2.3.1 Disadvantages

- A higher amount of hydrocarbon emissions.
- The purity of the H_2 produced is lower.
- Plants are expensive to construct.

3.2.4 Gasification method for hydrogen production

Fossil fuel is used by the US Department of Energy (DOE) for advanced coal-to-hydrogen technology which is the process of coal gasification with carbon capture, utilization, and storage [19]. The DOE expects that coal gasification for hydrogen generation with carbon capture, utilization, and storage can be achieved in a mid-term time frame.

3.2.4.1 Disadvantages

- A higher percentage of hydrocarbon is released in this process.
- The purity of the hydrogen is 80%.

3.3 Water electrolysis for hydrogen production

The theory of water electrolysis is basically the splitting of water molecules into oxygen (O_2) gas at the anode side, the oxygen evolution reaction (OER), and H_2 gas at the cathode side, the hydrogen evolution reaction (HER), using electrical energy as the input [20]. The equations are

$$\text{Attheanode: } 2H_2O \rightarrow O_2 \uparrow + 4H^+ + 4e^- \rightarrow E^0 = 1.23\text{ V}$$
$$\text{At thecathode: } 4H^+ + 4e^- \rightarrow 2H_2 \uparrow \rightarrow E^0 = 0.00\text{ V.}$$

3.3.1 Electrochemical water splitting

The thermodynamics reactions make clear that the water oxidation reaction is a non-spontaneous process. However, the water splitting reaction can be carried out by applying power to the system externally, which is known as the applied voltage or potential [21]. This reaction mechanism is called an endergonic transformation process; the device used for the approach is an electrolyzer. In a water electrolyzer, the potential is applied to the electrochemical water electrolyzer so that the water molecules can split into hydrogen and oxygen in the gaseous form.

3.3.1.1 In an acidic medium

A proton exchange membrane (PEM) is a semi-porous membrane made from ionomers and designed to conduct protons while producing gases such as oxygen or hydrogen. The porous membrane plays an important role in the separation of reactants and transport of protons when incorporated into a membrane electrode assembly (MEA) in the PEM electrolyzer. Figure 3.3 shows PEM water electrolysis based on the use of a polymeric PEM as the solid electrolyte [21]. The advantages of polymer electrolyte technology include:

1. Significant safety.
2. Reliability (high durability) is anticipated since no acid electrolyte is circulated in the cell stack.
3. High sustained rate for the membrane

The high various pressure without damage and were efficient in high purity oxygen and hydrogen gases. The reactions are as follows:

$$\text{Anode: } H_2O(l) \rightarrow 1/2O_2(g) + 2H^+ + 2e^-$$
$$\text{Cathode: } 2H^+ + 2e^- \rightarrow H_2(g)$$
$$\text{Full reaction: } H_2O(l) \rightarrow H_2(g) + 1/2O_2(g).$$

3.3.1.2 In an alkaline medium

Alkaline electrolysis shows the maximum efficiency in water splitting. Under current circumstances, the industrial application of water electrolysis systems is inconceivable without the use of steel [22]. Steel has a mass fraction of iron and stainless steel

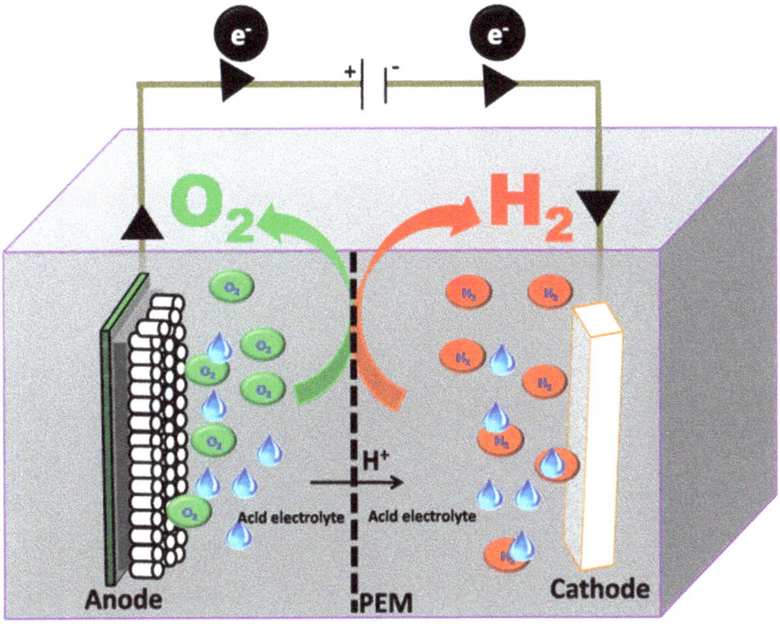

Figure 3.3. The principle of PEM electrolysis.

presents corrosion-resistance properties. The stainless steels are good electrocatalysts in alkaline membrane water electrolyzers and photoelectric devices.

The working principle of alkaline water electrolysis is shown schematically in figure 3.4. The water molecules are reduced to H_2 and two hydroxyl ions in the cathode part. The hydrogen escapes from the cathode (in a gaseous form), and the hydroxyl ions migrate through the porous diaphragm to the anode under the electrical field between the cathode and anode. Finally, the hydroxyl ions are split to a half molecule of oxygen and one molecule of water:

$$\text{Anode }(+): 2OH^- \rightarrow H_2O + 1/2O_2(g) + 2e^-$$
$$\text{Cathode }(-): 2H_2O + 2e^- \rightarrow H_2(g) + 2OH^-$$
$$\text{Full reaction}: H_2O(l) \rightarrow H_2(g) + 1/2O_2(g).$$

Water splitting technology is one of the best energy conversion systems for producing clean and sustainable energy using renewable sources. In this context, the OER and HER are involved in the water splitting process [23, 24]. The OER process is kinetically slow reaction, with a theoretical onset potential of 1.23 V and high energy loss [25]. The noble metal electrocatalysts such as RuO_2, IrO_2, and Pt are best for the OER and HER [26, 27], but are in low abundance and also high cost to deliver the technology. To improve H_2 production via the electrochemical route, understanding of the energy requirements, practical cell voltage, efficiency of the process, temperature and pressure effects on the potential, bubble mechanics and effects, kinetics of hydrogen production, and effect of electrode materials on conventional water electrolysis for the alkaline and proton exchange membrane

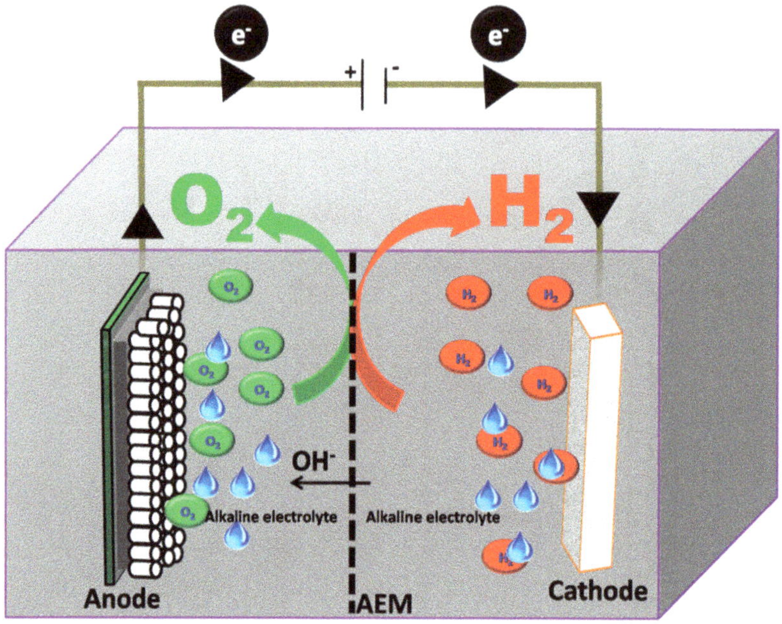

Figure 3.4. The principles of alkaline water electrolysis.

(PEC) water electrolyzer is crucial. Therefore, the major issues are the development of low-cost, Earth-abundant, environmentally friendly, and highly active electrocatalysts for the OER and HER [28–30]. Further, the synthesis and design of non-noble and carbon-based electrocatalysts for enhanced OER and HER to replace the noble metal catalysts such as Pt, IrO_2, and RuO_2 are essential [31]. Further, designing and developing robust bi-functional electrocatalysts (which perform OER and HER simultaneously) is the need of the hour for large-scale energy applications [32]. A variety of materials have been developed as OER and HER catalysts in the last decade. Among them, the overpotential of Ni, Fe, and Co metal/metal oxides has been well established in alkaline media and their operating potential is higher than the thermodynamic requirement value of 1.23 V [33]. This could be improved by support with carbon-based materials.

3.3.1.3 Non-noble electrocatalysts for OER and HER (alkaline media)

Observations of experiments show that introducing one metal/metal oxide onto another metal/metal oxide with carbon support can enhance the electrochemical activity resulting from the tailored electronic structure [34]. For such modifications, the prime focus is given for the Earth-abundant first-row transition metal/metal oxide-based materials such as Ni, Fe, Co, etc, to be used as bifunctional electrocatalysts. Among them, the Ni-based composite materials have proven to perform well in alkaline media due to their exceptional properties such as chemical stability, Earth-abundance, and excellent electrochemical properties [35]. Ni composite materials have gained intense scientific interest and the most studied materials are

Ni–Fe and Co–Ni [36]. In particular, the Ni composite metal/metal oxide electrocatalysts have promise and are gaining considerable interest for OER/HER activity with their controllable structure and active sites in alkaline media because of their electronic coordination [37]. A literature survey shows the large number of studies that give significant importance to the structure and active sites of Ni composites in the water oxidation process [38, 39]. In this direction, spike-like Ni–Fe(OOH) alloy nanostructures on toray carbon sheets have been developed via a simple and one-step electrochemical deposition method. The material exhibits decent electrochemical activity and high stability in alkaline media for hydrogen production.

Recent research activity on carbon-based catalysts such as graphene, carbon nanotubes (CNTs), and graphitic carbon nitride (g-C_3N_4), that contain Earth-abundant non-metallic elements (C, N, H, O, etc) were reported for the water splitting reaction [40]. Among the carbon catalysts, the g-C_3N_4 nanostructure is of much interest and has emerged as an efficient bi-functional catalyst. In addition, the *in situ* doping of N into the carbon lattice with a graphite-like 2D structure has been utilized for large-scale production of hydrogen [41–50]. However, the g-C_3N_4 nanostructures have some limitations such as poor electrical conductivity and a lack of redox sites that impede water splitting. Such issues of g-C_3N_4 can be overcome by composite fabrication with highly conductive and active counterparts. Some non-noble transition metals/metal oxides (Ni, Fe, Co, W, NiO, Fe_2O_3, Co_2O_4, etc) have been utilized and investigated as bi-functional catalysts for OER and HER in alkaline and acidic conditions [51–54]. Theoretical investigations have determined the energetics of the OER and HER intermediates that can be modulated by multi-metal/metal oxides composites for water splitting applications [55]. Among all the first-row transition metals/metal oxides, nickel (Ni) based multi-metal/metal oxides composite nanostructures are favored due to their superior bi-functional activity for OER and HER. Moreover, the theoretical analysis also shows that the activity of g-C_3N_4 can be tuned with Ni metal/metal oxides for OER/HER [56]. Consequently, tungsten oxide (WO_3), possessing the highest oxidation state and having structurally versatile organization and maintained with carbonaceous materials which showed a good electrocatalytic activity [57–60]. The modification of g-C_3N_4 via the integration of Ni metal/metal oxides and WO_3 nanoparticles supported with a Torray Carbon (TC) substrate is found to show an overall enhanced bi-functional activity in comparison with bare Ni metal under alkaline conditions. The design of such bi-functional catalysts for OER and HER with high activity and stability is widely favored for water splitting applications [61]. In order to develop the composite bi-functional catalyst, the modification of g-C_3N_4 with Ni metal/metal oxide nanoparticles, WO_3 nanoparticles, and TC was carried out using facile hydrothermal and thermal evaporation methods. This electrocatalyst is an excellent bifunctional catalyst for both OER and HER in 0.1 M NaOH aqueous electrolyte. Morphological and elemental analyses confirmed the formation of WO_3@g-C_3N_4@Ni-NiO on TC with multiple oxidation states. The TC@WO_3@g-C_3N_4@Ni-NiO matrix exhibits higher OER and HER activity in 0.1 M NaOH aqueous electrolyte, which is similar to bare Ni metal with low overpotentials of 385 and 450 mV at 10 mV cm^{-2} as shown figures 3.5 and 3.6.

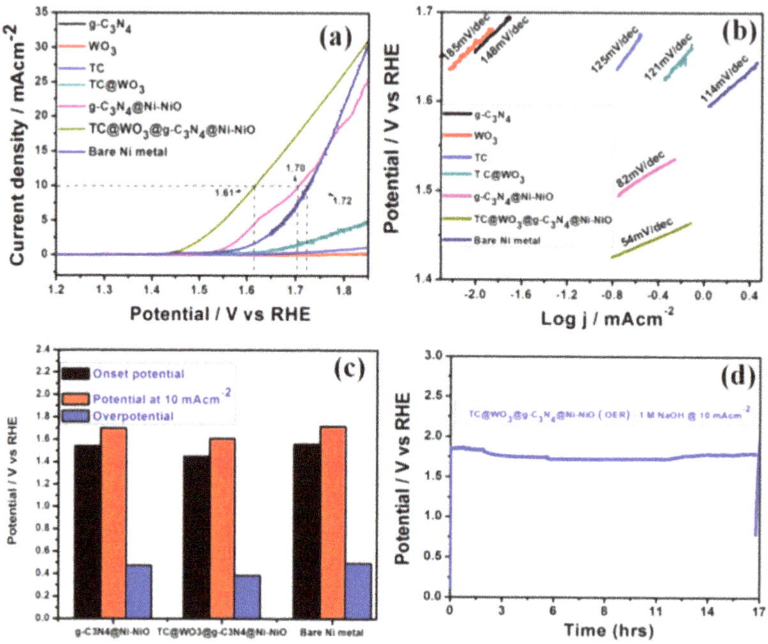

Figure 3.5. (a) LSV curve of g-C_3N_4, WO_3, TC, TC@WO_3, g-C_3N_4@Ni-NiO matrix, and TC@WO_3@g-C_3N_4@Ni-NiO matrix and bare Ni metal in 0.1 M NaOH aqueous electrolyte with a scan rate of 10 mA s^{-1} for OER. (b) Tafel plots for g-C_3N_4, WO_3, TC, TC@WO_3, g-C_3N_4@Ni-NiO matrix, and TC@WO_3@g-C_3N_4@Ni-NiO matrix, respectively. (c) Onset potential, potential required to reach 10 mA cm^{-2} current density, and overpotential of OER catalyzed by g-C_3N_4@Ni-NiO matrix, TC@WO_3@g-C_3N_4@Ni-NiO matrix, and bare Ni metal. (d) Stability performance of TC@WO_3@g-C_3N_4@Ni-NiO matrix for OER. (Reproduced with permission from [61]. Copyright 2017 Wiley.)

Nanostructured carbon-based hybrids, in particular Ni-based metal/metal oxides with nitrogen doped carbon materials, have high charge polarization and asymmetric electron spin density, and are currently considered to be promising OER electrocatalysts due to their tunable electronic structure and high electrical conductivity [62]. The application of Ni-based metals/metal oxides with nitrogen doped nanostructure carbons in OER electrocatalysis is a most excellent approach and their electrocatalytic activity is acceptable. To develop the overall water oxidation activity, more efforts should be concerned with the active sites of Ni-based metal/metal oxides with nitrogen doped carbon materials for the OER reaction [62–64].

The Cu-Ni composite metal/metal oxide electrocatalysts have potential and are attracting significant interest for OER activities with with their convenient structure and active sites in alkaline media because of the sympathetic electronic coordination of CuO [65]. Further, the CuO encourages numerous active sites in the Ni atomic lattice to improve the electrochemical activity of Ni in alkaline media for water oxidation processes [66]. A large number of reports in the literature report important mechanisms in the structure and active site of Ni composites in the water oxidation

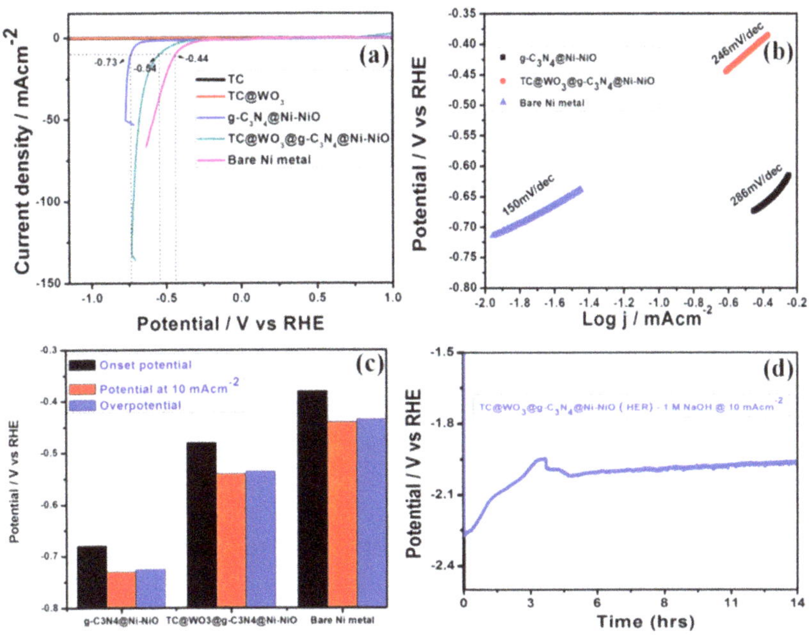

Figure 3.6. (a) HER studies of TC, TC@WO_3, g-C_3N_4@Ni-NiO matrix, TC@WO_3@g-C_3N_4@Ni-NiO matrix, and bare Ni metal in 0.1 M NaOH aqueous electrolyte with a scan rate of 10 mA s^{-1}. (b) Tafel plots and (c) comparison of onset potential, potential required reaching to 10 mA cm^{-2} current density and overpotential for g-C_3N_4@Ni-NiO matrix, TC@WO_3@g-C_3N_4@Ni-NiO matrix, and bare Ni metal. (d) Stability test of TC@WO_3@g-C_3N_4@Ni-NiO matrix for HER. (Reproduced with permission from [61]. Copyright 2017 Wiley.)

process [67, 68]. Recently, Pradhan *et al* established the classification of highly active sides in a Ni doped CuS nanocrystal electrocatalyst for an alkaline water oxidation process [66]. They clearly elucidated that the local electronic environment and structure modulations were obtained in the Ni lattice using CuS nanomaterial. Thus, current research is focusing on the development of CuO (nanosheets, nanoparticles, nanoplates, etc) composites with NiO for water splitting applications in alkaline media [69–71]. From this result, it is apparent that Cu based nanomaterials can support the enhancement of the electrocatalytic activity of Ni for OER in alkaline media. The Ni-based nanocomposite materials have been developed by different methods, such as hydrothermal methods [72], electrochemical deposition [73–75], simple chemical precipitation methods [76], thin film coating by chemical vapor deposition [77–79], and high temperature annealing processes [79, 80]. Among these methods, the simple chemical precipitation method allows the control of the growth of the nucleation process as well as tuning the structure and morphology, producing high purity materials [76].

A spike-like CuO–NiO nanocomposite on a toray carbon sheet has been reported, produced using a simple chemical method, where copper acetate and nickel chloride are used as the precursor materials [81]. The morphology and structure of the prepared nanomaterials were investigated using scanning electron microscopy (SEM) and high-resolution transmission electron microscopy (TEM), and the chemical compositions were observed using x-ray photoelectron spectroscopy (XPS). The fabricated CuO–NiO nanocomposite displayed an extraordinary OER performance and was highly stable in 1 M KOH electrolyte solution for the water splitting process. This electrocatalyst exhibited a lower overpotential of 1.37 V for OER applications. Additionally, OER stability was reported for 2 h in 1 M KOH electrolyte solution.

Similarly, copper oxide (Cu_2O) shows less OER activity in alkaline media [82, 83]. However, it is extremely stable in an alkaline medium, and can support improvements in OER activity through compositing with other suitable metal oxides [82, 83]. SnO_2 is recognized as an efficient component that makes it much easier to form the composites and is receiving much attention in current water splitting research activity [84, 85]. Hence, SnO_2 can be composited with Cu_2O to improve OER activity in alkaline media. Further, the Cu_2O-SnO_2 composite can be prepared using several methods, such as the solvothermal method, thermal treatment method, super critical method, electrodeposition method, and simple chemical methods.

The defect rich titania nanostructure is one of the best non-noble electrocatalysts, next to platinum, for electrochemical studies, and it has good electronic conductivity and is highly stable in both alkaline and acidic media [86–90]. The defect rich titania nanostructure is subjected to anodization [91, 92] and immediate cathodization [93, 94] in the acidic electrolyte (H_2SO_4) for several periods of time. In this process, oxygen vacancies are formed due to the lattice-strain in the titania nanostructure. The HER is observed in an acidic medium. The defect rich sites are created due to the oxygen vacancies formed, which structure the active sites for the HER.

A viable electrochemical cathodization method has been presented to design a highly efficient HER electrocatalyst using the defect rich titania nanostructure with outstanding features in terms of onset potential, Tafel slope, and stability. The developed electrocatalyst provided a lower onset potential of 0.17 V versus RHE and a current density of 150 mA cm^{-2} to drive a HER. Furthermore, the defect rich TiO_2 has revealed the HER activity in 1 M sulfuric acid by observing the Tafel slope of 0.0938 mV dec^{-1}, which is considerable. In addition, stability tests carried out on full cell characterizations showed that the defect rich TiO_2 is a tremendously stable electrocatalytic electrode material for HER in highly acidic conditions. This is mainly attributed to the high corrosion resistance. The promising and extremely optimistic results of the overall study suggest that defect rich TiO_2 can be chosen as a cost-efficient and stable electrocatalytic electrode material for bulk water electrolysis in the hydrogen generation. It can replace expensive and rare materials such as Pt, Ir, and Ru, which used in current water electrolyzers, to reduce the cost of hydrogen production [95].

Experimental and theoretical studies have revealed that N- and S-co-doped carbons are promising for enhanced OER performance [96–98]. A few systematic studies have focused on carbons with heteroatom-based catalysts for the OER. Consequently, most of the literature reported were mainly worked on the design of single or co-doped heteroatoms to attain the better effect on OER performance. The rationally designed heteroatom-doped carbons were analyzed for their combined effect on the production of electrocatalyst active sites, defect sites, and charge transportation. It is considered that controlled N and S contents along with hierarchical porous carbons enhance OER efficiency [96].

Sasaki *et al* [99] reported the synthesis of a superlattice of alternately stacked Ni–Fe hydroxide nanosheets and graphene using a homogeneous precipitation method with hexamethylenetetramine (HMT, $C_6H_{12}N_4$) and anthraquinone-2-sulfonate (AQS) anions ($C_{14}H_7O_5S\_$) as the source of $Ni_{2/3}Fe_{1/3}$-rGO superlattice composites. Figure 3.7 shows the x-ray diffraction (XRD) pattern for $Ni_{2/3}Fe_{1/3}$-GO and $Ni_{2/3}Fe_{1/3}$-rGO superlattice composites. The results confirmed that the graphene sheet was produced as layered double hydroxide (LDH) nanosheets without disturbing the crystallinity. TEM, high-resolution TEM (HRTEM), and selected area electron diffraction (SAED) patterns showed the presence of graphene in a Ni–Fe LDH superlattice composite, as shown figures 3.7(b) and 3.7(c). The electrocatalytic activity of the $Ni_{2/3}Fe_{1/3}$-rGO superlattice composites was studied for the OER reaction, which exhibited the best performance with a small overpotential of 0.21 V and a Tafel slope of 40 mV/decade, as shown figures 3.8(a) and 3.8(b). Furthermore, this electrocatalyst exhibited high stability for 3600 sec (figures 3.9(a) and 3.9(b)). This is higher activity than the LDH nanosheets alone. Further, the catalytic activity of Ni–Fe LDH catalyst can be enhanced with the increase of Fe content, which indicates the significant role of Fe in the bimetallic catalysts. Hence, the surface exposure of all transition metal atoms is very suitable to speed up the OER reaction due to a fast electron transfer process and rapid diffusion rate of the ions. Accordingly, the preparation of a hetero-assembly of hydroxide nanosheets and graphene has been a new approach to improve the OER activity in an alkaline medium. Graphene is a heterostructured sheet which has a high specific surface area

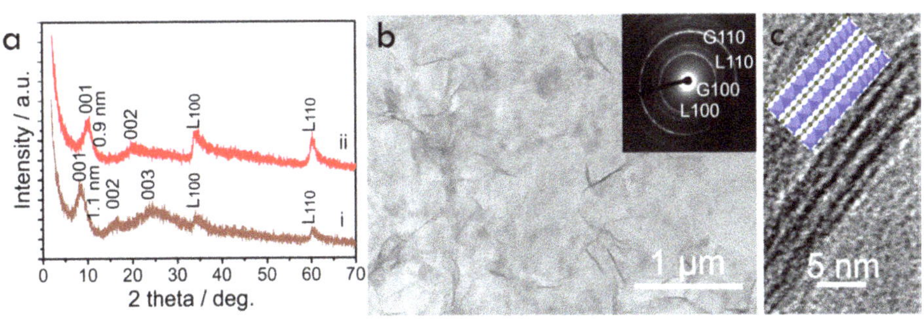

Figure 3.7. (a) The XRD patterns of superlattice composites of $Ni_{2/3}Fe_{1/3}$-NS and (i) GO and (ii) rGO. In the plots, the in-plane diffraction peaks are from the LDH nanosheet for L100 and L110. (b) TEM image of $Ni_{2/3}Fe_{1/3}$-NS and GO composites (inset) following the selected area diffraction pattern, which are, in order, the in-plane reflections of LDH for L100, L110, and graphite (G100, G110), respectively. (c) HRTEM and a schematic reveal the alternately stacked LDH nanosheets and graphene. (Reproduced with permission from [99]. Copyright 2015 American Chemical Society.)

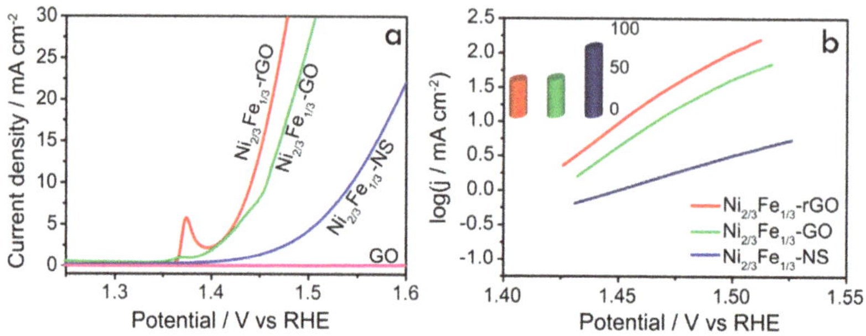

Figure 3.8. IR corrected (a) LSV study and (b) Tafel plots of $Ni_{2/3}Fe_{1/3}$-rGO, $Ni_{2/3}Fe_{1/3}$-GO, $Ni_{3/4}Fe_{1/4}$-rGO, and $Ni_{3/4}Fe_{1/4}$-GO (inset: histogram of corresponding Tafel slopes). (Reproduced with permission from [99]. Copyright 2015 American Chemical Society.)

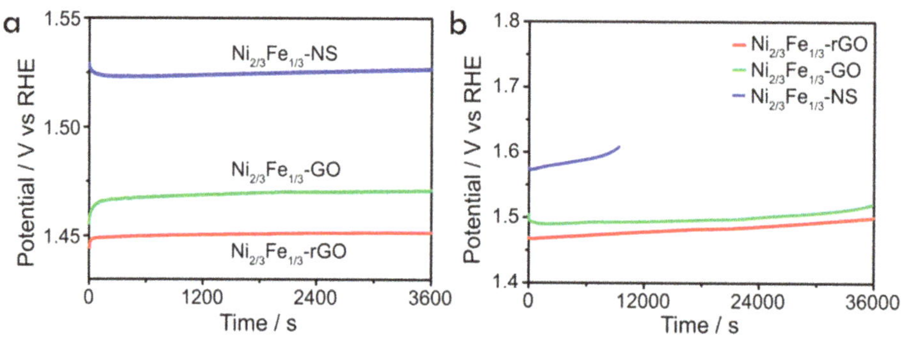

Figure 3.9. Chronopotentiometry study of $Ni_{2/3}Fe_{1/3}$-rGO, $Ni_{2/3}Fe_{1/3}$-GO, and $Ni_{2/3}Fe_{1/3}$-NS at a constant current density of (a) 5 mA cm^{-2} and (b) 10 mA cm^{-2}, respectively. (Reproduced with permission from [99]. Copyright 2015 American Chemical Society.)

of 2600 m^2 g^{-1} and good electrical conductivity of ~10^6 S cm^{-1} (theoretical value). It may provide an electron carrier lane to increase charge transfer and mass transport throughout the OER reaction in the alkaline electrolyte solution. In particular, the electrostatic stacking contact between positively charged LDH nanosheets and negatively charged graphene (oxide) at the molecular scale, which permits direct interfacial contact between the 3D transition metals and carbon. Likewise, this interaction can notably restrict the diffusion distance.

Polyvinylpyrrolidone (PVP)-assisted nanomaterial preparation was achieved by utilizing a controlled solvothermal method to fabricate suitable electrode materials for electrochemical water splitting processes [100]. Bi_2WO_6 and $FeWO_4$ nanomaterials have been produced using the well-established signature of the (113) and (111) crystal planes, respectively. The self-supporting nanoplate array surface morphology of Bi_2WO_6 and the well-ordered nanosphere particles of $FeWO_4$ nanomaterials were examined using SEM images. Using these results, the preparation of highly stable and long-term electrodes for electrochemical water splitting has been achieved for the efficient oxidation of water. Electrochemical studies confirm this, obtaining 2.79 and 1.96 mA g^{-1} for 0.5 g PVP-assisted Bi_2WO_6 and $FeWO_4$ nanomaterials, respectively. The water oxidation process mechanisms of both nanomaterials have been found to be highly stable for 24 h for continuous water oxidation and faster charge transfer. Smaller Tafel slope values of 75 and 78 mV dec^{-1} have been achieved for Bi_2WO_6 and $FeWO_4$ nanomaterials, respectively. Usually, these nanomaterials are applied for photoelectrochemical water splitting applications. The current investigation discovered the PVP-assisted synthesis to construct electrocatalytically active nanomaterials and their electrochemical water splitting mechanism, which will tender a lane for research benefit with observe to the production of multi-functional nanomaterials for both electro- and photoelectrochemical catalytic applications in the future.

Members of Hou's group introduced a new hybrid of Co_3O_4 embedded in tubular nanostructures of graphitic carbon nitride (GCN) [101]. They prepared this via a simple, large-scale chemical method at low temperature. The well-established synergistic property between Co_3O_4 and GCN resulted in outstanding performance as a bifunctional electrocatalyst for overall water splitting (OER and HER). Bifunctional electrocatalysis benefits from the high surface area, unique tubular nanostructure, and composition of the hybrid, which makes all redox sites easily available, and supplies faster ionic and electronic conduction. The Co_3O_4@GCN tubular nanostructured (TNS) hybrid displayed a lower overpotential (0.12 V) and tremendous current density (147 mA cm^{-2}) in OER with high stability in an alkaline medium, better than the standard electrocatalyst of IrO_2 and RuO_2. In addition, the Co_3O_4@GCN TNS hybrid showed tremendous performance with the lowest onset and overpotential in HER, and a high current density. It is predictable that the Co_3O_4@GCN TNS hybrid designed in this study will be a good alternative to precious metal electrocatalysts in large-scale water splitting and fuel cell applications.

The growth of a NiO/Ni catalyst directly on a carbon fiber paper sheet was achieved by atomic layer deposition and subsequent reduction and oxidation annealing treatments were reported [102]. The 10–20 nm NiO/Ni nanoparticle as a bi-functional electrocatalyst could reach a current density of 10 mA cm^{-2} at an overpotential of 189 mV for HER and 257 mV for OER with high stability in an alkaline electrolyte solution. This bi-functional electrocatalyst was used in a typical two-electrode set-up for water splitting, which showed the same performance with a current density of 10 mA·cm^{-2} at 1.78 V. The results suggest that nanoparticulate NiO/Ni is as an active, stable, and noble-metal-free bi-functional electrocatalyst, which makes it a possible method for future water splitting processes.

Electrocatalysts were prepared easily by pyrolyzing a source composed of nickel and iron salts with urea under inert atmospheres without any post-treatment [103]. The synthesized nanomaterials were composed of metallic nickel–iron alloy nanoparticles and are isolated on nitrogen doped bamboo-like carbon nanotubes (CNTs). This facile preparation method can simultaneously achieve nanostructuring, doping, and hybridizing with nanocarbon, which establishes it as an efficient approach to optimize the electrocatalytic activity of bi-functional electrocatalysts. The *in situ* created hybrid bi-functional electrocatalyst showed excellent catalytic performance for overall water splitting (OER and HER) in alkaline solutions, and the doping of iron considerably influenced the activities. A symmetric two-electrode cell could carry out overall water splitting at a current density of 10 mA cm^{-2} with an onset potential of 1.58 V, and shows insignificant degradation after 24 h of operation, when the best bifunctional electrocatalyst is loaded on nickel foam with a loading of 2 mg cm^{-2}. The exceptional electrocatalytic activity and simple synthesis route facilitate this hybrid bi-functional electrocatalyst to be a promising candidate for future large-scale overall water splitting applications.

From literature reports, the topotactic conversion of $NiCo_2O_4$ nanowire arrays on carbon cloth ($NiCo_2O_4$ NA/CC) into $NiCo_2S_4$ NA/CC was achieved via a facile route, and it was used as a bi-functional electrocatalyst for efficient water splitting with highly stable and excellent electrocatalytic activity in 1.0 M KOH electrolyte solution [104]. This bi-functional $NiCo_2S_4$ NA/CC electrode generates an HER current density of 100 mA cm^{-2} at an overpotential of 305 mV, and 100 mA cm^{-2} at an overpotential of 340 mV for OER. To give a 10 mA cm^{-2} overall water splitting current, the alkaline water electrolyzer constructed from $NiCo_2S_4$ NA/CC requires a cell voltage of 1.68 V with high stability.

A new electrochemical bifunctional electrocatalyst consisting of CoO@Co nanoparticles with a core–shell structure immobilized on N-doped reduced graphene oxide (rGO) (CoO@Co/N-rGO) was prepared via a convenient and controllable technique combining a rapid microwave-polyol route with a vacuum thermal treatment [105]. The outstanding surface appearance with the Co/N-doping effect, the introduction of CoO@Co particles with a core–shell structure, and excellent contact between the CoO@Co particles and the N-doped graphene result in highly multi-functional and efficient electrocatalytic activity. The bi-functional electrocatalyst shows extraordinary electrocatalytic activity and great stability towards the HER with a lower overpotential of 140 mV at 10 mA cm^{-2} in 0.5 M H_2SO_4 and 237 mV in 0.1 M KOH. Additionally, excellent activity for the OER was observed for the same catalyst, which exhibits a current density of 10 mA cm^{-2} with a low overpotential of 1.67 V in 0.1 M KOH, which is like that of a commercial RuO_2 electrocatalyst. Extraordinary electrocatalytic activity occurs from the synergetic chemical coupling effects of metallic Co, cobalt oxides, and Co/N-doped graphene. This investigation provided a new smart multi-functional electrocatalyst material for future regenerative fuel cells and overall water splitting applications.

The non-noble metals/metal oxides with some carbon-based nanomaterials are popular materials with improved electrocatalytic potential in several applications. The exclusive physicochemical properties (shape/size selectivity, tunable solid acidity, large surface area, and high conductivity) of metal/metal oxide-based composites could provide overall water splitting for hydrogen generation in acidic and alkaline media.

3.4 Conclusion

In conclusion, the global demand for energy is increasing, and the major source of energy remains the Earth's reserves of fossil fuels such as oil, coal, and natural gas. However, the commonly used steam reforming of natural gas and water splitting are emerging as promising solutions to generate hydrogen toward carbon-free and environmentally friendly green energy. Water is a renewable energy sources which is abundant in nature, and it is used to produce the hydrogen. The hydrogen can be harvested by splitting or breaking of water molecules into hydrogen and oxygen using the water splitting process. Currently, the water electrolysis process accounts for 4% of the world's hydrogen generation. To improve the hydrogen generation efficiency, catalytic water splitting is a highlight of current investigations on various areas such as electrocatalysis, the thermo-chemical cycle, and enzymatic reactions. The varieties of sustainable hydrogen generation pathways are not only limited to thermolysis and electrolysis, but also focus on their combination, including the thermo-electrolysis of water. In this chapter we have reviewed hydrogen production using electrochemical water splitting routes with special emphasis on their advantages and limitations. Looking at prominent electrochemical water splitting more generally, Earth-abundant HER and OER catalysts that operate at very high and very low pH will certainly also be of great utility (for example, to produce cheaper conventional electrolyzers that run using other forms of renewable energy). This chapter also presents various designs and development of different nanomaterials as electrocatalysts using simple chemical and electrochemical methods for water oxidation processes. Nanomaterials were investigated for their electrochemical and photoelectrochemical activities in alkaline and acid media using linear sweep voltammetry (LSV), chronoamperometry, and impedance spectroscopy.

Acknowledgments

The authors MPK and RVM duly acknowledge the FONDECYT postdoctoral project No. 3220475, Government of Chile and Universidad Adolfo Ibanez, Santiago, Chile, for financial support.

References

[1] Perera F 2018 Pollution from fossil-fuel combustion is the leading environmental threat to global pediatric health and equity: solutions exist *Int. J. Environ. Res. Public Health* **15** 16

[2] Rehman S, Hadhrami L M A and Alam M M 2015 Pumped hydro energy storage system: a technological review *Renew. Sustain. Energy Rev.* **44** 586–98

[3] Yu Q, Wang Q, Tan X, Fang G and Meng J 2019 A review of compressed-air energy storage *J. Renew. Sustain. Energy* **11** 042702–12

[4] Sarbu I and Sebarchievici C 2018 A comprehensive review of thermal energy storage *Sustainability* **10** 191–224

[5] Hesse H C, Schimpe M, Kucevic D and Jossen A 2017 Lithium-ion battery storage for the grid—a review of stationary battery storage system design tailored for applications in modern power grids *Energies* **10** 2107–49

[6] Mubarak S, Dhamodharan D, Ghoderao P N P and Byun H S 2022 A systematic review on recent advances of metal–organic frameworks-based nanomaterials for electrochemical energy storage and conversion *Coord. Chem. Rev.* **471** 214741
[7] Dhamodharan D, Ghoderao P P, Dhinakaran V, Mubarak S, Divakaran N and Byun H S 2021 A review on graphene oxide effect in energy storage devices *J. Ind. Eng. Chem.* **106** 20–36
[8] The Royal Society 2018 Options for producing low-carbon hydrogen at scale *The Royal Society* https://royalsociety.org/topics-policy/projects/low-carbon-energy-programme/
[9] Hosseini S E and Wahid M A 2016 Hydrogen production from renewable and sustainable energy resources: promising green energy carrier for clean development *Renew. Sustain. Energy Rev.* **57** 850–66
[10] Mubarak S, Dhamodharan D, Byun H S, Arya S B and Pattanayak D K 2022 Effective photoelectrocatalytic reduction of CO_2 to formic acid using controllably annealed TiO_2 nanoparticles derived from porous structured Ti foil *J. CO_2 Util.* **63** 102152
[11] Arulraj A, Thachnatharen N, Rajkumar C and Mangalaraja R V 2022 Engineering 2D chalcogenides for energy and environmental remediation *Nanostructured Mater. Sustain. Energy Environ. Remediat.* (Bristol: IOP Publishing) pp 2-1–2-28
[12] Vineesh T V, Mubarak S, Hahm M G, Prabu V, Alwarappan S and Narayanan T N 2016 Controllably alloyed, low density, free-standing Ni-Co and Ni-graphene sponges for electrocatalytic water splitting *Sci. Rep.* **6** 31202
[13] Dufour J, Serrano D P, Galvez J L, Gonzalez A, Soria E and Fierro J L G 2012 Life cycle assessment of alternatives for hydrogen production from renewable and fossil sources *Int. J. Hydrogen Energy* **37** 1173–83
[14] Ramachandran R and Menont R K 1998 An overview of industrial uses of hydrogen *Int. J. Hydrogen Energy* **23** 593–8
[15] Silvera I F and Cole J W 2010 Metallic hydrogen: the most powerful rocket fuel yet to exist *J. Phys. Conf. Ser.* **215** 012194
[16] Zhang L, Wen X, Kong T, Zhang L, Gao L, Miao L and Li Y H 2019 Preparation and mechanism research of Ni-Co supported catalyst on hydrogen production from coal pyrolysis *Sci. Rep.* **9** 9818
[17] Melis A and Happe T 2001 Hydrogen production. Green algae as a source of energy *Plant Physiol.* **127** 740–8
[18] Vagia E C and Lemonidou A A 2008 Hydrogen production via steam reforming of bio-oil components over calcium aluminate supported nickel and noble metal catalysts *Appl. Catal.* A **351** 111–21
[19] Sun Y, Nakano J, Liu L, Wang X and Zhang Z 2015 Achieving waste to energy through sewage sludge gasification using hot slags: syngas production *Sci. Rep.* **5** 11436
[20] Buscema M, Island J O, Groenendijk D J, Blanter S I, Steele G A, van der Zant H S J and Castellanos-Gomez A 2015 Photocurrent generation with two-dimensional van der Waals semiconductors *Chem. Soc. Rev.* **44** 3691–718
[21] Naimi Y and Antar A 2018 Hydrogen generation by water electrolysis *Advances In Hydrogen Generation Technologies* ed M Eyvaz (London: InTech Open)
[22] Wang Z, Liu H, Ge R, Ren X, Ren J, Yang D, Zhang L and Sun X 2018 Phosphorus-doped Co_3O_4 nanowire array: a highly efficient bifunctional electrocatalyst for overall water splitting *ACS Catal.* **8** 2236–41

[23] Smith R D L, Prévot M S, Fagan R D, Zhang Z, Sedach P A, Siu M K J, Trudel S and Berlinguette C P 2013 Photochemical route for accessing amorphous metal oxide materials for water oxidation catalysis *Science* **340** 60–3

[24] Arulraj A, Praveen Kumar M, Alejandra T Z and Mangalaraja R V 2023 Nanoarchitectonics of layered metal chalcogenides-based ternary electrocatalyst for water splitting *Energies* **16** 1669

[25] Trotochaud L, Young S L, Ranney J K and Boettcher S W 2014 Nickel–iron oxyhydroxide oxygen-evolution electrocatalysts: the role of intentional and incidental iron incorporation *J. Am. Chem. Soc.* **136** 6744–53

[26] Koza J A, He Z, Miller A S and Switze J A 2012 Electrodeposition of crystalline Co_3O_4—a catalyst for the oxygen evolution reaction *Chem. Mater.* **24** 3567

[27] Praveen Kumar M, Sasikumar M, Arulraj A, Rajasudha V, Murugadoss G, Rajesh Kumar M, Gouse Peera S and Mangalaraja R V 2022 NiFe layered double hydroxide electrocatalyst prepared via an electrochemical deposition method for the oxygen evolution reaction *Energies* **12** 1470

[28] Park H W, Lee D U, Zamani P, Seo M H, Nazar L F and Chen Z 2014 Electrospun porous nanorod perovskite oxide/nitrogen-doped graphene composite as a bi-functional catalyst for metal air batteries *Nano Energy* **10** 192–200

[29] Haber J A, Cai Y, Jung S, Xiang C, Mitrovic S, Jin J, Bellbd A T and Gregoire J M 2014 Discovering Ce-rich oxygen evolution catalysts, from high throughput screening to water electrolysis *Energy Environ. Sci.* **7** 682–8

[30] Haber J A, Anzenburg E, Yano J, Kisielowski C and Gregoire J M 2015 Multiphase nanostructure of a quinary metal oxide electrocatalyst reveals a new direction for OER electrocatalyst design *Adv. Energy Mater.* **5** 1402307–117

[31] Hu W, Wang Y, Hu X, Zhou Y and Chen S 2012 Three-dimensional ordered macroporous IrO_2 as electrocatalyst for oxygen evolution reaction in acidic medium *J. Mater. Chem.* **22** 6010–6

[32] Lewerenz H J, Stucki S and Kötz R 1983 Oxygen evolution and corrosion: XPS investigation on Ru and RuO_2 electrodes *Surf. Sci. Lett.* **126** 463–8

[33] Xu H, Ci S, Ding Y, Wang G and Wen Z 2019 Recent advances in precious metal-free bifunctional catalysts for electrochemical conversion systems *J. Mater. Chem.* A **7** 8006–29

[34] Sarnoa M and Ponticorvo E 2019 Metal–metal oxide nanostructure supported on graphene oxide as a multifunctional electro-catalyst for simultaneous detection of hydrazine and hydroxylamine *Electrochem. Commun.* **107** 106510–7

[35] Li B, Zheng M, Xue H and Pang H 2016 High performance electrochemical capacitor materials focusing on nickel based materials *Inorg. Chem. Front* **3** 175–202

[36] Zheng S, Li X, Yan B, Hu Q, Xu Y, Xiao X, Xue H and Pang H 2017 Transition-metal (Fe, Co, Ni) based metal–organic frameworks for electrochemical energy storage *Adv. Energy Mater.* **7** 1602733

[37] Yu M, Moon G, Bill E and Tuysuz H 2019 Optimizing Ni–Fe oxide electrocatalysts for oxygen evolution reaction by using hard templating as a toolbox *ACS Appl. Energy Mater.* **2** 1199–209

[38] Hassan M H, Soliman A B, Elmehelmey W A, Bugable A A A, Karakalos S G, Elbahri M, Hassanien A and Alkordi M H 2019 A Ni-loaded, metal–organic framework–graphene composite as a precursor for *in situ* electrochemical deposition of a highly active and durable water oxidation nanocatalyst *Chem. Commun.* **55** 31–4

[39] Louie M W C and Bell A T 2013 An investigation of thin-film Ni–Fe oxide catalysts for the electrochemical evolution of oxygen *J. Am. Chem. Soc.* **135** 12329–37
[40] Kumar M P, Vineesh T V, Takahashi C, Kalita G, Alwarappan S, Pattanayak D K and Narayanan T N 2015 Bifunctional electrocatalytic activity of boron-doped graphene derived from boron carbide *Adv. Energy Mater.* **5** 1500658
[41] Zhao Y, Zhao F, Wang X P, Xu C Y, Zhang Z P, Shi G Q and Qu L T 2014 Graphitic carbon nitride nanoribbons: graphene-assisted formation and synergic function for highly efficient hydrogen evolution *Angew. Chem. Int. Ed.* **53** 13934
[42] Tahir M, Cao C B, Butt F K, Idrees F, Mahmood N, Ali Z, Aslam I, Tanveer M, Rizwan M and Mahmood T 2013 Tubular graphitic-C_3N_4: a prospective material for energy storage and green photocatalysis *J. Mater. Chem.* A **1** 13949
[43] Ma T Y, Dai S, Jaroniec M and Qiao S Z 2014 Graphitic carbon nitride nanosheet–carbon nanotube three-dimensional porous composites as high-performance oxygen evolution electrocatalysts *Angew. Chem. Int. Ed.* **53** 7281
[44] Martin D J, Qiu K P, Shevlin S A, Handoko A D, Chen X W, Guo Z X and Tang J W 2014 Highly efficient photocatalytic H_2 evolution from water using visible light and structure-controlled graphitic carbon nitride *Angew. Chem. Int. Ed.* **53** 9240
[45] Li Q, Yang J P, Feng D, Wu Z X, Wu Q L, Park S S, Ha C S and Zhao D Y 2010 Facile synthesis of porous carbon nitride spheres with hierarchical three-dimensional mesostructures for CO_2 capture *Nano Res.* **3** 632
[46] Han C, Wang Y D, Lei Y P, Wang B, Wu N, Shi Q and Li Q 2015 *In situ* synthesis of graphitic-C_3N_4 nanosheet hybridized N-doped TiO_2 nanofibers for efficient photocatalytic H_2 production and degradation *Nano Res.* **8** 1199
[47] Han Q, Zhao F, Hu C G, Lv L X, Zhang Z P, Chen N and Qu L T 2015 Facile production of ultrathin graphitic carbon nitride nanoplatelets for efficient visible-light water splitting *Nano Res.* **8** 1718
[48] Tahir M, Cao C B, Mahmood N, Butt F K, Mahmood A, Idrees F, Hussain S, Tanveer M, Ali Z and Aslam I 2014 Multifunctional g-C_3N_4 nanofibers: a template-free fabrication and enhanced optical, electrochemical, and photocatalyst properties *ACS Appl. Mater. Interfaces* **6** 1258
[49] Tahir M, Cao C B, Butt F K, Butt S, Idrees F, Ali Z, Aslam I, Tanveer M, Mahmood A and Mahmood N 2014 Large scale production of novel g-C_3N_4 micro strings with high surface area and versatile photodegradation ability *Cryst. Eng. Comm.* **16** 1825
[50] Tahir M, Mahmood N, Zhu J H, Mahmood A, Butt F K, Rizwan S, Aslam I, Tanveer M, Idrees F and Shakir I 2015 One dimensional graphitic carbon nitrides as effective metal-free oxygen reduction catalysts *Sci. Rep.* **5** 12389
[51] Sivanantham A, Ganesan P and Shanmugam S 2016 Hierarchical $NiCo_2S_4$ nanowire arrays supported on Ni foam: an efficient and durable bifunctional electrocatalyst for oxygen and hydrogen evolution reactions *Adv. Funct. Mater.* **26** 4661–72
[52] Li J, Li J, Zhou X, Xia Z, Gao W, Ma Y and Qu Y 2016 Highly efficient and robust nickel phosphides as bifunctional electrocatalysts for overall water-splitting *Appl. Mater. Interfaces* **8** 10826
[53] Liu D, Lu Q, Luo Y, Su X and Asiri A M 2015 $NiCo_2S_4$ nanowires array as an efficient bifunctional electrocatalyst for full water splitting with superior activity *Nanoscale* **7** 15122

[54] Fang W, Liu D, Lu Q, Sun X and Asiri A M 2016 Nickel promoted cobalt disulfide nanowire array supported on carbon cloth: an efficient and stable bifunctional electrocatalyst for full water splitting *Electrochem. Commun.* **63** 60
[55] Zhang B *et al* 2016 Homogeneously dispersed multimetal oxygen-evolving catalysts *Science* **352** 333–7
[56] Gogotsi Y 2014 What nano can do for energy storage *ACS Nano* **8** 5369
[57] Yin Q, Tan J M, Besson C, Geletii Y V, Musaev D G, Kuznetsov A E, Luo Z, Hardcastle K I and Hill C L 2010 A fast soluble carbon-free molecular water oxidation catalyst based on abundant metals *Science* **328** 342
[58] Toma F M *et al* 2010 Efficient water oxidation at carbon nanotube–polyoxometalate electrocatalytic interfaces *Nat. Chem.* **2** 826
[59] Liu Q, Jin J and Zhang J 2013 $NiCo_2S_4$@graphene as a bifunctional electrocatalyst for oxygen reduction and evolution reactions *Appl. Mater. Interfaces* **5** 5002
[60] Zhang J, Xia Z and Dai L 2015 Carbon-based electrocatalysts for advanced energy conversion and storage *Sci. Adv.* **1** e1500564
[61] Praveen Kumar M, Murugesan P, Vivek S and Ravichandran S 2017 Nanoparticles NWO_3 grown on graphitic carbon nitride (g-C_3N_4) supported toray carbon as an efficient bifunctional electrocatalyst for oxygen and hydrogen evolution reactions *Part. Part. Syst. Charact.* **34** 1700043
[62] Hou Y *et al* 2017 Efficient electrochemical and photoelectrochemical water splitting by a 3D nanostructured carbon supported on flexible exfoliated graphene foil *Adv. Mater.* **29** 1604480
[63] Zhang J, Zhao Z, Xia Z and Dai L 2015 A metal-free bifunctional electrocatalyst for oxygen reduction and oxygen evolution reactions *Nat. Nanotechnol.* **10** 444
[64] Fei H *et al* 2018 General synthesis and definitive structural identification of MN_4C_4 single-atom catalysts with tunable electrocatalytic activities *Nat. Catal.* **1** 63–72
[65] Wang H, Lee H W, Deng Y, Lu Z, Hsu P C, Liu Y, Lin D and Cui Y 2015 Bifunctional non-noble metal oxide nanoparticle electrocatalysts through lithium-induced conversion for overall water splitting *Nat. Commun.* **6** 7261
[66] Kundu J, Khilari S, Bhunia K and Pradhan D 2019 Ni-doped CuS as an efficient electrocatalyst for the oxygen evolution reaction *Catal. Sci. Technol.* **9** 406–17
[67] Li Y F and Selloni A 2014 Mechanism and activity of water oxidation on selected surfaces of pure and Fe-doped NiO_x *ACS Catal.* **4** 1148–53
[68] Liao P, Keith J A and Carter E A 2012 Water oxidation on pure and doped hematite (0001) surfaces: prediction of Co and Ni as effective dopants for electrocatalysis *J. Am. Chem. Soc.* **134** 13296–309
[69] Li C, Zhang B, Li Y, Hao S, Cao X, Yang G, Wu J and Huang Y 2019 Self-assembled Cu-Ni bimetal oxide 3D in-plane epitaxial structures for highly efficient oxygen evolution reaction *Appl. Catalysis* B **244** 56–62
[70] Kim J S, Kim B, Kim H and Kang K 2018 Recent progress on multimetal oxide catalystsfor theoxygen evolution reaction *Adv. Energy Mater.* **8** 1702774
[71] Qian M, Liu X, Cui S, Jia H and Du P 2018 Copper oxide nanosheets prepared by molten salt method for efficient electrocatalytic oxygen evolution reaction with low catalyst loading *Electrochim. Acta* **263** 318–27

[72] Zhou J P, Lv L, Liu Q, Zhang Y X and Liu P 2012 Hydrothermal synthesis and properties of $NiFe_2O_4$@$BaTiO_3$ composites with well-matched interface *Sci. Technol. Adv. Mater.* **13** 045001

[73] Merrill M D and Dougherty R C 2008 Metal oxide catalysts for the evolution of O_2 from H_2O *J. Phys. Chem.* C **112** 3655–66

[74] Kim K H, Zheng J Y, Shin W and Kang Y S 2012 Preparation of dendritic NiFe films by electrodeposition for oxygen evolution *RSC Adv.* **2** 4759–67

[75] Gangasingh D and Talbot J B 1991 Anomalous electrodeposition of nickel–iron *J. Electrochem. Soc.* **138** 3605–11

[76] Diva T N, Zare K, Taleshi F and Yousefi M 2017 Synthesis, characterization, and application of nickel oxide/CNT nanocomposites to remove Pb^{2+} from aqueous solution *J. Nanostruct Chem.* **7** 273–81

[77] Kang J X, Zhao W Z and Zhang G F 2009 Influence of electrodeposition parameters on the deposition rate and microhardness of nanocrystalline Ni coatings *Surf. Coat. Technol.* **203** 1815–8

[78] Sahoo P and Das S K 2011 Tribology of electroless nickel coatings—a review *Mater. Design* **32** 1760–75

[79] Fanicchia F, Axinte D, Kell J, McIntyre R, Brewster G and Norton A D 2017 Combustion flame spray of CoNiCrAlY& YSZ coatings *Surf. Coat. Technol.* **315** 546–57

[80] Li B Q, Tang C, Wang H F, Zhu X L and Zhang Q 2016 An aqueous preoxidation method for monolithic perovskite electrocatalysts with enhanced water oxidation performance *Sci. Adv.* **2** e1600495

[81] Praveen Kumar M, Murugadoss G and Rajesh Kumar M 2020 Synthesis and characterization of CuO–NiO nanocomposite: highly active electrocatalyst for oxygen evolution reaction application *J. Mater. Sci., Mater. Electron.* **31** 11286–94

[82] Travitzky N, Kumar P, Sandhage K H, Janssen R and Claussen N 2003 Rapid synthesis of Al_2O_3 reinforced Fe–Cr–Ni composites *Mater. Sci. Eng.* A **344** 245–52

[83] Wang H, Lee H W, Deng Y, Lu Z, Hsu P C, Liu Y, Lin D and Cui Y 2015 Bifunctional non-noble metal oxide nanoparticle electrocatalysts through lithium-induced conversion for overall water splitting *Nat. Commun.* **6** 7261–8

[84] Kundu J, Khilari S, Bhunia K and Pradhan D 2019 Ni-doped CuS as an efficient electrocatalyst for the oxygen evolution reaction *Catal. Sci. Technol.* **9** 406–17

[85] Hutchings R, Müller K and Stucki S 1984 A structural investigation of stabilized oxygen evolution catalysts *J. Mater. Sci.* **19** 3987–94

[86] Park M S, Kang Y M, Wang G X, Dou S X and Liu H K 2008 The effect of morphological modification on the electrochemical properties of SnO_2 nanomaterials *Adv. Funct. Mater.* **18** 455–61

[87] Wang R, Hashimoto K, Fujishima A, Chikuni M, Kojima E and Kitamura A 1997 Light-induced amphiphilic surfaces *Nature* **388** 431–2

[88] Su Z and Zhou W 2009 Formation, microstructures and crystallization of anodic titanium oxide tubular arrays *J. Mater. Chem.* **19** 2301–9

[89] Allam N K, Shankar K and Grimes C A 2008 A general method for the anodic formation of crystalline metal oxide nanotube arrays without the use of thermal annealing *Adv. Mater.* **20** 3942–6

[90] Feng H *et al* 2018 Activating titania for efficient electrocatalysis by vacancy engineering *ACS Catal.* **8** 4288–93

[91] Robin A, Ribeiro de A, Rosa M B, Nakazato J L and Silva R Z 2014 Formation of TiO_2 nanotube layer by anodization of titanium in ethylene glycol–H_2O electrolyte *J. Surf. Eng. Mater. Adv. Technol* **4** 123–30
[92] Kavan L, Stoto T, Graetzel M, Fitzmaurice D and Shklover V 1993 Quantum size effects in nanocrystalline semiconducting titania layers prepared by anodic oxidative hydrolysis of titanium trichloride *J. Phy. Chem.* **97** 9493–8
[93] Swaminathan J, Ravichandran S and Vengatesan S 2016 Defect-rich metallic titania ($TiO_{1.23}$)—an efficient hydrogen evolution catalyst for electrochemical water splitting *ACS Catal.* **6** 2222–9
[94] Swaminathan J and Ravichandran S 2018 Insights into the electrocatalytic behavior of defect-centered reduced titania ($TiO_{1.23}$) *J. Phys. Chem.* C **122** 1670–80
[95] Praveen Kumar M, Murugadoss b G, Mangalaraja R V, Arunachalam P, Rajesh Kumar M, Hartley U W, Salla S, Rajabathar J, ALOthman Z A and ALTalhi T 2021 Design and development of defect rich titania nanostructure for efficient electrocatalyst for hydrogen evolution reaction in an acidic electrolyte *J. Mater. Res. Technol.* **14** 2739–50
[96] Zhou J, Wang Z, Yang D, Zhang W, Chen Y and Freestanding S 2019 N co-doped graphene/Ni foam as highly efficient and stable electrocatalyst for oxygen evolution reaction *Electrochim. Acta* **317** 408–15
[97] Murdachaew G and Laasonen K 2018 Oxygen evolution reaction on nitrogen-doped defective carbon nanotubes and graphene *J. Phys. Chem.* C **122** 25882–92
[98] Kong X, Liu Q, Chen D and Chen G 2017 Identifying the active sites on N-doped graphene toward oxygen evolution reaction *Chem. Cat. Chem.* **9** 846–52
[99] Ma W, Ma R, Wang C, Liang J, Liu X, Zhou K and Sasaki T 2015 A superlattice of alternately stacked Ni–Fe hydroxide nanosheets and graphene for efficient splitting of water *ACS Nano* **9** 1977–84
[100] Jansi Rani B, Ravi G, Yuvakkumar R, Praveenkumar M, Ravichandran S, Mareeswaran P M and Hong S I 2019 Bi_2WO_6 and $FeWO_4$ nanocatalysts for the electrochemical water oxidation process *ACS Omega* **4** 5241–53
[101] Tahir M *et al* 2015 Bifunctional catalysts of Co_3O_4@GCN tubular nanostructured (TNS) hybrids for oxygen and hydrogen evolution reactions *Nano Res.* **8** 3725–36
[102] Zhang R, Wei H, Si W, Ou G, Zhao C, Song M, Zhang C and Wu H 2017 Enhanced electrocatalytic activity for water splitting on NiO/Ni/carbon fiber paper *Materials* **10** 15
[103] Zhang X, Xu H, Li X, Li Y, Yang T and Liang Y 2016 Facile synthesis of nickel–iron/nanocarbon hybrids as advanced electrocatalysts for efficient water splitting *ACS Catal.* **6** 580–8
[104] Liu D, Lu Q, Luo Y, Sun X and Asiri A M 2015 $NiCo_2S_4$ nanowires array as an efficient bifunctional electrocatalyst for full water splitting with superior activity *Nanoscale* **7** 15122–6
[105] Liu X X, Zang J B, Chen L, Chen L B, Chen X, Wu P, Zhoua S Y and Wang Y H 2017 A microwave-assisted synthesis of CoO@Co core–shell structures coupled with N-doped reduced graphene oxide used as a superior multi-functional electrocatalyst for hydrogen evolution, oxygen reduction and oxygen evolution reactions *J. Mater. Chem.* A **5** 5865–72

Chapter 4

The state-of-the-art of solid sorbents for carbon capture applications: opportunities and future perspectives

Venkadeshkumar Ramar, N B Karthik Babu and Karthikeyan Balasubramanian

Carbon dioxide (CO_2) has been recognized to be the biggest driver of global warming and climate change, which is one of the most serious problems that our world is facing at present. Therefore, carbon capture and storage (CCS) techniques have received wide attention from the research community. For an effective CO_2 capture process, several features of the CO_2 capture stage must be considered carefully. Currently, the widely used CO_2 capture technology is liquid amine scrubbing. This technology is associated with some disadvantages, such as solvent loss, its corrosive nature, and energy costs. CO_2 capture using solid sorbents has been accepted as a promising alternative method that has attracted great attention from industrial and academic researchers in the last decade. Nanomaterials, in particular, have been considered as a favorable adsorbents for carbon capture owing to their wide availability, low cost, and high adsorption capacity. This chapter mentions the key technologies used in CO_2 capture, highlights the recent research status of nanomaterials for CO_2 capture, and explores the impact of the surface modification and microstructure of materials on CO_2 capture capacity. We focus on the recent advances in developing synthetic sorbents, and the correlation between their design and their cyclic CO_2 capture performance, which is described by cyclic stability, rate of carbonation, and CO_2 uptake capacity. The state-of-the-art, challenges, opportunities, and future research directions for these nanomaterial-based sorbents are discussed. Future research should focus on designing low-cost, high-availability nanomaterials to reduce CO_2 emission from large-scale power plants.

doi:10.1088/978-0-7503-5724-1ch4

4.1 Introduction

The fast growth of global industrialization and the rise of anthropogenic activities has led to increased consumption of energy in the twenty-first century [1]. Almost 80% of energy consumption relies on fossil fuels, owing to their high energy density and wide availability [1]. The combustion of fossil fuels results in the greenhouse effect through the excessive emission of carbon dioxide (CO_2) gas into the atmosphere. This issue is becoming more serious and immediate action must be taken in order to mitigate climate change and global warming [2, 3]. Consequently, there are ongoing developments in CO_2 capture technology and CO_2 capture materials, which are receiving much more attention in the current context. Among the several CO_2 capture technologies, post-combustion carbon capture is considered to be a highly suitable method and it has gained a great deal of attention during the past decade [4]. Membranes and sorbents are the two types of materials which are widely used for carbon capture [5]. Aqueous amine-based solvents are considered to be state-of-the-art for CO_2 capture and are widely used for the large-scale separation of CO_2 from flue gas. For its regeneration there are high energy costs, amine degradation leads to severe corrosion problems, and the net solvent consumption is very high, which are the major drawbacks for its application. Thus solid sorbent based CO_2 removal is considered to be a promising and competitive alternative approach to the conventional aqueous amine scrubbing technique [6].

Recently, several adsorption-based separation methods have been reported for cost-effective and energy-efficient CO_2 remediation. The objectives of these techniques are increasing the adsorption capacity of CO_2, high selectivity of CO_2 compared to nitrogen (N_2), fast kinetics of CO_2 adsorption/desorption, adequate enthalpy heat, and excellent mechanical, chemical, and thermal stability [7]. In order to satisfy the above conditions, there are different solid adsorbents such as zeolites [8], porous organic polymers (POP) [9], porous carbons [10], metal oxides [11], silica [12], metal–organic frameworks (MOFs) [13], and covalent-organic frameworks (COFa) [14], which have been used due to their lack of equipment corrosion, ease of operation, lower energy consumption, and low capital investment [9, 15, 16]. Solid sorbent based CO_2 capture typically depends on the physical adsorption mechanism and/or weak interaction between the sorbent surface and CO_2, which results in the requirement for lower regeneration energy [17]. The solid sorbents should have desirable properties such as good reusability, ease of regeneration, good CO_2 selectivity, low cost, moisture insensitivity, and high adsorption capacity [11]. When assessing the physical solid sorbents, it is essential to consider their sustainability, abundance, stability, production method, regeneration feasibility, pore volume, pore size, density, and surface area [18].

Several parameters such as kinetic factors and thermodynamic properties need to be considered to design a high-performance CO_2 adsorbent. Thermodynamically favorable adsorbents interact moderately with CO_2 molecules to achieve strong and reversible adsorption and desorption. In physical binding, the energy required to release the gas is lower than the chemisorption. In ambient conditions, they are not able to form strong adsorbent–adsorbate interactions to hold CO_2 molecules which results in the rapid decrement of CO_2 adsorption capacity with a temperature increase [19]. In the widely used amine-based chemisorption technique, the captured linear and symmetric CO_2 molecules form strong C–N covalent bonds, and requires high input energy to desorb the CO_2 molecules [20].

A kinetically favorable adsorbent needs rapid capture and reversible release of CO_2, which depends on three different hierarchical structural length scales. The first scale involves the micropore structure which provides the CO_2 adsorption at low pressure and high CO_2 selectivity. The second scale corresponds to the mesopore structures, which are attached strongly to the adsorption sites and facilitate the paths to the micropores. The third scale corresponds to the macroporous structures which have a pore size of hundreds of nanometers, which are responsible for the effective transport of CO_2 [4]. The CO_2 molecule has a kinetic diameter of 3.30 Å [3]. Thus micropores smaller than 1 nm are required for the high kinetic selectivity of CO_2 adsorption at ambient conditions. Hence, materials possessing micropores, mesopores, and macropores will be suitable candidates for the effective capture, storage, and utilization of carbon.

The remarkable advancements in 2D materials in the past two decades have provided tunable spaces for carbon capture and storage (CCS). The layer-by-layer assembly of 2D materials makes abundant vacant space and tunable interlayer spacing. Graphene is a rising star of materials science, and has gained much attention in the research community owing to its distinctive molecular structure and many other superior properties such as large surface area, outstanding chemical stability, mechanical strength, super-flexible, light weight, and tunable porosity, and it also has mesoporous and 3D interconnected macroporous structures [4, 7]. Because of its outstanding properties, graphene and its hybrid or composite materials has been used widely in various applications for catalysis, energy storage, supercapacitors, sensors, and many more.

It is postulated that when CO_2 molecules penetrate through nanofluids, both the nanochannels between the graphene sheets and the structural defects within the graphene sheets act as selective gas barriers. Meanwhile, the CO_2 molecules react with amine groups according to the transport mechanism [1]. The existence of carboxyl, carbonyl, epoxy, and hydroxyl groups makes the graphene oxide (GO) sheets compatible with amines, with good hydrophilicity and stable suspension [21]. First, the GO is a 2D material with a high surface area to volume ratio and high flexibility. Thus it is easy to make a porous network by random crosslinking of graphene layers. Second, the planar geometry of GO makes it easy to functionalize, modify, and adjust the surface chemistry properties and porosity which makes GO an outstanding candidate for the large-scale industrial CO_2 capture applications. More significantly GO is chemically inert, therefore it can endure in the harsh conditions of flue gas [7].

In this chapter, we discuss critically the novel and widely used CO_2 adsorbents, such as carbonaceous materials, zeolites, and MOFs, including their unique features, such as durability, tunable microporosities, hierarchical porosities, and high surface area. Moreover, the effects of different functionalizations on the CO_2 capture performance of carbonaceous materials, zeolites, and MOFs are also discussed in detail.

4.2 CO_2 capture technologies

There are three main technologies used in the carbon capture and utilization process: (i) pre-combustion carbon capture, (ii) post-combustion carbon capture, and (iii) oxy-fuel combustion carbon capture.

4.2.1 Pre-combustion carbon capture

In this process, traditional fuels such as natural gas or coal will react with oxygen or air in the presence or absence of steam to produce synthetic gas or syngas, which is a combination of hydrogen (H_2) and carbon monoxide (CO). During the production of H_2 and CO, there are several processes such as gasification, reforming, and downstream distillation. The manufactured hydrogen is called blue hydrogen. The complete process of syngas production through a pre-combustion carbon capture technique is shown in figure 4.1. The synthetic gas is produced from the reaction between fuel and oxygen/air, and is then sent to a shift reactor to yield CO_2 and hydrogen. The yielded hydrogen can be used for electricity generation through a gas turbine and used as a fuel for electric cars.

There are two main processes (partial oxidation and stream reforming reactions) involved in syngas production, which are

$$C_nH_m + \frac{n}{2}O_2 \rightarrow nCO + \left(\frac{m}{2}\right)H_2 \ \Delta H_{CH_4} = -36 \text{ kJ mol}^{-1} \tag{4.1}$$

$$C_nH_m + nH_2O \rightarrow nCO + \left(\frac{n+m}{2}\right)H_2 \ \Delta H_{CH_4} = 206 \text{ kJ mol}^{-1}. \tag{4.2}$$

The typical reforming products such as 6% CO_2, 21% H_2O, 11% CO, and 43% H_2 are commonly used for steam reforming [23]. When steam reforming and partial oxidation are used simultaneously in pre-combustion carbon capture it is known as auto-thermal reforming, where the heat released from the partial oxidation process by its exothermic nature can drive the endothermic steam reforming reaction.

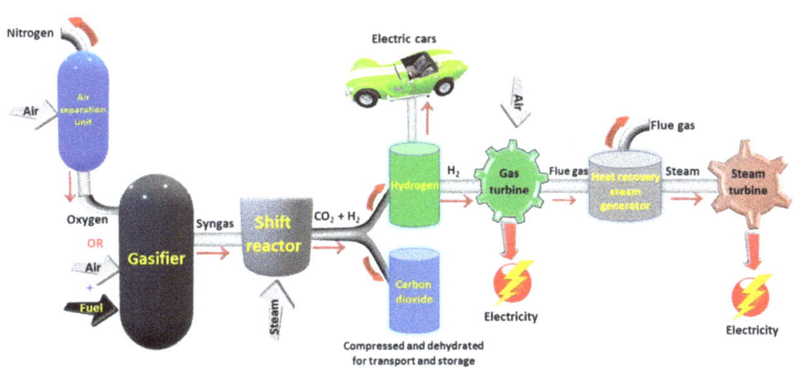

Figure 4.1. Pre-combustion carbon capture technology for the syngas production, hydrogen fuel production, and electricity generation. (Reproduced from [22]. CC BY 4.0.)

The syngas combination is then cooled down and cleaned of impurities such as carbonyl sulfide, mercury, hydrochloric acid, and hydrogen sulphide [24]. The sanitized syngas is then allowed to react CO with H_2O steam through the water-gas shift reaction (WGSR) (as shown in the following equation) to enhance the CO_2 percentage level and enable CO_2 separation along with the production of H_2 fuel in the later steps:

$$CO + H_2O \rightarrow CO_2 + H_2 \quad \Delta H = -41 \text{ kJ mol}^{-1}. \tag{4.3}$$

In industry, the pre-combustion carbon capture technique has been used for more than 95 years. Even though the CO_2 separation process is much easier and requires less energy than other methods such as post-combustion and oxy-fuel combustion, it still requires developments in energy-efficient recovery, air separation, and extra energy for reforming. In addition to that, further purification steps are essential to remove the impurities, such as sulfur and ash containing compounds. The main reason for the loss in efficiency is WGSR stage, which is responsible for almost 44% of the total efficiency loss. This is due to the required energy for the generation of steam along with the heat released during the WGSR process as it is an exothermic process and has limited equilibrium.

4.2.2 Oxy-fuel combustion carbon capture

In the oxy-fuel combustion method, the CO_2 capture and separation process is easy. A schematic representation of oxy-fuel combustion technology with an air separation unit is shown in figure 4.2. The fuel is burned in the presence of pure oxygen and flue gas in a boiler. Then the flue gas is sent to the particle remover unit to remove microparticles. Then it is fed into a cooler and condenser unit to remove the water followed by a sulfur removal unit. Again, it sent to the cooler and condenser unit, then the CO_2 is compressed and dehydrated for transport and storage purposes. This oxy-fuel combustion method is assumed to be the most promising energy-efficient

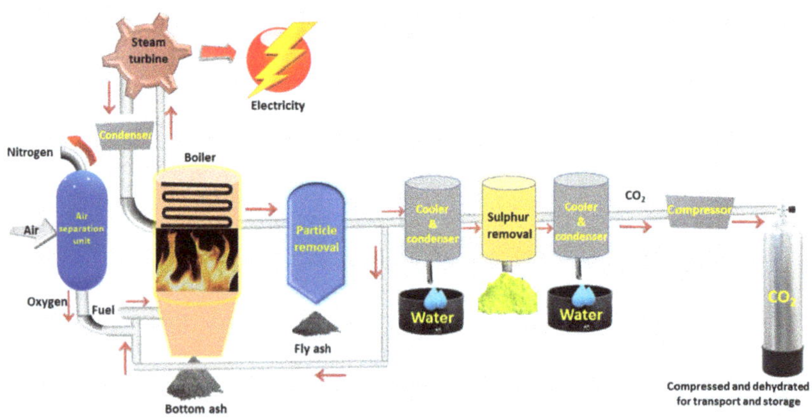

Figure 4.2. Schematic representation of oxy-fuel combustion technology with air separation unit. (Reproduced from [22]. CC BY 4.0.)

method compared to the other two methods (pre- and post-combustion), with a lower efficiency penalty of 4% compared to the post-combustion method (8%–12%). The reduction of nitrogen gas emission and exhausted flue gas volume along with the boiler efficiency increment can be attained by executing the oxy-fuel combustion method in power systems. An air separation unit is used supply the pure oxygen and it is a costly and energy-intensive process, which is considered as a big challenge in this method. Cryogenic distillation is used to produce a large quantity of oxygen with high purity for large-scale production. Oxygen-transport membranes were introduced to reduce the energy related difficulties associated with cryogenic air separation. Carbo *et al* stated that the introduction of oxygen-transport membranes in oxy-fuel combustion can increase the economic savings around 19%–50%, compared to post-combustion carbon capture techniques [25].

Instead of an air separation unit, the utilization of chemical looping can greatly increase the net efficiency of a power plant by 3%, when compared to oxyfired along with an integrated gasification combined cycle (IGCC). However, the cost of electricity and power plant capital costs will decrease by 7%–12% and 10%–18%, respectively. This oxy-fuel combustion technique can be applied in existing or new power plants along with the use of different types of fuels, such as lignocellulosic biomass and municipal solid waste, which is the main advantage of this technique. The integration between bioenergy and carbon capture and storage is referred as bioenergy carbon capture and storage (BECCS) or BioCCS, and aims to decrease carbon for the mitigation of climate change. Twenty BECCS projects are running including biogas plants, biomass gasification, combustion of biomass and co-firing, ethanol production, and waste to energy. However, there are still some challenges in the BECCS method, such as lower efficiency, a high cost of electricity, and a higher cost of biomass compared to fossil fuel. When including compression units, CO_2 purification and air separation units in the oxy-fuel combustion method lead to a decrease of 9%–13% in cycle efficiency. Hence, it is important to integrate the heat and processing to making the overall process more attractive for commercialization. Several pilot-scale plants and projects have been demonstrated in the last few decades which show the increase of worldwide attention on the viability of oxy-fuel combustion technology.

4.2.3 Post-combustion carbon capture

CO_2 capture and separation from flue gas combustion in an oxidant environment is known as post-combustion carbon capture. A schematic representation of the post-combustion carbon capture technique is shown in figure 4.3. Initially, the exhausted flue gas undergoes a desulphurization and denitrification process along with the removal of dust and is cooled to avoid the degradation of the solvent. At this stage, the flue gas consists of N_2, H_2O, and CO_2, which is then fed into an absorber unit containing the solvent (as shown in figure 4.3). The cleaned gas is then washed with water, followed by the regeneration of CO_2. Generally, the captured CO_2 is compressed into supercritical fluid and transported for storage in saline aquifers or geological reservoirs.

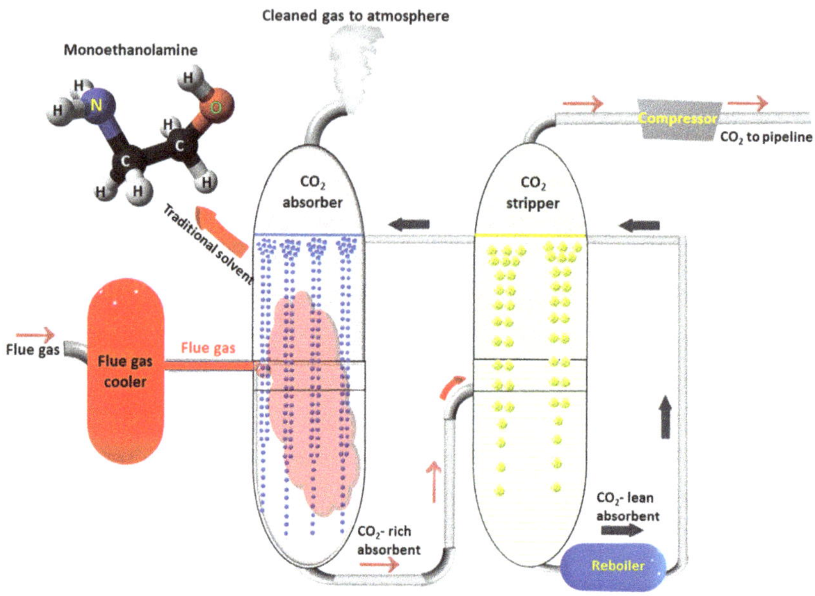

Figure 4.3. Schematic representation of post-combustion carbon capture technology. (Reproduced from [22]. CC BY 4.0.)

Amine solutions are used as the most common solvents for CO_2 capture owing to their good selectivity and high adsorption capacity towards acidic gases. In particular, monoethanolamine based absorption is commercially available and is the most common method in the post-combustion carbon capture approach. However, the amine-based solutions have several drawbacks such as evaporation, solvent loss, degradation, a high energy penalty during regeneration, and corrosion. Although, the monoethanolamine based sorbent is the only commercially existing method, its capital costs are expensive. To reduce the capital costs that accompany the post-combustion carbon capture method, membrane-based separation seems to be an appropriate technology as it has low operational costs, a low carbon footprint, low energy requirements, and ease of retrofitting and scaling up with existing power plants. Nevertheless, there are several drawbacks related to membrane separation, such as a rapid decrease of selectivity, emission of SO_x and NO_x after passing through the membrane, and water condensation on the membrane. A few membranes are also affected by humidity fluctuation and temperature adjustment difficulties, which results in severe changes to the transport properties in the membrane.

Although the pre-combustion carbon capture technique provides greater efficiency than the post-combustion carbon capture technique, it is more expensive. Thus, it is essential to reduce the cost associated with the pre-combustion method by finding a better solvent for the absorption. As stated earlier, amine solvent based chemisorption is a readily available technology for retrofitting existing power plants. Based on that technology, pilot scale power plants have been established showing

80 tonnes per day of CO_2 absorption capacity [26]. It is expected that before 2025, the first integrated commercial carbon capture and storage will be established alongside coal-fired power plants. Afterwards, it will be employed in the rest of the commercial scale carbon capture processes. Vega *et al*, in 2020, reported a comparison between traditional (post-combustion) and novel (partial oxy-combustion) carbon capture and storage methods. In the pilot-scale power plants, novel techniques along with blend solvents were used to decrease the energy penalty in the overall process. There are several desirable characteristics in novel solvents such as lower degradation, low corrosiveness, low production cost, lower by-products, and high cyclic capacities. Based on the amine solvent technology, the company Shell has established a new carbon capture method (Shell Cansolv™) [27]. The developed carbon capture technology is suitable for different industries such as the chemical industry, mining, energy production, and refineries, etc. This Cansolv™ technology requires a regeneration energy of 2.5–2.9 GJ per tonne of CO_2, which is lower than the most common monoethanolamine solvent regeneration energy of 3.5 GJ per tonne of CO_2 [28]. The industrial sector decarbonization needs an assessment of the technology readiness level (TRL) for the different carbon capture, storage, and utilization techniques. Pre-combustion is the only capture technology which has touched the highest commercial scale of TRL9 [29]. The other capture technologies, such as direct air capture, BECCS technology, membrane polymeric (natural gas), pre-combustion (IGCC), oxy-fuel combustion (coal power plants), post-combustion, and adsorption are at the demonstration scale of TRL7. The pilot-scale level capture techniques such as chemical looping combustion, carbonate looping combustion, post-combustion (biphasic solvents), and membrane polymeric (power plants) are at the TRL6 level. The other capture technologies, ranging from laboratory scale plants (TRL5) to proof of concept (TRL3), are pre-combustion low-temperature, ionic liquid post-combustion, oxy-fuel combustion (gas turbine), and inorganic dense membrane technologies. In the case of carbon storage technology, post-combustion in CO_2-enhanced oil recovery, saline formations, and power plants are the three technologies touching the commercial scale of TRL9 [30]. There are some technologies such as depleted gas and oil fields and CO_2-enhanced gas recovery that are in the demonstration level of TRL7. Other storage technologies, such as mineral storage and ocean storage, are in the initial formulation stage (TRL2) and proof of concept in laboratory tests in TRL3. On the other hand, the transport technologies using either offshore or onshore pipelines along with shipping transport are in the commercial scale of TRL9.

4.3 Solid sorbents for carbon capture

The different solid sorbents, such as carbonaceous material, zeolites, and MOFs, and the functionalization of these materials for efficient CO_2 capture performance were discussed in this section. The recent advancements in developing solid sorbents and the correlation between the synthesis approaches and designs, and the cyclic stability, carbonation rate, and CO_2 uptake capacities, are discussed.

4.3.1 Carbon-based sorbents

Carbon is an important element and is used in several scientific areas. It is the sixth most abundant element on Earth with an average abundance range of 180–270 ppm. All the fabricated allotropic carbons have a different crystal structures. Based on the surface functionalities, preparation methods, and structural features, carbon materials are classified into various types such as carbon aerogels (CA), carbon nanofibers (CNF), template based carbons, carbon molecular sieves, biomass derived carbons, pyrogenic carbons, porous carbon, amorphous carbon, fullerenes, activated carbons (AC), carbon nanotubes (CNTs), and graphene, etc [31]. These carbon materials have the following excellent desirable properties: high porosity, ease of morphological tunability, high surface area, good mechanical and electronic properties, excellent gas storage capacity, and high regenerative ability [32]. These carbon-based nanostructures play a significant role in nanoscience and nanotechnology.

4.3.1.1 Fullerene based carbon capture

In 1985 Kroto *et al* successfully developed fullerenes at Rice University through laser ablation of graphite under helium flow. Fullerenes are caged spherical molecule with 20 or more atomic carbons positioned at polyhedral structured corners and contain pentagonal and hexagonal rings as the basis of an icosahedral symmetry closed cage structure. The spherical form of fullerenes is the most stable, when the pentagons are divided by no more than one hexagon. The properties of fullerene depend on the number of carbon atomic clusters (C_n). The fullerenes can be prepared by various methods such as flash vacuum pyrolysis, electron beam ablation, bottom-up chemical methods, and electric arc discharge methods. The prepared fullerenes and their derivatives are used effectively in photonic, electronic, magnetic, and biomedical applications. Fullerenes are used, in particular, in capturing CO_2 from flue gas. Numerous theoretical and experimental results have been reported for fullerene-based carbon capture. In particular, in 2015 Dong *et al* established B_{40} fullerene theoretically for carbon capture applications, which provides excellent charge transfer, kinetic feasibility, and high adsorption capacity. The developed B_{40} fullerene demonstrates good CO_2 selectivity with 13.87 mmol g^{-1} of CO_2 adsorption capacity [33]. In 2014 Andreoli *et al* synthesized polyethylenimine cross-linked with C_{60} fullerene, where the prepared material shows a CO_2 adsorption capacity of 140 mg g^{-1} under a temperature and pressure of 363 K and 0.1 bar, respectively [34]. The same research group synthesized polypropylene cross-linked with C_{60} fullerene, where they achieved a CO_2 adsorption capacity of 17.78 mmol g^{-1} [35]. Recently, in 2022, Mehdi *et al* examined B_{38} fullerene for carbon capture via adensity functional theory approach. This resulted in CO_2 uptake capacities of 2.34 and 17.04 mmol g^{-1} under temperatures of 300 and 200 K at 1 bar of pressure, respectively [36].

4.3.1.2 Graphene based carbon capture

Geim and Novoselov discovered graphene, which consist of a one atom thick sheet of carbon atoms organized in a honeycomb pattern and has excellent electronic and mechanical properties [37]. In 2010 they were awarded the Nobel Prize in Physics for their discovery. The main reasons for the delay in the discovery of graphene were the following: (i) before its discovery, graphene was anticipated to be unstable in the free state, and (ii) no experimental tools existed to detect the one atom thickness of graphene [38]. Graphene is a monolayer thick atomic sheet formed by sp^2 bonded carbon atoms of hexa atomic cyclic molecules. Its distinctive properties, such as high thermal conductivity, mechanical strength, large surface area, and good chemical stability, make it a potential material for several applications such as gas sensors, DNA sequencing, touch panels, solar cells, transistors, supercapacitors, catalysis, hazardous material adsorption, organic pollutant removal, and nonlinear optical applications, etc [39–46]. Graphene and graphene-based materials can be prepared via different techniques such as epitaxial growth, oxidation techniques, solvothermal methods, liquid phase exfoliation, and microwave assisted exfoliation, etc [47] Graphene and graphene-based composite materials have increased interest in CO_2 capture applications, because of its sheet-like morphology providing a large surface area and its porous nature. Several research groups are working on modifying graphene and graphite structures to improve their CO_2 capture performance [48]. Graphene and its derivatives show better CO_2 capture efficiency by KOH activation, making hybrid composites and amine modification, etc. Different types of amine modified graphene-based sorbents were examined for CO_2 capture applications. In 2016, Shin *et al*, synthesized PEI functionalized graphite oxide for CO_2 capture. The prepared PEI-GO 60 sample shows the highest CO_2 adsorption capacity of 84 mg g^{-1} under 1 bar of pressure and 298 K temperature. The presence of a high amount of PEI (60) in the graphite oxide leads to the increase the interaction between the electron-donor and acceptor among the CO_2 species and amine group, which results in enhancement of the adsorption capacity [49]. Moreover, at the GO surface, CO_2 molecules react with two amine functionalities, leading to the formation of a carbamate complex. The increase of PEI impregnation into the GO surface increases CO_2 adsorption capacity owing to the development of chemical linkages. The highest CO_2/N_2 selectivity was observed to be 37.13 for PEI-GO 60 adsorbent, which is due to the strong acid–base interaction between nitrogen moieties and CO_2 molecules [49]. In another report, by Liu *et al* in 2019, TEPA functionalized GO was developed through an ultrasound mediated amination method. The dynamic CO_2 uptake was done at 0.1 bar of pressure and 338 K temperature. The results show that the ultrasonic activated TEPA-GO exhibits a better CO_2 uptake capacity of 1.2 mmol g^{-1} at 338 K, while the bare GO shows 0.3 mmol g^{-1} at 338 K. Ultrasonication treatment plays a significant role in CO_2 adsorption enhancement, where ultrasonic waves are used to exfoliate graphene layers and increase the number of active sites. Through longer period ultrasonication, the specific surface area increases as well as a large number of amine functionalities attaching to the surface of GO owing to the longer reaction between GO and TEPA. As a result of the interaction between electrophilic CO_2 molecules and nucleophilic amine, the CO_2 adsorption capacity is increased [50]. In 2018 Alghamdi *et al* prepared nitrogen (N) doped GO sheets through the N-rich polymeric compounds polyaniline (PANI), polypyrrole (PPy), and polymer blends (PANI-PPy) doped with different acids. (benzenesulphonic acid, sulfuric acid, and hydrochloric acid). These acid doped polymers were chemically activated by KOH. The GO sheets doped with PPy-Ar-1 exhibited a CO_2 uptake capacity of 1.28 mmol g^{-1} under 1 bar pressure and 50°C temperature. The existence of pyridinic-N and pyrrolic-N at high concentrations and broad pore size probably encourages the CO_2 adsorption capacities. Therefore, the preparation of N-doped GO is a promising idea for future CO_2 capture processes.

The development of GO based aerogels has also been done for the CO_2 adsorption process. For example, in 2019 Hsan *et al* fabricated eco-friendly chitosan incorporated GO aerogels for CO_2 capture. CS was attached to GO through a crosslinking and freeze-drying method, where N-hydroxy succinimide (NHS) and (1-ethyl-3-(3-dimethylaminopropyl) carbodiimide hydrochloride (EDC) were used as crosslinkers [51]. The CO_2 adsorption was measured through the Brunauer–Emmett–Teller (BET) method. The CO_2 uptake capacity was found to be 0.25 mmol g^{-1} at 1 bar and 298 K, which was higher than the bare CS (0.05 mmol g^{-1}) at the same temperature and pressure. The improved CO_2 capture performance is due to the presence of nitrogen containing functional groups (exhibiting a lone pair of electrons) within the polymer chain. Also, the presence of a large number of amine functional groups in the chitosan causes a strong interaction between CO_2 molecules and chitosan graphene oxide (CSGO), which leads to an increase in CO_2 capture performance. Composite based GO adsorbents were developed by Pokhrel *et al* in 2018, where they developed a novel composite adsorbent consisting of GO with a zeolite imidazolate framework (ZIF-8) and functionalization of amines (ethylene diamine (ED), polyethyleneimine (PEI), and 3-aminopropyl triethoxysilane (APTES)) [52]. The amine groups were successfully introduced into GO sheets, which led to an increase in the distance of the interlayer. The modification of amine greatly influences the CO_2 adsorption performance of GO, which is determined by the thermogravimetric analyzer at 1 bar of pressure and a 30 °C temperature. The GO-APTES sorbent shows 36% higher CO_2 uptake capacity compared to bare GO. In the presence of pre-adsorbed moisture (10% RH), the GO-APTES adsorbent shows 33% higher adsorption capacity than dry adsorption. Because the bicarbonate generation in the presence of water molecules offers higher active sites for the CO_2 adsorption, the amine modified ZIF-8 sorbent shows 43% higher CO_2 adsorption than the bare ZIF-8. This study may assist in regulating the optimal modifications for CO_2 adsorption enhancement and help to develop extremely efficient sorbents for CO_2 capture in the future.

4.3.1.3 Carbon nanotube-based carbon capture

Carbon nanotubes (CNTs) are another important carbon allotrope and have a cylindrical morphology with a diameter range of 100–1000 nm. While fullerene is a sealed spherical shell consisting of sp^2 atomic carbon, CNTs consist of sp^2 atomic carbons as well, but as single or multiple coaxial tubes made up of graphite sheets. CNTs exist in the form of single-walled CNTs (SWCNTs) and multi-walled CNTs (MWCNTs). A single layer of graphitic carbon rolled up into a tube is called an SWCNT, and two or more layers of graphitic carbon rolled up into two or more tubes is called an MWCNT. CNTs have outstanding physical, chemical, electronic, and mechanical properties. They have a high specific surface area (200–900 m^2 J^{-1}), good electrical conductivity (0.002–0.25 S cm^{-1}), good thermal conductivity (6600 W m^{-1} K^{-1}), and good specific gravity (0.2 g^{-1} cm^{-2}). These properties make CNTs as an outstanding material for CO_2 capture applications. In order to increase the CO_2 uptake capacity, researchers have carried out several modifications in CNT structures, such as acidic pretreatment, amine impregnation, wet impregnation, chemical activation by KOH, amine activation grafting, the co-precipitation or intermixing method, and urea melt method, etc [10]. In 2017 Hu *et al* fabricated phenylenediamine (PDA) functionalized MWCNTs for carbon capture applications. The carboxylated MWCNTs were obtained through pretreatment of acid and then activation grafting methodology was used to functionalize the carboxylated MWCNTs with PDA through covalent bonding. The CO_2 adsorption was measured through a static volumetric analyzer. It was found that the PDA functionalized MWCNT shows a CO_2 uptake capacity of 0.6 mmol g^{-1} under 200 kPa pressure and 303 K temperature, which are higher than for bare MWCNTs (0.17 mmol g^{-1}). Because of PDA functionalization, a large number of primary ($-NH_2$) and secondary (–NH–) amines are attached over the surface of the MWCNTs. These amine groups capture the CO_2, resulting the formation carbamate anions. In addition to the nitrogen moieties, amino functionalities also attract CO_2 molecules though electrostatic forces owing to their larger electronegativity. Moreover, the hydrogen bonding between amino functionality and CO_2 molecules is also responsible for the increased CO_2 adsorption process. In another report, Khalili *et al* also prepared PDA functionalized MWCNTs, which showed much improved CO_2 adsorption efficiency [53]. A two-step route was followed for the preparation of PDA functionalized MWCNTs. First, H_2SO_4/HNO_3 was used to oxidize the MWCNTs and then treated with PDA to obtain the MWCNT–NH_2 adsorbent. A volumetric approach was used to measure the CO_2 adsorption performance and the results show 18 mmol g^{-1} of CO_2 adsorption capacity at 40 bar and 298 K. The strong interaction between amine active sites and gas molecules is responsible for the highest adsorption capacity. The adsorption mechanism was proposed through two steps. In the first step, amine-CO_2 zwitterion generation takes place through the interaction between CO_2 molecules and amine nitrogen. In the second step, deprotonation of zwitterion occurs, which leads to the formation of carbamate ionic species ($RNHCO_2$-), see figure 4.4(a). Furthermore, the isosteric heat of adsorption (Q_{st}) was calculated using Clausius–Clapeyron equation and was found to be 0.5 mmol g^{-1} for the MWCNT–NH_2 sample.

Several studies report increasing the affinity of CO_2 molecules and the number of active sites based on PDA grafting of CNTs. In particular, in 2019 Rahimi *et al* followed a two-step method to prepare (as shown in figure 4.4(b)) 1,3-diaminopropane (DAP) modified MWCNTs for CO_2 adsorption [54]. In the first step, acid treatment took place for MWCNTs in order to carboxylate the MWCNTs through oxidation and to remove catalytic metal species. In the second step, oxidized MWCNTs were functionalized with diamine DAP solution. A static volumetric adsorption method was used to examine the CO_2 uptake performance, where the DAP/MWCNT showed a maximum CO_2 uptake capacity of 2.10 mmol g^{-1} (93 mg g^{-1})

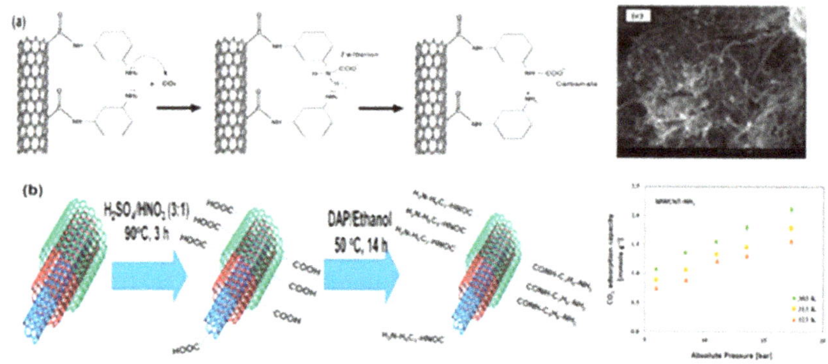

Figure 4.4. (a) Mechanistic scheme for CO_2 adsorption over $-NH_2$–MWCNT. (b) Schematic representation of the acidic pre-treatment and amine modification of MWCNTs. (c) SEM image of MWCNT–NH_2. (d) CO_2 adsorption capacities of MWCNT–NH_2. (Reproduced with permission from [10]. Copyright 2021 Elsevier.)

at 17.3 bar and 303 K, which is higher compared to bare MWCNTs (48.5 mg g^{-1}) (shown in figure 4.4(d)). The Q_{st} values were calculated for the DAP/MWCNTs and was found to be in the range of 20–26 kJ mol^{-1}.

Instead of diamine functionalization, the amine modified CNTs have also been investigated widely for carbon capture applications, where urea, tetraethylenepentamine (TEPA), and polyethyleneimine (PEI) were used as an amine source. In particular, in 2015 Lee *et al* prepared PEI functionalized MWCNTs [55]. Initially, H_2SO_4/HNO_3 acids were used for the pretreatment of MWCNTs and then a simple wet impregnation method was followed for functionalization of PEI. The CO_2 adsorption measurement was carried out using a thermogravimetric analyzer, which showed the highest CO_2 adsorption capacity of 94 mg g^{-1} at 1 bar and 298 K. During the studies, it was observed that more PEI impregnation into MWCNTs makes the nature of the MWCNTs surface more basic, which results in strong interactions between acidic CO_2 species and basic amine functionality. Moreover, these functionalized sorbents exhibit an excellent ability of regeneration for ten adsorption–desorption cycles. In another investigation, commercial, industrial grade, and low-cost purified MWCNTs modified with TEPA have been reported by Liu *et al* [56]. The TEPA functionalized industrial grade-MWCNTs (IG-MWCNTs) show the highest CO_2 adsorption capacity of 3.08 mmol g^{-1} at 343 K. The amine impregnated different diameters of SWCNTs and MWCNTs were prepared through the urea-melt method, where urea was used as an amine source [57]. The urea modified MWCNTs with large diameter show the highest CO_2 adsorption capacity of 64 mg g^{-1} at 70 °C and 1 bar. The urea modification induces new active adsorption sites of the prevailing amino groups over the MWCNTs, which bind with the CO_2 molecules leading to carbamates generation.

The CNT based nanohybrids and composites have also been established and demonstrated as suitable adsorbents for CO_2 capture processes. For example, APTES functionalized, ZnO decorated MWCNT nanohybrid adsorbents were developed by Jena *et al* in 2019 [58]. Acid treatment was carried out to carboxylate the MWCNTs and then ZnO-APTES was added into the carboxylated MWCNTs via a combination of the sol–gel and hydrothermal methods. The prepared nanohybrids exhibit the highest CO_2 uptake capacity of 1.32 mmol g^{-1} at 1 bar and 273 K. The enhanced electrostatic interactions involve coordinate bonding, dipole–dipole interactions, and hydrogen bonding towards the attraction of CO_2 molecules. Amine modified MWCNTs with hollow silica spherical particles were prepared by Zohdi *et al* in 2019 [59]. A simple co-precipitation route was followed to prepare the composite adsorbents. The CO_2 adsorption performance was assessed through a systematic volumetric method. The results show 13.8 mmol g^{-1} of CO_2 adsorption capacities under 20 bar and 298 K. However, the prepared composites possess better CO_2 gas selectivity than N_2 and CH_4. The quadruple moment is the reason for the higher selectivity of CO_2. The CO_2 molecule's quadruple moment is three times higher than the N_2 quadruple moment and more CO_2 adsorption is due to the facilitation of more polarizability. The prepared composite adsorbents exhibit good stability and recyclability by maintain the adsorption of CO_2 even after four adsorption–desorption cycles.

4.3.1.4 Activated carbon-based carbon capture

Activated carbon (AC) is another type of carbon nanomaterial and results from biomass through pyrolysis. It must be activated through different physical and chemical processes. It has high porosity, high adsorption capacity, large surface reactivity, and good mechanical stability, which makes it a potential candidate for gas sorption applications. AC can be manufactured from several natural products such as lignin, coffee grounds, coal, coconut shells, chitosan, synthetic organic polymers, and chicken feathers, etc. The addition of a base, acid, and/or salt is essential for chemically activated carbon preparation. In order to improve AC's surface area, different functional groups can be added over the surface of the AC. Sometimes ammonia is used as an activation agent at high temperatures, where the addition of nitrogen groups and acidic oxygen group removal takes place simultaneously, which leads to improving the basicity. Several functional groups, such as amide, pyridinic, lactam, imide, and pyrrolic, can be formed during the reaction with nitrogen containing functional groups such as nitric acids, amines, and ammonia [60]. The activated carbon's chemical characteristics are mostly examined by the existence of heteroatoms such as phosphorous, sulfur, hydrogen, nitrogen, and oxygen. The quantity and type of these atoms are derived from incorporation through activation processes or the nature of starting material. The acidic property of AC is associated with oxygen containing functional groups. These functional groups are present on the basal plane edge or outer surface of the carbon. Since the outer surface contributes the majority of the adsorption process, the oxygen functional group concentration on the surface impacts the adsorption capacities of the carbon greatly. There are several oxygen containing functional groups such as ether, quinone, chromene, pyrone, carbonyl, phenol, lactone, and carboxylic groups present in the AC surface [61]. The oxidation of the carbon surface results the formation of oxygen containing functional groups. Aqueous and gas oxidants are widely used activation methods to incorporate oxygen contain acidic functional groups. Steam, CO_2, air, and oxygen can be used as the gas oxidant. In these methods, strong acids (carboxylic) are formed by the low-temperature oxidations, while weak acids (phenolic) are formed during high temperature oxidation [62]. When compared to gas phase oxidation, a large amount oxygen can be incorporated on the activated carbon surface in liquid phase oxidation. It has been confirmed that the gas phase oxidation of activated carbon increases because of the concentration of carbonyl and hydroxyl functional groups.

The activated carbon's basicity is related to (i) basic surface functionalities and (ii) the resonating π-electrons of carbon aromatic rings, and both are able to bind with protons. The basic properties of activated carbon mainly arise from the delocalized π-electrons of the graphene layer. It was also noted that the π-electrons of the carbon layer can act as Lewis bases. Leon *et al* reported the surface basicity of two carbon sequences and proved that the oxygen-free carbon sites can adsorb protons from the solution. Lewis sites are located in π-electron rich areas on the activated carbon basal plane.

Activated carbons are considered as good candidates for CO_2 capture because no separate process is required for moisture removal, their ease of regeneration, and their high adsorption capacity and high selectivity at ambient pressure. The adsorption capacity is affected strongly by activated carbon's surface chemistry [63]. Since CO_2 is an acidic gas (weak Lewis acid), the incorporation of Lewis base on the surface of AC may increase the CO_2 adsorption performance of the AC [64]. There are several methods followed for the fabrication of AC with higher basicity. Among these, the replacement of acidic groups with basic functional groups and the neutralizing or removal of acidic functional groups are the most popular methods. Nitric acids, amines, and ammonia are the main nitrogen containing functional groups used to increase the basicity. The purpose of ammonia treatment is to increase the basicity of AC. During the ammonia treatment at high temperatures, ammonia will decompose into free radicals such as atomic nitrogen, hydrogen, NH, and NH_2. These free radicals are attached to the AC's surface to form nitrogen functionalized AC.

4.3.2 Zeolite-based sorbents

Zeolites are categorized as microporous crystal structures and are classified into two types: synthetic and natural aluminosilicates. The natural and synthetic zeolites were discovered in 1756 and 1940, respectively [65, 66]. More than 40 natural and 200 synthetic zeolites are reported by the International Zeolite Association (IZA). It has a TO_4 tetrahedra periodic array, where T denotes either silicon or aluminum, and it has a porosity in the order of 0.5–1.2 nm. Zeolites are highly crystalline materials with high stability, strong adsorption sites, and high surface area. Because of these good qualities, zeolites have been used extensively in several applications, such as separation, chemical sensing, detergency, and catalysis, etc [67]. Some of the widely used zeolites accepted by the IZA are shown in figure 4.5.

Synthetic zeolites can be prepared by the hydrothermal sol–gel method in the presence of aqueous solution under moderate temperature (>300 °C) and low pressure. The basic character of the zeolites can be increased by incorporating aluminum atoms in the silicate-based framework. This is due to the electronegativity of aluminum, when compared to silicon. The inclusion of cations such as K, Na, and Li into the zeolite framework can compensate the negative framework charges from aluminum [68]. The basicity of the zeolites can be increased by adding cations with decreased electronegativity. The increase in basicity of the zeolites results the

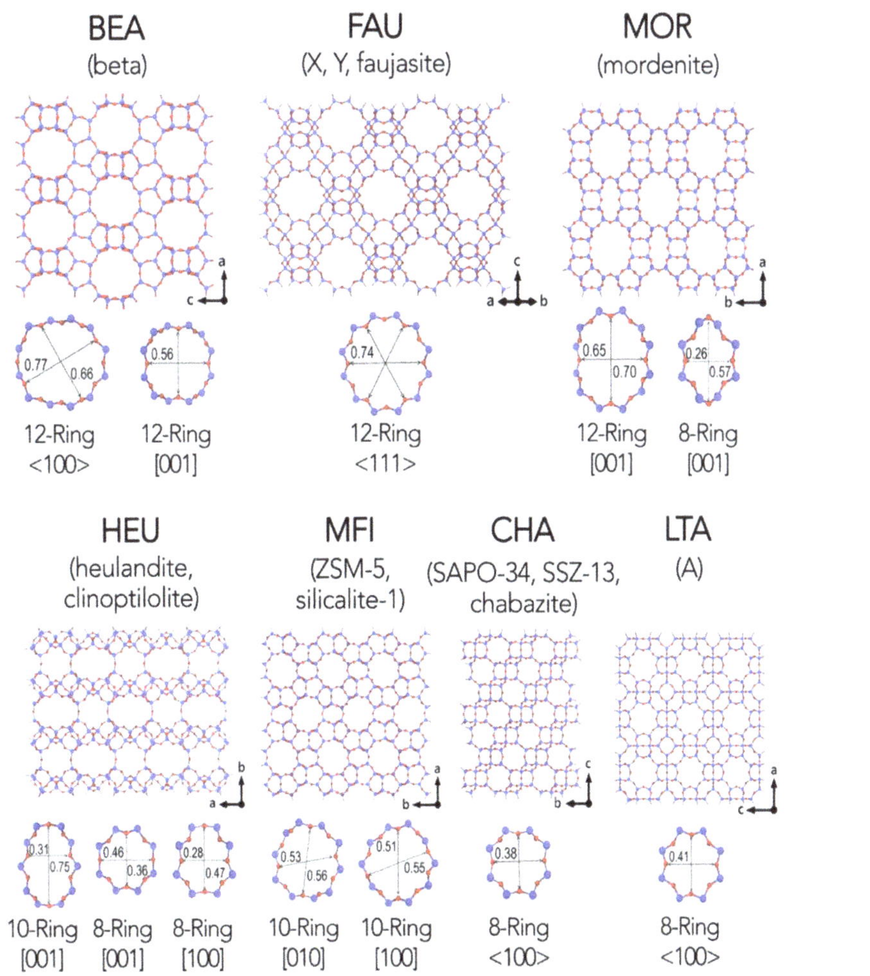

Figure 4.5. Some of the widely used zeolite frameworks. Blue spheres denote T atoms and red spheres represents the oxygen atoms. The three letter code is assigned by the International Zeolite Association (IZA). (Reproduced with permission from [121]. Copyright 2017 Elsevier.)

increase of CO_2 adsorption capacity [69]. The adsorptive properties of zeolites are affected by the number of neutralizing cations inside the pores and Si/Al ratios.

Carbon gating is a special phenomenon of zeolites, where CO_2 can be allowed into the framework but other compounds are prohibited. Generally, cationic zeolites have a carbon gating property, where the additional cationic framework blocks entry to the narrow pores. This carbon gating effect provides the high selectivity for CO_2 adsorption over additional small adsorbates, which is useful during CO_2 adsorption from flue gases.

Traditionally, zeolites have been used for the adsorption of gases. For the past few years, zeolites have been applied widely in continuous separation processes, because the zeolites exhibit an excellent trade-off property under relatively low temperature (57 °C) and low CO_2 partial pressure (0.1–0.2 atm.) of post-combustion CO_2 capture. Moreover, N_2 and CO_2 molecules have kinetic diameters of 3.64 and 0.33 nm (almost similar), which makes it difficult to separate them. The polarizability and quadrupole moment of the CO_2 molecule are two and three times higher than for the N_2 molecules [70]. In addition, the N_2 molecule is less condensable than CO_2 molecules (the critical temperatures are 126.2 and 304.1 respectively). The zeolites can interact more strongly with CO_2 molecules than N_2 resulting in high CO_2/N_2 selectivity [71]. In order to obtain good selectivity, a continuous polycrystalline structure of zeolites must be prepared by interconnecting them to each other. A secondary growth method is widely used to prepare a dense polycrystalline structure of zeolites [72]. In this method, a zeolite seed layer will be deposited on a solid support followed by densification via hydrothermal preparation of the polymer cast or zeolite membrane on top of the zeolite layer.

Faujasite (FAU) is a type of zeolite which is a familiar framework for CO_2 capture. FAU has a relatively low Si/Al ratio and it has been demonstrated by several simulations and experimental studies to have the highest CO_2 adsorption selectivity [69, 73]. Based on the Si/Al ratio, FAU is divided into two types: zeolite Y (Si/Al 1.5–3.8) and zeolite X (Si/Al < 1.5) [67]. Krishna *et al* reported a barrier permeability of 10 000 and selectivity of 500 for FAU zeolite membrane [70]. In the aluminum–silicon frameworks, negative charges are created by aluminum, where the metal cations are used to neutralize it. Na^+ is a widely used cation, hence it is known as NaY and NaX. CO_2 interacts strongly with the micropore electric field. The strength of interaction depends on the type of cations and number of cations. NaY has a higher Si/Al ratio than NaX and it has more cations leading to higher CO_2 adsorption capacity [74]. However, when it comes to membrane separation, the strong interaction causes the molecules to slow down. The zeolite Y shows the higher CO_2 permeance and a better trade-off between permeance and selectivity. Some other zeolites are also used in the CO_2 capture applications such as zeolites A, β, Y, 13X, ZSM, and chabazites, etc [75]. The CO_2 adsorption capacities of widely used zeolites are shown in table 4.1.

Zeolites are more effective at ambient temperatures and less effective at high temperatures. In order to make them more effective at high temperatures, zeolites modified with amines have been proposed, where physisorption and chemisorption phenomena take place. In particular, various amines such as monoethanalamine

Table 4.1. The widely used zeolites and their CO_2 uptake capacities.

S. No.	Adsorbents	Pressure (bar)	Temperature (°C)	CO_2 uptake (mmol g^{-1})	Reference
1.	Zeolite 13X	1.6	25	5.1	[76]
2.	Zeolite 13X (13X-C)	1	6.2	25	[73]
3.	Zeolite 13X (13X-B)	8–20	4.9	25	[77]
4.	Zeolite 13X (13X-C)	8–20	6.5	25	[77]
5.	Zeolite 13X (13X-F)	8–20	3.9	25	[77]
6.	Zeolite 13X (13X-K)	12–20	6.3	35	[77]
7.	Na-zeolite	1	2.8	25	[78]
8.	Na RTH zeolite	1	3.2	25	[79]
9.	Zeolite	1	1.76	30	[80]
10.	Basalt based zeolite 4A	15	5.9	20	[81]
11.	Merlinoite zeolite	0.1	3.5	25	[82]
12.	Zeolite A	1	0.2	20	[83]

(MEA), polyethanalamine (PEA), and tetraethylenepentamine (TEPA) are used to impregnate the zeolites [84]. These amine impregnated zeolites show better adsorption capacity, owing to the strong interaction between basic amine molecules and acidic CO_2 molecules. The nitrogen atom basicity and density of electrons plays a significant role in choosing suitable amines for CO_2 capture. Sometimes ammonium carbamate will be produced during the interaction between CO_2 and amines. Diffusion interruption of amine impregnated molecules leads to pore blockages which reduce the CO_2 adsorption capacity. For example, zeolite 4A has small pores which are not suitable for CO_2 capture. Therefore, large pore sized zeolites such as 13X and NaY impregnated with amines are used in CO_2 capture applications [85–87]. In 2007 Jadhav *et al* prepared MEA impregnated zeolite 13X and assessed its CO_2 capture efficiency in temperatures between 30 °C and 120 °C. It was found that the CO_2 capture efficiency was increased by 1.6 times at 30 °C and 3.5 times at 120 °C [88]. Recently, in 2019, Karka *et al* reported PEI impregnated zeolite 13X, where they achieved the highest CO_2 adsorption capacity of 1.22 mmol g^{-1} with 60 wt% at 75 °C [84]. Bezerra *et al* synthesized methanolic MEA grafted zeolite 13X for CO_2 uptake. The adsorption capacity of pure zeolites 13X is higher than for the amine grafted zeolites. The adsorption capacity is found to increase with increasing temperature. Even though the adsorption capacities of amine grafted zeolites are low, they provide the superior properties which could lead to new research directions. The observed results from the reports revealed the appearance of strong adsorption sites on the amine grafted zeolite surfaces, indicating the contribution of the sites into the CO_2 selectivity [85]. The newly synthesized zeolites from fly ash through the hydrothermal method have been used successfully for CO_2 adsorption. Although fly ash has been recognized as an environmental hazard, it has great advantages for the preparation of such a useful material. The acid activated zeolite

13X prepared from palm oil mill fly ash through an alkaline fusion route demonstrated a CO_2 uptake capacity of 6.42 mmol g^{-1} [89].

Chen *et al* developed a molecular simulation which simulates the cation exchange zeolite 13X with different quantities of Ca^{2+}, K^+, and Li^+, allowing the design and development of high-performance cation exchange zeolites. The cation exchanged zeolites' CO_2 separation performance was assessed using its N_2/CO_2 selectivity, isosteric heat, adsorption isotherm, energy, and pore volume [90]. Girimonte *et al* developed a new LiX-80, which is a polymetallic cation exchanged zeolite, where Ag^{2+} and Pd^{2+} metals have been incorporated. The designed LiPdAgX zeolites showed good CO_2/N_2 selectivity and higher CO_2 loading than LiX-80 and zeolite 13X [91]. Kim *et al* prepared ethylene diamine (ED) altered zeolite Y through gas phase titration. The free amine group in the modified zeolite capture the CO_2 efficiently from wet flue gas. The adsorption capacity of the prepared material was 1.1 mmol g^{-1}. However, urea was not formed in this process. The ED altered zeolite has a higher H_2O tolerance than pure inorganic zeolites during the CO_2 uptake process. The prepared crystalline adsorbent was hydrothermally stable when compared to conventional amine supports such as amorphous silica, where in the amorphous silica, pore structural collapse takes place during the H_2O vapor exposure comprising CO_2 streams [92, 93].

4.3.3 Metal–organic framework (MOF)-based sorbents for carbon capture

A metal–organic framework is a type of porous crystalline material introduced in 1995 [94]. MOF materials consist of metal-containing nodes linked with organic ligands joined with clusters or inorganic ions through strong coordination bonds. MOFs have well-defined three-dimensional geometric and crystallographic microporous structures. By controlling the geometry and size, more than 12 million MOFs have been designed and synthesized to date (15 October 2022) in the Cambridge structural database (https://www.ccdc.cam.ac.uk/Community/blog/june-2022-csd-update/). These structures are durable and strong, which allows the guest species to be removed, resulting in permanent porosity in MOFs. MOFs can be tuned, synthesize, and designed easily. Compared to the other porous materials such as activated carbon and zeolites, simple optimization of its surface functions and pores make MOFs potential materials for several precise and specific applications such as catalysis [95], adsorption [96], separation [97], and purification [98], etc. MOFs can be classified into two types: rigid and dynamic/flexible. The flexible MOFs consist of a soft and dynamic framework with quick responses to external stimuli such as pressure, temperature, and guest molecules. This outstanding property of external stimulus sensitivity allows MOFs to act as a pressure/temperature-dependent molecular sieve, which makes them traditional adsorbents. On the other hand, rigid MOFs have a strong and stable porous framework with permanent porosity. Additionally, rigid MOFs have been used widely in selective gas adsorption processes, which are based on the molecular sieving effect. This is maybe achieved through the adsorbate–adsorbent and adsorbate–adsorbate interaction strengths. Li *et al* revealed that the adsorption selectivity of rigid MOFs depends on the following three key factors: shape/size exclusion, surface–adsorbate interactions, and concurrent incorporation of both factors.

The chemical structures and chemical formulas of some organic ligands used often in MOFs are provided in table 4.2. In the porous material category, MOFs have excelled due to the following properties: high surface area, good physical adsorption, methane storage, and very high CO_2 uptake. Along with these properties, the intertwined/bridged porous structure of MOFs makes them welcoming hosts of liquid or gas molecules. These particular features of MOFs are the key reasons for studying MOFs in carbon capture applications.

In 2016 Chen *et al* fabricated a 3D MOF with zirconium metals and $-NH_2$ functionalized tetrazole was used as pillars, where a pore volume of 0.17 $cm^3\ g^{-1}$ and an S_{BET} value of 312 m^2/g were observed and they achieved the CO_2 uptake capacity of 50 $cm^3\ g^{-1}$ under 1 bar pressure and 298 K temperature [100]. In 2018 Li *et al* reported the performance and structure of porphyrinic Zr-MOFs by integrating additional open metal sites of Al^{3+} and Fe^{3+}, which showed 112.2% and 22.1% increases of CO_2 uptake capacity when compared to PCN-X at 298 K and 1 bar [101]. In 2016 Ethiraj *et al* developed three cerium-based MOFs, such as MOF-76-Ce-fs, MOF-76-Ce-hs, and MOF-76-Ce-ds, and tested its CO_2 adsorption capacity. They found that the MOF-76-Ce-ds sample showed a 4 mmol g^{-1} sorption capacity at 1.1 bar and 25 °C [102]. Nandi *et al* prepared MOFs with a porous structure of 4.8 and 3.5 Å, which can be suitable for pre-combustion carbon capture applications. They reported a CO_2 adsorption capacity of 3.95 mmol g^{-1} under a pressure of 10 bar and a temperature of 40 °C [103]. A summary of the CO_2 adsorption capacities of some MOF based materials are represented in table 4.3.

Some computational investigations have reported molecular level insights into the adsorption process. The design of MOF structures with a variety of choices makes them suitable for the CO_2 adsorption which can be achieved through molecular simulations. The grand canonical Monte Carlo (GCMS) simulation is widely used to evaluate the gas uptake on MOFs through open access software packages such as RASPA and MuSiC [117, 118]. The examination of CO_2 capture among other molecules and molecular exchange sites with the host are considered to be the advantages of the molecular simulation process. For example, in 2015 Tan *et al* examined the adsorption mechanisms of a variety of gas molecules such as CO_2, O_2, N_2, NO_2, NO, NH_3, and H_2O. The molecular simulations revealed that the CO_2 adsorption was affected by the interaction of guest molecules with the host structure along with the open metal sites. Such interactions lead to CO_2 molecule displacement with other gas molecules [119]. The molecular simulations also opened a door to realize the carbon capture performance of amine functionalized MOFs. In 2018 Lee *et al* studied the required binding energies between CO_2 molecules and amine functionalized X_2 type MOFs (X = Zn, Co, Fe, Mn, Mg). Additionally, it is also used to identify the adsorption sites and electrostatic interactions of the molecules. The results show that the calculated adsorption enthalpy of functionalized MOFs is more effective than for non-functionalized MOFs, even in the presence of water [120].

Table 4.2. Chemical structure and chemical formulas of some organic ligands used very often in MOFs. (Adapted from [99]. CC BY 4.0.)

S. No.	MOF name	Organic ligand	Organic ligand structure
01.	MOF-177	BTB: 4,4′,4"-benzene-1,3,5-triyl-tribenzonate.	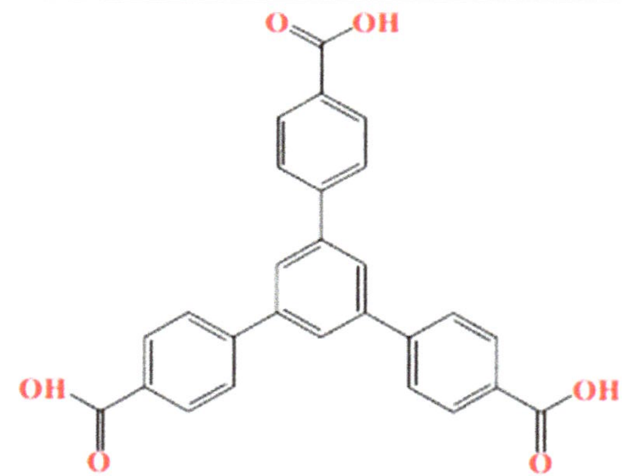
02.	MOF-180	BTE: 4,4′,4"-benzene-1,3,5-triyl-tris (ethyne-2,1-diyl) benzoate.	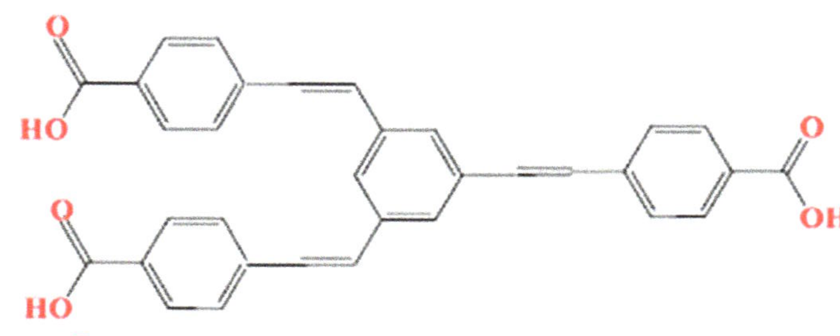

(*Continued*)

Table 4.2. *(Continued)*

S. No.	MOF name	Organic ligand	Organic ligand structure
03.	MOF-200	BBC: 4,4′,4"-benzene-1,3,5-triyl-tris (benzene-4,1-diyl) tribenzoate	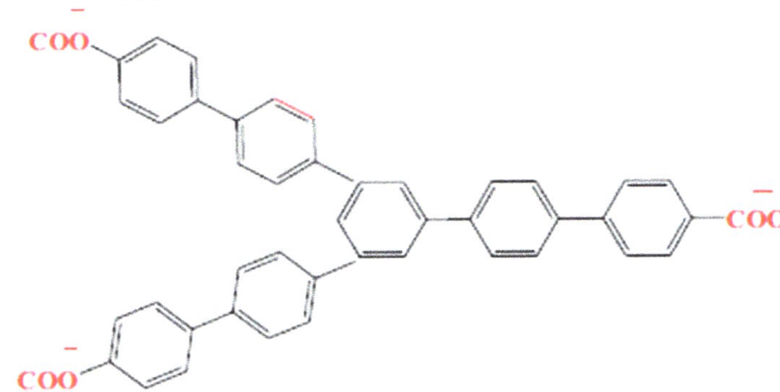
04.	MOF-205	BTB+ 2,6-napthalenedicarboxylate (NDC)	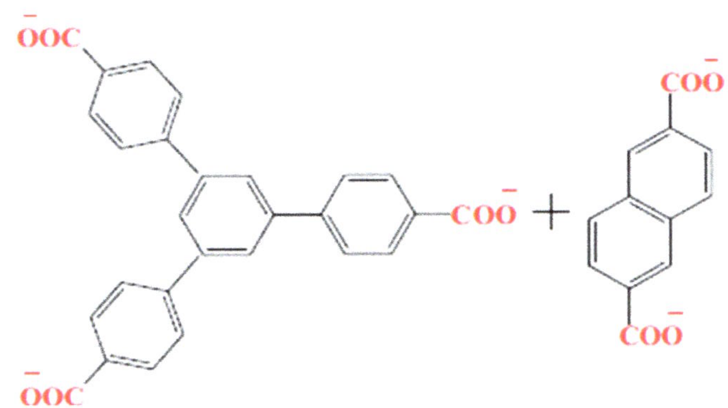

05.	MOF-210	BTE+ biphenyl-4,40-dicarboxilate (BPDC)	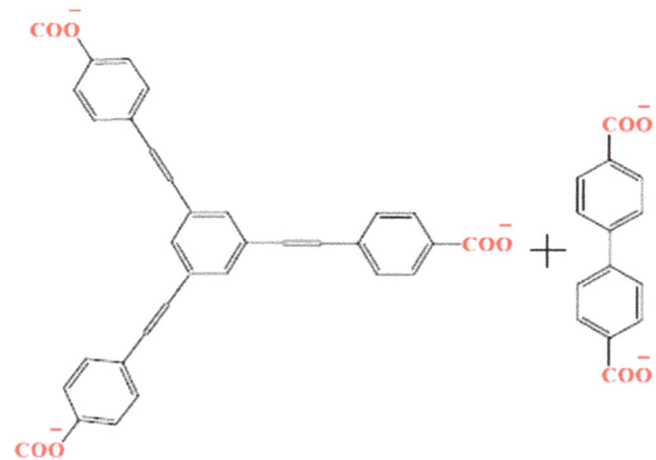

Table 4.3. The CO_2 adsorption capacity of MOF based materials.

MOF	Pressure (bar)	Temperature (°C)	CO_2 uptake (mmol g^{-1})	Reference
UMCM-1-NH_2-MA	18	25	19.8	[104]
Mg-MOF-74	1	23	8	[105]
MIL-53 with BNH_x	1	0	4.5	[106]
MOF with PEI	0.15	78	4.2	[107]
Zn(II)-based MOFs	1	25	9.2	[108]
Zn doped Ni-ZIF-8	1	0	4.3	[109]
HCM-$Cu_3(BTC)_2$-3	1	25	2.8	[110]
MOF-5/graphite oxide	4	25	1.1	[111]
PCN-124	1	0	9.1	[112]
NJU-Bai12	20	0	23.8	[113]
A core–shell MOF	1	0	41	[114]
2D-MOF	1	0	2.9	[115]
1D-MOF	1	78	4.0	[106]
SNU-110	1	78	6	[116]

4.4 Summary of adsorbent materials

In summary, various kinds of sorbent materials have been used for carbon capture applications. The suitability of a sorbent material depends on the following properties, such as production cost, moisture tolerance, cycling stability, mechanical and thermal stability, adsorption/desorption temperature, rate of adsorption/desorption, selectivity, and adsorption capacity. The pros and cons of the different adsorbents are listed in table 4.4.

4.5 Conclusion

This chapter analyzes the recent progresses in the solid (particularly carbon-based, zeolite based, and MOF based) sorbents for carbon capture applications. The pre-combustion carbon capture method is promising for CO_2 capture, but it is facing several challenges to improve its overall efficiency. In particular, the solvent regeneration must be performed at low temperatures to protect from solvent degradation. In the oxy-fuel combustion method, novel methods such as oxygen-transport membranes and ion-transport along with chemical looping have to be investigated to reduce the air separation unit cost. In the post-combustion carbon capture method, the solvent should have low degradation, low corrosiveness, low production costs, high cyclic capacities, and lower by-product formation. Although the pre-combustion carbon capture method provides more efficiency than post-combustion carbon capture, it is highly expensive. To decrease the production costs, it is essential to find a novel absorption solvent. The captured CO_2 can be utilized for two main purposes: (i) physical utilization of CO_2 and (ii) conversion to useful chemicals and fuels. In addition to that, it can also used in carbonated beverages, a welding medium, a refrigerant, and in fire extinguishers, etc. The direct applications

Table 4.4. Advantages and disadvantages of the different types of adsorbents.

Sorbents	Advantages	Disadvantages
MOFs	The surface properties and pore sizes are easily tunable. It has high porosity and a high surface area. Its CO_2 uptake capacity is high.	The synthesis process is expensive and complex. There are impurities and the moisture tolerance level is low. The thermal stability is limited.
Natural zeolites	Low cost. High availability.	The separation performance is limited. As a result of different chemical compositions, its purity is low.
Synthetic zeolites	The thermal stability is high. The CO_2 uptake capacity is high. It has high porosity and a high surface area. The surface properties and pore sizes are easily tunable.	The synthesis process is expensive. The selectivity is low. There are impurities and the moisture tolerance level is low.
Carbon nanomaterials	Quick adsorption kinetics. Excellent physio-chemical stability. Good moisture tolerance. CO_2 uptake capacity is high. Large specific surface area.	The selectivity is limited. The synthesis process is expensive and complex.
Activated carbon	Availability of precursor is wide. Ease of regeneration. Moisture tolerance is good. Large specific surface area. Thermal stability is good. Low cost.	High friability. The selectivity is low. The CO_2 uptake capacity is low.

of CO_2 are limited, but indirect applications of CO_2 in large-scale industries are used to improve the performance of different processes. Different types of solid sorbents such as carbonaceous materials, zeolites, and MOFs, and their CO_2 adsorption capacities are discussed in detail herein due to their unique features such as durability, tunable microporosities, hierarchical porosities, and high surface area. It is clearly understood that the capture efficiency was greatly improved when we functionalize the materials. There are various types of functionalization, such as mono-/co-doping, heteroatom doping, chemical and physical activation, imine and amine functionalization, composite and hybrid formation, acidic treatment, metal oxide impregnation, etc. Furthermore, the functionalized materials exhibit higher stability, high isostatic heats of adsorption, higher CO_2 selectivity, and good regenerative abilities.

References

[1] Dong G, Zhang Y, Hou J, Shen J and Chen V 2016 Graphene oxide nanosheets based novel facilitated transport membranes for efficient CO_2 capture *Ind. Eng. Chem. Res.* **55** 5403–14

[2] Arulraj A, Thachnatharen N, Rajkumar C and Mangalaraja R V 2022 Engineering 2D chalcogenides for energy and environmental remediation Nanostructured Materials for Sustainable Energy and Environmental Remediation (Bristol: IOP Publishing) pp 2-1–2-28

[3] Wang R, Xi S C, Wang D Y, Dou M and Dong B 2021 Defluorinated porous carbon nanomaterials for CO_2 capture *ACS Appl. Nano Mater.* **4** 10148–54

[4] Liu F Q, Wang L L, Li G H, Li W and Li C Q 2017 Hierarchically structured graphene coupled microporous organic polymers for superior CO_2 capture *ACS Appl. Mater. Interfaces* **9** 33997–4004

[5] Wang S, Mahurin S M, Dai S and Jiang D E 2021 Design of graphene/ionic liquid composites for carbon capture *ACS Appl. Mater. Interfaces* **13** 17511–6

[6] Wang J, Huang L, Yang R, Zhang Z, Wu J, Gao Y, Wang Q, O'Hare D and Zhong Z 2014 Recent advances in solid sorbents for CO_2 capture and new development trends *Energy Environ. Sci.* **7** 3478–518

[7] An L, Liu S, Wang L, Wu J, Wu Z, Ma C, Yu Q and Hu X 2019 Novel nitrogen-doped porous carbons derived from graphene for effective CO_2 capture *Ind. Eng. Chem. Res.* **58** 3349–58

[8] Kumar S, Srivastava R and Koh J 2020 Utilization of zeolites as CO_2 capturing agents: advances and future perspectives *J. CO_2 Util.* **41** 101251

[9] Zou L, Sun Y, Che S, Yang X, Wang X, Bosch M, Wang Q, Li H, Smith M and Yuan S 2017 Porous organic polymers for post-combustion carbon capture *Adv. Mater.* **29** 1700229

[10] Kamran U and Park S J 2021 Chemically modified carbonaceous adsorbents for enhanced CO_2 capture: a review *J. Clean. Prod.* **290** 125776

[11] Li P and Zeng H C 2017 Hierarchical nanocomposite by the integration of reduced graphene oxide and amorphous carbon with ultrafine MgO nanocrystallites for enhanced CO_2 capture *Environ. Sci. Technol.* **51** 12998–3007

[12] Chen C, Zhang S, Row K H and Ahn W S 2017 Amine–silica composites for CO_2 capture: a short review *J. Energy Chem.* **26** 868–80

[13] Ghanbari T, Abnisa F and Daud W M A W 2020 A review on production of metal organic frameworks (MOF) for CO_2 adsorption *Sci. Total Environ.* **707** 135090

[14] Zeng Y, Zou R and Zhao Y 2016 Covalent organic frameworks for CO_2 capture *Adv. Mater.* **28** 2855–73

[15] Dutcher B, Fan M and Russell A G 2015 Amine-based CO_2 capture technology development from the beginning of 2013: a review *ACS Appl. Mater. Interfaces* **7** 2137–48

[16] Ma L, Bai R, Hu G, Chen R, Hu X, Dai W, Dacosta H F and Fan M 2013 Capturing CO_2 with amine-impregnated titanium oxides *Energy Fuels* **27** 5433–9

[17] Creamer A E and Gao B 2016 Carbon-based adsorbents for postcombustion CO_2 capture: a critical review *Environ. Sci. Technol.* **50** 7276–89

[18] Nelson T O, Coleman L J, Kataria A, Lail M, Soukri M, Quang D V and Zahra M R A 2014 Advanced solid sorbent-based CO_2 capture process *Energy Procedia* **63** 2216–29

[19] Lee K B and Sircar S 2008 Removal and recovery of compressed CO_2 from flue gas by a novel thermal swing chemisorption process *AIChE J.* **54** 2293–302

[20] Yang Z Z, He L N, Gao J, Liu A H and Yu B 2012 Carbon dioxide utilization with C–N bond formation: carbon dioxide capture and subsequent conversion *Energy Environ. Sci.* **5** 6602–39

[21] Nagarajan L, Sampath Kumar P, Arulraj A, Senguttuvan G and Kumaraguru K 2020 Solar assisted reduced graphene oxide as adsorbent for carbon dioxide and its kinetic studies *Physica E: Low Dimens. Sys. and Nanostructures* **116** 113739

[22] Osman A I, Hefny M, Abdel Maksoud M, Elgarahy A M and Rooney D W 2021 Recent advances in carbon capture storage and utilisation technologies: a review *Environ. Chem. Lett.* **19** 797–849

[23] Osman A I, Abu-Dahrieh J K, Cherkasov N, Fernandez-Garcia J, Walker D, Walton R I, Rooney D W and Rebrov E 2018 A highly active and synergistic $Pt/Mo_2C/Al_2O_3$ catalyst for water-gas shift reaction *Mol. Catal.* **455** 38–47

[24] Cao M, Zhao L, Xu D, Ciora R, Liu P K, Manousiouthakis V I and Tsotsis T T 2020 A carbon molecular sieve membrane-based reactive separation process for pre-combustion CO_2 capture *J. Membr. Sci.* **605** 118028

[25] Carbo M C, Jansen D, Hendriks C, de Visser E, Ruijg G J and Davison J 2009 Opportunities for CO_2 capture through oxygen conducting membranes at medium-scale oxyfuel coal boilers *Energy Procedia* **1** 487–94

[26] Xu Y, Ren Q, Zheng Z J and He Y L 2017 Evaluation and optimization of melting performance for a latent heat thermal energy storage unit partially filled with porous media *Appl. Energy* **193** 84–95

[27] Stéphenne K 2014 Start-up of world's first commercial post-combustion coal fired CCS project: contribution of Shell Cansolv to Sask Power Boundary Dam ICCS project *Energy Procedia* **63** 6106–10

[28] Yun S, Oh S Y and Kim J K 2020 Techno-economic assessment of absorption-based CO_2 capture process based on novel solvent for coal-fired power plant *Appl. Energy* **268** 114933

[29] Bui M, Adjiman C S, Bardow A, Anthony E J, Boston A, Brown S, Fennell P S, Fuss S, Galindo A and Hackett L A 2018 Carbon capture and storage (CCS): the way forward *Energy Environ. Sci.* **11** 1062–176

[30] Campbell M 2014 Technology innovation and advancements for Shell Cansolv CO_2 capture solvents *Energy Procedia* **63** 801–7

[31] Karthik P, Himaja A and Singh S P 2014 Carbon-allotropes: synthesis methods, applications and future perspectives *Carbon Lett.* **15** 219–37

[32] Zhang Y and Park S J 2019 Imidazolium-optimized conductive interfaces in multilayer graphene nanoplatelet/epoxy composites for thermal management applications and electro-active devices *Polymer* **168** 53–60

[33] Dong H, Lin B, Gilmore K, Hou T, Lee S T and Li Y 2015 B40 fullerene: an efficient material for CO_2 capture, storage and separation *Curr. Appl. Phys.* **15** 1084–9

[34] Andreoli E, Dillon E P, Cullum L, Alemany L B and Barron A R 2014 Cross-linking amine-rich compounds into high performing selective CO_2 absorbents *Sci. Rep.* **4** 1–5

[35] Andreoli E and Barron A R 2015 Activation effect of fullerene C60 on the carbon dioxide absorption performance of amine-rich polypropylenimine dendrimers *Chem. Sus. Chem.* **8** 2635–44

[36] Esrafili M D and Abolghasemzadeh S 2022 Carbon dioxide storage and separation using all-boron B38 fullerene: DFT calculations *Chem. Phys. Lett.* **790** 139361

[37] Novoselov K S, Geim A K, Morozov S V, Jiang D e, Zhang Y, Dubonos S V, Grigorieva I V and Firsov A A 2004 Electric field effect in atomically thin carbon films *Science* **306** 666–9
[38] Neto A C, Guinea F, Peres N M, Novoselov K S and Geim A K 2009 The electronic properties of graphene *Rev. Mod. Phys.* **81** 109
[39] Ghosh A, Subrahmanyam K, Krishna K S, Datta S, Govindaraj A, Pati S K and Rao C 2008 Uptake of H_2 and CO_2 by graphene *J. Phys. Chem.* C **112** 15704–7
[40] Ramar V and Balasubramanian K 2021 Reduced graphene oxide/WO_3 nanorod composites for photocatalytic degradation of methylene blue under sunlight irradiation *ACS Appl. Nano Mater.* **4** 5512–21
[41] Ramar V and Karthikeyan B 2021 Effect of reduced graphene oxide on the sunlight-driven photocatalytic activity of rGO/h-MoO_3 nanocomposites *J. Phys.* D: Appl. Phys. **54** 155502
[42] Ramar V and Karthikeyan B 2020 Unravelling the synergistic effect of reduced graphene oxide on optical, phonon and optical power limiting properties of rGO/α-MoO_3 nanohybrids *Appl. Phys.* A **126** 1–9
[43] Ramar V and Balasubramanian K 2020 Strong violet emission and optical power limiting properties of reduced graphene oxide/MoO_3 synergistic composites *J. Appl. Phys.* **127** 193102
[44] Ramar V and Balasubramanian K 2019 Optical and highly enhanced solar light-driven photocatalytic activity of reduced graphene oxide wrapped α-MoO_3 nanoplates *Sol. Energy* **194** 1–10
[45] Ramar V and Balasubramanian K 2018 Charge transfer induced tunable bandgap and enhanced saturable absorption behavior in rGO/WO_3 composites *Appl. Phys.* A **124** 1–11
[46] Swaminathan H, Ramar V and Balasubramanian K 2017 Excited-state electron and energy transfer dynamics between 2D MoS_2 and GO/RGO for turn ON BSA/HSA sensing *J. Phys. Chem.* C **121** 12585–92
[47] Ramar V and Balraj A 2022 Critical review on carbon-based nanomaterial for carbon capture: technical challenges, opportunities, and future perspectives *Energy Fuels* **36** 13479–505
[48] Vinodh R, Babu C M, Abidov A, Palanichamy M and Jang H T 2019 Facile synthesis of amine modified silica/reduced graphene oxide composite sorbent for CO_2 adsorption *Mater. Lett.* **247** 44–7
[49] Shin G J, Rhee K and Park S J 2016 Improvement of CO_2 capture by graphite oxide in presence of polyethylenimine *Int. J. Hydrogen Energy* **41** 14351–9
[50] Liu L, Su S, Xu K, Li H, Qing M, Hu S, Wang Y and Xiang J 2019 Insights into the highly efficient Co modified MnSm/Ti catalyst for selective catalytic reduction of NO_x with NH_3 at low temperature *Fuel* **255** 115798
[51] Hsan N, Dutta P, Kumar S, Bera R and Das N 2019 Chitosan grafted graphene oxide aerogel: synthesis, characterization and carbon dioxide capture study *Int. J. Biol. Macromol.* **125** 300–6
[52] Pokhrel J, Bhoria N, Anastasiou S, Tsoufis T, Gournis D, Romanos G and Karanikolos G N 2018 CO_2 adsorption behavior of amine-functionalized ZIF-8, graphene oxide, and ZIF-8/graphene oxide composites under dry and wet conditions *Microporous Mesoporous Mater.* **267** 53–67
[53] Khalili S, Ghoreyshi A A, Jahanshahi M and Pirzadeh K 2013 Enhancement of carbon dioxide capture by amine-functionalized multi-walled carbon nanotube. *Clean–Soil, Air, Water* **41** 939–48

[54] Rahimi K, Riahi S, Abbasi M and Fakhroueian Z 2019 Modification of multi-walled carbon nanotubes by 1,3-diaminopropane to increase CO_2 adsorption capacity *J. Environ. Manage.* **242** 81–9
[55] Lee M S, Lee S Y and Park S J 2015 Preparation and characterization of multi-walled carbon nanotubes impregnated with polyethyleneimine for carbon dioxide capture *Int. J. Hydrogen Energy* **40** 3415–21
[56] Liu Q, Shi Y, Zheng S, Ning L, Ye Q, Tao M and He Y 2014 Amine-functionalized low-cost industrial grade multi-walled carbon nanotubes for the capture of carbon dioxide *J. Energy Chem.* **23** 111–8
[57] Omidfar N, Mohamadalizadeh A and Mousavi S H 2015 Carbon dioxide adsorption by modified carbon nanotubes *Asia-Pacific J. Chem. Eng.* **10** 885–92
[58] Jena K K, Panda A P, Verma S, Mani G K, Swain S and Alhassan S M 2019 MWCNTs-ZnO-SiO_2 mesoporous nano-hybrid materials for CO_2 capture *J. Alloys Compd.* **800** 279–85
[59] Zohdi S, Anbia M and Salehi S 2019 Improved CO_2 adsorption capacity and CO_2/CH_4 and CO_2/N_2 selectivity in novel hollow silica particles by modification with multi-walled carbon nanotubes containing amine groups *Polyhedron* **166** 175–85
[60] González Plaza M, Pevida García C, Arias Rozada B, Casal Banciella M D, Fernández Martín C, Fermoso Domínguez J, Pis Martínez J J and Rubiera González F 2009 Different approaches for the development of low-cost CO2 adsorbents *J. Env. Eng.* **135** 426–32
[61] Dubinin M 1966 *Chemistry and Physics of Carbon* (New York: Marcel Dekker) p 51
[62] Mangun C L, Benak K R, Daley M A and Economy J 1999 Oxidation of activated carbon fibers: effect on pore size, surface chemistry, and adsorption properties *Chem. Mater.* **11** 3476–83
[63] Pevida C, Plaza M G, Arias B, Fermoso J, Rubiera F and Pis J 2008 Surface modification of activated carbons for CO_2 capture *Appl. Surf. Sci.* **254** 7165–72
[64] Plaza M, Pevida C, Arias B, Fermoso J, Rubiera F and Pis J 2009 A comparison of two methods for producing CO_2 capture adsorbents *Energy Procedia* **1** 1107–13
[65] Xu R, Pang W, Yu J, Huo Q and Chen J 2009 *Chemistry of Zeolites and Related Porous Materials: Synthesis and Structure* (New York: Wiley)
[66] Storck S, Bretinger H and Maier W F 1998 Characterization of micro- and mesoporous solids by physisorption methods and pore-size analysis *Appl. Catal.* A **174** 137–46
[67] Auerbach S M, Carrado K A and Dutta P K 2003 *Handbook of Zeolite Science and Technology* (Boca Raton, FL: CRC Press)
[68] Lee S Y and Park S J 2015 A review on solid adsorbents for carbon dioxide capture *J. Ind. Eng. Chem.* **23** 1–11
[69] Choi S, Drese J H and Jones C W 2009 Adsorbent materials for carbon dioxide capture from large anthropogenic point sources *Chem. Sus. Chem.* **2** 796–854
[70] Krishna R and van Baten J M 2010 In silico screening of zeolite membranes for CO_2 capture *J. Membr. Sci.* **360** 323–33
[71] Ramasubramanian K, Severance M A, Dutta P K and Ho W W 2015 Fabrication of zeolite/polymer multilayer composite membranes for carbon dioxide capture: deposition of zeolite particles on polymer supports *J. Colloid Interface Sci.* **452** 203–14
[72] Ge Q, Shao J, Wang Z and Yan Y 2012 Effects of the synthesis hydrogel on the formation of zeolite LTA membranes *Microporous Mesoporous Mater.* **151** 303–10
[73] Chen C, Park D W and Ahn W S 2014 CO_2 capture using zeolite 13X prepared from bentonite *Appl. Surf. Sci.* **292** 63–7

[74] Maurin G, Llewellyn P and Bell R 2005 Adsorption mechanism of carbon dioxide in faujasites: grand canonical Monte Carlo simulations and microcalorimetry measurements *J. Phys. Chem.* B **109** 16084–91
[75] Akhtar F, Liu Q, Hedin N and Bergström L 2012 Strong and binder free structured zeolite sorbents with very high CO_2-over-N_2 selectivities and high capacities to adsorb CO_2 rapidly *Energy Environ. Sci.* **5** 7664–73
[76] Cavenati S, Grande C A and Rodrigues A E 2004 Adsorption equilibrium of methane, carbon dioxide, and nitrogen on zeolite 13X at high pressures *J. Chem. Eng. Data* **49** 1095–101
[77] Garshasbi V, Jahangiri M and Anbia M 2017 Equilibrium CO_2 adsorption on zeolite 13X prepared from natural clays *Appl. Surf. Sci.* **393** 225–33
[78] Yang S T, Kim J and Ahn W S 2010 CO_2 adsorption over ion-exchanged zeolite beta with alkali and alkaline Earth metal ions *Microporous Mesoporous Mater.* **135** 90–4
[79] Jo D, Lim J B, Ryu T, Nam I S, Camblor M A and Hong S B 2015 Unseeded hydroxide-mediated synthesis and CO_2 adsorption properties of an aluminosilicate zeolite with the RTH topology *J. Mater. Chem.* A **3** 19322–9
[80] Xu X, Zhao X, Sun L and Liu X 2009 Adsorption separation of carbon dioxide, methane and nitrogen on monoethanol amine modified β-zeolite *J. Nat. Gas Chem.* **18** 167–72
[81] Hwang K J, Choi W S, Jung S H, Kwon Y J, Hong S, Choi C, Lee J W and Shim W G 2018 Synthesis of zeolitic material from basalt rock and its adsorption properties for carbon dioxide *RSC Adv.* **8** 9524–9
[82] Georgieva V M, Bruce E L, Verbraeken M C, Scott A R, Casteel W J, Brandani S and Wright P A 2019 Triggered gate opening and breathing effects during selective CO_2 adsorption by merlinoite zeolite *J. Am. Chem. Soc.* **141** 12744–59
[83] Rzepka P, Bacsik Z, Pell A J, Hedin N and Jaworski A 2019 Nature of chemisorbed CO_2 in zeolite A *J. Phys. Chem.* C **123** 21497–503
[84] Karka S, Kodukula S, Nandury S V and Pal U 2019 Polyethylenimine-modified zeolite 13X for CO_2 capture: adsorption and kinetic studies *ACS Omega* **4** 16441–9
[85] Bezerra D P, da Silva F W, de Moura P A, Sousa A G, Vieira R S, Rodriguez-Castellon E and Azevedo D C 2014 CO_2 adsorption in amine-grafted zeolite 13X *Appl. Surf. Sci.* **314** 314–21
[86] Su F, Lu C, Kuo S C and Zeng W 2010 Adsorption of CO_2 on amine-functionalized Y-type zeolites *Energy Fuels* **24** 1441–8
[87] Bezerra D P, Oliveira R S, Vieira R S, Cavalcante C L and Azevedo D 2011 Adsorption of CO_2 on nitrogen-enriched activated carbon and zeolite 13X *Adsorption* **17** 235–46
[88] Jadhav P, Chatti R, Biniwale R, Labhsetwar N, Devotta S and Rayalu S 2007 Monoethanol amine modified zeolite 13X for CO_2 adsorption at different temperatures *Energy Fuels* **21** 3555–9
[89] Kongnoo A, Tontisirin S, Worathanakul P and Phalakornkule C 2017 Surface characteristics and CO_2 adsorption capacities of acid-activated zeolite 13X prepared from palm oil mill fly ash *Fuel* **193** 385–94
[90] Chen S, Zhu M, Tang Y, Fu Y, Li W and Xiao B 2018 Molecular simulation and experimental investigation of CO_2 capture in a polymetallic cation-exchanged 13X zeolite *J. Mater. Chem.* A **6** 19570–83
[91] Girimonte R, Formisani B and Testa F 2017 Adsorption of CO_2 on a confined fluidized bed of pelletized 13X zeolite *Powder Technol.* **311** 9–17

[92] Kim C, Cho H S, Chang S, Cho S J and Choi M 2016 An ethylenediamine-grafted Y zeolite: a highly regenerable carbon dioxide adsorbent via temperature swing adsorption without urea formation *Energy Environ. Sci.* **9** 1803–11

[93] Li W, Bollini P, Didas S A, Choi S, Drese J H and Jones C W 2010 Structural changes of silica mesocellular foam supported amine-functionalized CO_2 adsorbents upon exposure to steam *ACS Appl. Mater. Interfaces* **2** 3363–72

[94] Yaghi O M, Li G and Li H 1995 Selective binding and removal of guests in a microporous metal–organic framework *Nature* **378** 703–6

[95] Zhao M, Yuan K, Wang Y, Li G, Guo J, Gu L, Hu W, Zhao H and Tang Z 2016 Metal–organic frameworks as selectivity regulators for hydrogenation reactions *Nature* **539** 76–80

[96] Mason J A, Oktawiec J, Taylor M K, Hudson M R, Rodriguez J, Bachman J E, Gonzalez M I, Cervellino A, Guagliardi A and Brown C M 2015 Methane storage in flexible metal–organic frameworks with intrinsic thermal management *Nature* **527** 357–61

[97] Cadiau A, Adil K, Bhatt P, Belmabkhout Y and Eddaoudi M 2016 A metal–organic framework-based splitter for separating propylene from propane *Science* **353** 137–40

[98] Chen C X, Zheng S P, Wei Z W, Cao C C, Wang H P, Wang D, Jiang J J, Fenske D and Su C Y 2017 A robust metal–organic framework combining open metal sites and polar groups for methane purification and CO_2/fluorocarbon capture *Chem. Eur. J.* **23** 4060–4

[99] Elhenawy S E M, Khraisheh M, AlMomani F and Walker G 2020 Metal–organic frameworks as a platform for CO_2 capture and chemical processes: adsorption, membrane separation, catalytic-conversion, and electrochemical reduction of CO_2 *Catalysts* **10** 1293

[100] Chen D M, Tian J Y, Chen M, Liu C S and Du M 2016 Moisture-stable Zn (II) metal–organic framework as a multifunctional platform for highly efficient CO_2 capture and nitro pollutant vapor detection *ACS Appl. Mater. Interfaces* **8** 18043–50

[101] Li Z, Li X, Chen C, Zhou L, Guo Q, Yuan D, Wan H, Ding J and Guan G 2018 Extra unsaturated metal centers of zirconium-based MOFs: a facile approach towards increasing CO_2 uptake capacity at low pressure *Eur. J. Inorg. Chem.* **2018** 194–202

[102] Ethiraj J, Bonino F, Vitillo J G, Lomachenko K A, Lamberti C, Reinsch H, Lillerud K P and Bordiga S 2016 Solvent-driven gate opening in MOF-76-Ce: effect on CO_2 adsorption *Chem. Sus. Chem.* **9** 713–9

[103] Nandi S, De Luna P, Daff T D, Rother J, Liu M, Buchanan W, Hawari A I, Woo T K and Vaidhyanathan R 2015 A single-ligand ultra-microporous MOF for precombustion CO_2 capture and hydrogen purification *Sci. Adv.* **1** e1500421

[104] Xiang Z, Peng X, Cheng X, Li X and Cao D 2011 CNT@ Cu_3 (BTC) 2 and metal–organic frameworks for separation of CO_2/CH_4 mixture *J. Phys. Chem.* C **115** 19864–71

[105] Caskey S R, Wong-Foy A G and Matzger A J 2008 Dramatic tuning of carbon dioxide uptake via metal substitution in a coordination polymer with cylindrical pores *J. Am. Chem. Soc.* **130** 10870–1

[106] Bataille T, Bracco S, Comotti A, Costantino F, Guerri A, Ienco A and Marmottini F 2012 Solvent dependent synthesis of micro-and nano-crystalline phosphinate based 1D tubular MOF: structure and CO_2 adsorption selectivity *CrystEngComm* **14** 7170–3

[107] Lin Y, Yan Q, Kong C and Chen L 2013 Polyethyleneimine incorporated metal–organic frameworks adsorbent for highly selective CO_2 capture *Sci. Rep.* **3** 1–7

[108] Masoomi M Y, Stylianou K C, Morsali A, Retailleau P and Maspoch D 2014 Selective CO_2 capture in metal–organic frameworks with azine-functionalized pores generated by mechanosynthesis *Cryst. Growth Des.* **14** 2092–6

[109] Yu J and Balbuena P B 2013 Water effects on postcombustion CO_2 capture in Mg-MOF-74 *J. Phys. Chem.* C **117** 3383–8
[110] Qian D, Lei C, Hao G P, Li W C and Lu A H 2012 Synthesis of hierarchical porous carbon monoliths with incorporated metal–organic frameworks for enhancing volumetric based CO_2 capture capability *ACS Appl. Mater. Interfaces* **4** 6125–32
[111] Zhao Y, Ding H and Zhong Q 2013 Synthesis and characterization of MOF-aminated graphite oxide composites for CO_2 capture *Appl. Surf. Sci.* **284** 138–44
[112] Park J, Li J R, Chen Y P, Yu J, Yakovenko A A, Wang Z U, Sun L B, Balbuena P B and Zhou H C 2012 A versatile metal–organic framework for carbon dioxide capture and cooperative catalysis *Chem. Commun.* **48** 9995–7
[113] Zheng B, Yun R, Bai J, Lu Z, Du L and Li Y 2013 Expanded porous MOF-505 analogue exhibiting large hydrogen storage capacity and selective carbon dioxide adsorption *Inorg. Chem.* **52** 2823–9
[114] Li T, Sullivan J E and Rosi N L 2013 Design and preparation of a core–shell metal–organic framework for selective CO_2 capture *J. Am. Chem. Soc.* **135** 9984–7
[115] Yan Q, Lin Y, Wu P, Zhao L, Cao L, Peng L, Kong C and Chen L 2013 Designed synthesis of functionalized two-dimensional metal–organic frameworks with preferential CO_2 capture *Chem. Plus. Chem.* **78** 86–91
[116] Hong D H and Suh M P 2012 Selective CO_2 adsorption in a metal–organic framework constructed from an organic ligand with flexible joints *Chem. Commun.* **48** 9168–70
[117] Sweatman M B 2013 Preface to the special issue on 'Monte Carlo Codes, Tools and Algorithms' *Mol. Simul.* **39** 1123–4
[118] Dubbeldam D, Calero S, Ellis D E and Snurr R Q 2016 RASPA: molecular simulation software for adsorption and diffusion in flexible nanoporous materials *Mol. Simul.* **42** 81–101
[119] Tan K, Zuluaga S, Gong Q, Gao Y, Nijem N, Li J, Thonhauser T and Chabal Y J 2015 Competitive coadsorption of CO_2 with H_2O, NH_3, SO_2, NO, NO_2, N_2, O_2, and CH_4 in M-MOF-74 (M = Mg, Co, Ni): the role of hydrogen bonding *Chem. Mater.* **27** 2203–17
[120] Lee J H, Siegelman R L, Maserati L, Rangel T, Helms B A, Long J R and Neaton J B 2018 Enhancement of CO_2 binding and mechanical properties upon diamine functionalization of M_2(dobpdc) metal–organic frameworks *Chem. Sci.* **9** 5197–206
[121] Li Y and Yu J 2017 Applications of zeolites in sustainable chemistry *Chem* **3** 928–49

IOP Publishing

Materials Technology for the Energy and Environmental Nexus, Volume 1

R V Mangalaraja, Arunachalam Arulraj, Nalandhiran Pugazhenthiran and Chan Siew Hwa

Chapter 5

MOF-based electrocatalysts for oxygen evolution reactions

Sam Sankar Selvasundarasekar and Subrata Kundu

Metal organic framework (MOF) based materials are highly anticipated in the field of electrocatalysis. The physical and mechanical stability of MOFs make them a topic of enormous interest in electrochemistry. MOFs are usually assembled with metal ions and organic linkers. They have received attention owing to their properties such as large surface area and flexible tuning of their functionalities. The oxygen evolution reaction (OER) is a bottleneck for the electrocatalytic water splitting reaction which effectively hinders the overall energy required for total water splitting, specifically the hydrogen evolution reaction (HER). Hence, finding Earth-abundant metal catalysts is highly desirable. Adding to the above-mentioned advantages, MOFs have been considered as suitable candidates for the electrocatalytic OER with their vast exposed active surface area. This chapter highlights the synthesis of MOFs and MOF derived materials and their utilization in electrocatalytic OER studies.

5.1 Introduction

The global energy demand has been increasing rapidly, owing to the fast expansion of the human population and the growth of industrialization. This scenario has led to the over-use of traditional fossil fuels. A great disadvantage associated with conventional fossil fuels is the emission of CO_2 as a by-product [1, 2]. It causes huge environmental problems such as global temperature increases, climatic changes, the greenhouse effect, air pollution, water pollution, etc. In addition, the limited resources of fossil fuels will also struggle to meet future energy needs. Hence, finding an alternative to the fossil fuels is highly necessary in the current scenario. For sustainable and clean energy development, the alternative fuel should meet the

doi:10.1088/978-0-7503-5724-1ch5

criteria of environmental friendliness, cost-effectiveness, and efficiency similar to fossil fuels. Among the different energy carriers, hydrogen fuel is considered as one of the best fuels for a sustainable future owing to its high combustion efficiency, non-toxicity, non-polluting nature, and, most importantly, renewability [2–4].

Many processes are available for the production of hydrogen fuel. Among these electrocatalytic water splitting and steam reforming of fossil fuels are the most efficient. However, our requirement is to fulfill the aforementioned criteria. Electrocatalytic water splitting is one of the pioneering ways to produce high purity hydrogen in an efficient way and is commonly understood to undergo two basic reactions, the hydrogen evolution reaction (HER) occurs at the cathode and the oxygen evolution reaction (OER) normally occurs at the anode [5–7]. Of these reactions, OER is a bottleneck for the production of hydrogen. Since it is a demanding four proton–electron coupled electron transfer process, the formation of the oxygen–oxygen bond is not favored kinetically under the reaction pathway. Thus the catalyst is highly recommended to expedite the OER process. Noble metal-based catalysts such as IrO_2 and RuO_2 are utilized as electrocatalysts to reduce the overpotential and Tafel slope value of the OER [8]. Due to the scarcity and high cost of the noble metal catalysts, they are highly restricted in large scale production. Hence, many efforts have been made to produce cost-effective transition metal based oxide [9, 10], hydroxide [11, 12], nitride [13, 14], carbide [15, 16], and phosphide [17, 18] materials to catalyze the OER process under lower overpotential regions.

Recently, metal organic framework (MOF) based materials have been gaining interest, owing to their novel physio-chemical properties. These are new types of materials which show unique optoelectronic and catalytic properties. They can offer several advantages such as flexible control of their functionalities through ligands as well as the metal centers, tunable pore size, a large surface area to volume ratio, and rich combinations of metal composites [19]. Normally, the metal centers are separated by organic linkers which are considered to be quantum dots. A specific surface area and the band gap of the MOFs are achieved mainly through the tailoring of organic ligands as linkers and/or metal centers. These phenomena maximize the electrocatalytic performance of MOFs significantly. Thus MOF based materials have gained notable attention to replace the noble metal electrocatalysts for large scale hydrogen production [20].

In this chapter we have elaborated on the various synthesis methods available for the preparation MOF based materials and their utilization as electrocatalysts for the oxygen evolution reaction. In the analysis, we have only focused on pure MOF and MOF derived material synthesis and their improvement towards electrocatalytic active sites in the OER process. Further, current perspectives and challenges associated with the utilization of MOF based materials toward the electrocatalytic OER process are also discussed.

5.2 Synthesis of MOF materials

MOF based materials are considered as unique category of porous materials, and are known for their specific functionalities and notable surface properties. These materials are constructed by the bridging of organic ligands with suitable metal ions. Many synthesis routes and parameters are decisive for the feasible formation of MOF materials. Parameters such as temperature, pressure, pH, reaction time, and suitable solvents, should be considered for MOF preparation. MOF based materials can be synthesized using many methods, among these the main synthesis methods are the hydrothermal method, microwave assisted method, electrochemical method, diffusion method, mechanochemical method, and sonochemical method. Through these synthesis methods, the structure and properties can be highly tailored and altered as per the material requirements and the synthesized materials have also been utilized for various applications.

5.2.1 Hydrothermal method

The hydro/solvothermal approach is used to achieve self-assembly of the product from soluble precursors. An autoclave (sealed container) is used for the synthesis under optimum pressure with an operational temperature range from 80 °C to 260 °C, which can affect the reaction rate and modulate the product formation pathway (Figure 5.1) [25]. Yaqoob *et al* synthesized a $NiNH_2BDC$ MOF/5 wt% rGO hybrid via a facile solvothermal method and used it for the methanol oxidation reaction. The synthesized material was crystalline in nature and also exhibited greater stability and high conductivity. These improvements in electrocatalytic activity obtained from the solvothermal method were applied for material synthesis [21]. Arul *et al* developed surfactant-free Cu-MOF preparation through a solvothermal method along with the protonation and deprotonation approach. The synthesized material showed better electrocatalytic activity in a therapeutic drug application [22].

Xu *et al* developed two-dimensional Co-MOF nanosheets via a simple surfactant-assisted hydrothermal method and the synthesized material was subjected to the electrocatalytic oxygen evolution reaction in an alkaline medium. The resultant material exhibited excellent OER performance with a lower overpotential value.

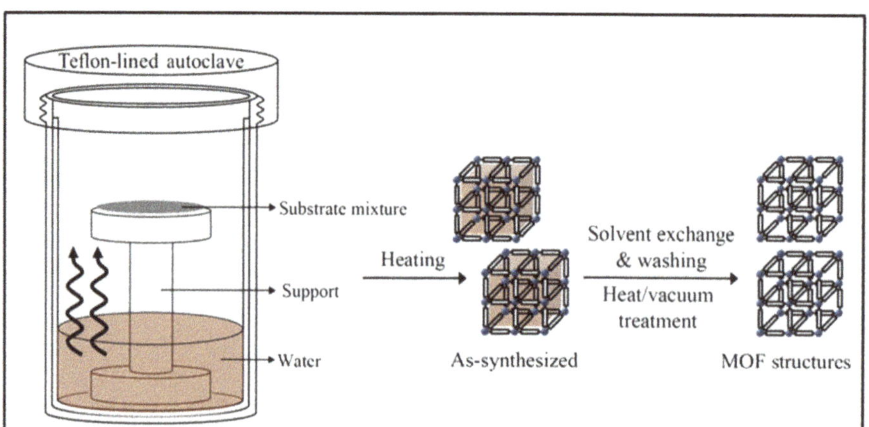

Figure 5.1. The hydro/solvothermal method of synthesizing MOF materials. (Reproduced with permission from [25]. Copyright 2013 Korean Institute of Chemical Engineers, Seoul, Korea.)

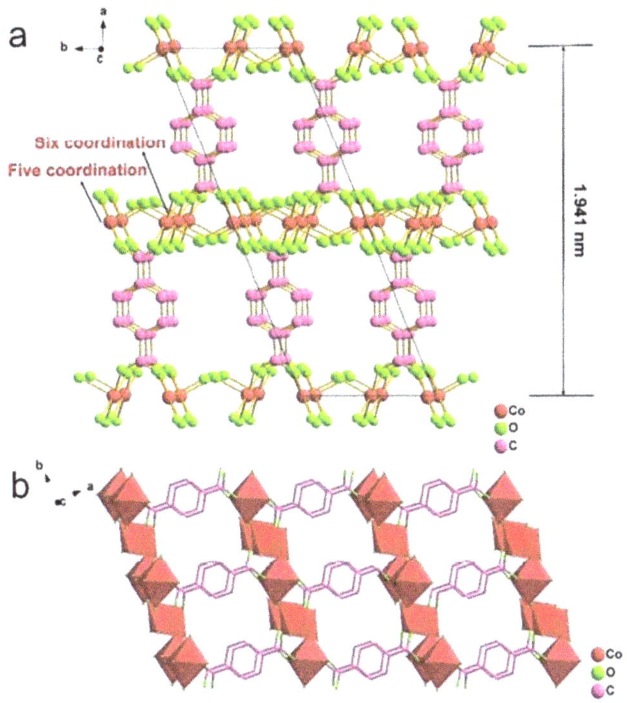

Figure 5.2. (a) Molecular structure and (b) crystal structure of 2D Co-MOF nanosheets. (Reproduced with permission from [23]. Copyright 2018 Royal Society of Chemistry.)

Here, the Co^{II} active center on the 2D nanosheets enhanced the electrocatalytic OER performance and the coordination was ensured by the synthesis method [23]. The crystal structure of the 2D Co-MOF nanosheets is given in figure 5.2. Normally, the reaction is carried out over a long period of time. However, sometimes many days and even weeks are needed for feasible MOF formation via the hydrothermal as well as solvothermal approach. Noor *et al* synthesized a novel bimetallic nickel oxide/copper oxide MOF based nanocomposite by hydrothermal treatment and the prepared MOF material exhibited excellent electrochemical properties towards the electrocatalytic methanol oxidation reaction. The synthesis process enables the coordination between NiO and CuO with BTC and pyrazine ligands, which directly influences the electrocatalytic activity which leads to a better outcome [24].

5.2.2 Microwave method

Metals and metal oxides with smaller dimensions can be synthesized via a microwave assisted method of synthesis. In this process, the solution temperature can be increased by the assistance of microwaves which leads to nanosized particle formation. It is not an appropriate approach for the routine synthesis of crystalline MOF materials [25]. It is considered one of the vital methods to achieve the product in faster way, which means that the reaction time is fast (figure 5.3). The resultant

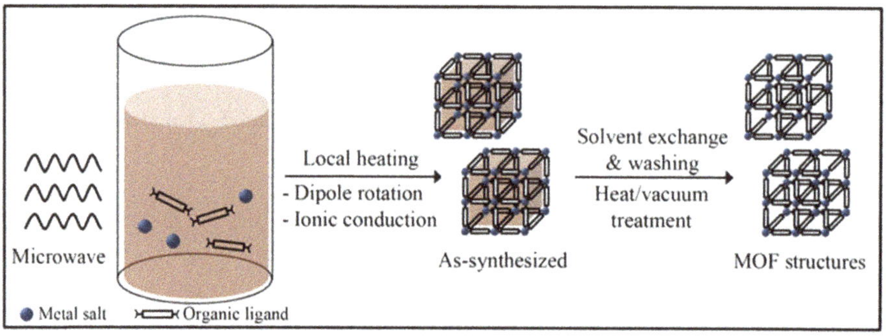

Figure 5.3. MOF materials synthesized via the microwave assisted method. (Reproduced with permission from [25]. Copyright 2013 Korean Institute of Chemical Engineers, Seoul, Korea.)

particles can be controlled properly for size and shape using this approach. As mentioned, it is difficult to produce crystals for single x-ray analysis. Certain conditions are strongly required for solvent evaporation so as to ensure crystal growth in the saturated solutions and solvent solubility at a given temperature range, which can directly influence the formation of crystals in the cooling phase.

Wu *et al* prepared MOF-74 (M = Ni, Mg) material by a hydrothermal method followed by a microwave assisted method and utilized it for gas separation. This study revealed that the microwave assisted method provides a smaller particle size and relatively uniform distribution of the particles. This method also helps to improve the large surface area and micro-pore volume [26]. Further, Cho *et al* developed the Co-MOF-74 material via a microwave heating method and the structure of the Co-MOF-74 was as given in figure 5.4. The synthesized material was utilized for CO_2 adsorption studies. Through this method the catalytic ability was improved significantly [27].

5.2.3 Diffusion method

The diffusion method is considered one of the conventional ways to synthesize MOF based materials with enhanced catalytic properties. The best known diffusion method is solvent liquid diffusion where it forms two layers of different density media. One medium is the precipitant in solvent and the other one includes the product in solvent.

At the interface, crystal growth can appear after the diffusion of precipitant solvent from the separated layers. In the second method of diffusion, reactants are regularly diffused by the physical barriers which comprise two different sized vials. In some cases, gels are applied in the crystallization process to reduce the diffusion and also prevent the formation of bulk precipitation. Al-Ghoula *et al* have synthesized single crystals of MOF-199 via a reaction–diffusion method at room temperature and studied the various sizes and structural benefits of the observed crystals, according to the exposed properties which can also be utilized for electro-catalytic applications [28].

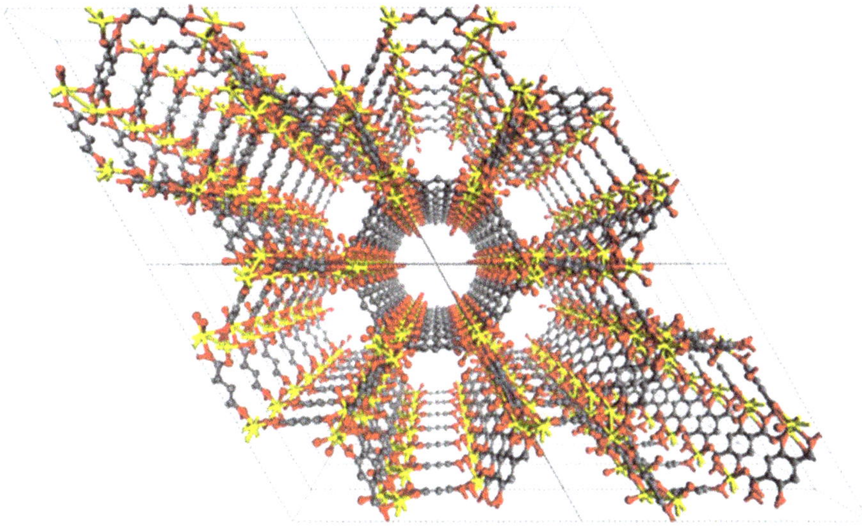

Figure 5.4. Structure of Co-MOF-74, where carbon, oxygen, and cobalt atoms are shown in gray, red, and yellow, respectively. (Reproduced with permission from [27]. Copyright 2012 Elsevier.)

5.2.4 Electrochemical method

The electrochemical method has been used to manufacture MOF based materials at an industrial scale. The major advantage of this method is avoiding the participation of anions such as nitrate, chloride, etc. It is considered as a quick synthesis method compared to the hydro/solvothermal method. In crystallization, it has shown some limitations in the bulk step synthesis where the production of metal ions was observed nearer to the supported material (figure 5.5). Although it is an *in situ* process, some unavoidable accumulations of crystals are observed during membrane formation. In addition, there is the possibility of thermally induced cracking of the membrane during the cooling process. High temperatures are needed to carry out this process. The thermal growth coefficient depends on the structural formation, however, in many cases, due to a lack of proper handling, it may lead to MOF cracking. Thus careful observation is highly recommended. Compared to other available synthetic methods, the electrochemical method can create more parameters for fine-tuning of structures through applying simple voltage and pulses (signals). Liu *et al* synthesized $Cu_3(MTCP)_2$ films from Cu (100) foil using a controllable electrochemical method and measured the conductivity of the $Cu_3(MTCP)_2$ film. It showed superior conductivity because of the electrochemical growth strategy over the Cu foil [29].

5.2.5 Mechanochemical method

In the mechanochemical method, a chemical reaction is influenced by the mechanical breakage of the inter-molecular bonds. In synthetic chemistry, mechanochemistry has a long history, where it has been exploited in multicomponent reactions to

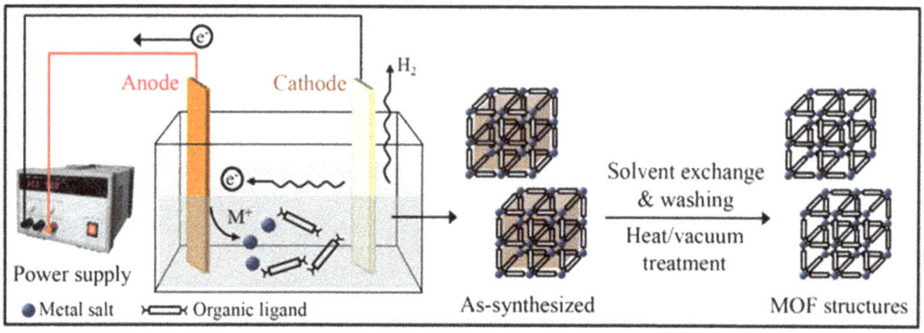

Figure 5.5. MOF material synthesized through the electrochemical method. (Reproduced with permission from [25]. Copyright 2013 Korean Institute of Chemical Engineers, Seoul, Korea.)

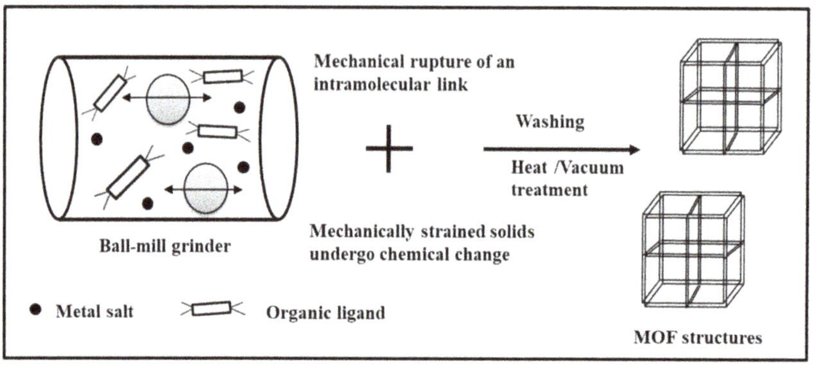

Figure 5.6. MOF materials synthesized through the mechanochemical method.

establish co-crystals. This method has attracted many other fields such as inorganic solid-state chemistry, polymer science, organic chemistry, etc. A detailed schematic representation of the synthesis process is given in figure 5.6. Based on environmental concerns, the mechanochemical method is a highly promising way to synthesis MOF materials because it does not involve organic solvents and the reactions are carried out in ambient temperature conditions without the aid of a solvent. The time needed for the reaction to convert the product yield is normally 10–15 min, and low solubility metal oxides show poor product formation towards solvent based reactions. Applying organic reactants and hydrated metal sources is essential to form preferable anions. Metal acetates and carbonate are utilized to form well-crystallized structures. As an improvement of this technique, ion/liquid assisted grinding was developed for selective production of metal compounds. Wang *et al* developed MOF-74 analogues through sustainable mechanochemical synthesis by the use of Hünig's organic base. In the same way, Cook *et al* made an attempt to develop ZIF-74 analogues from ZnO and 2,5-dihydroxyterephthalic acid with DMF as a liquid additive. After a prolonged milling process, highly crystalline MOF-74 was observed [30].

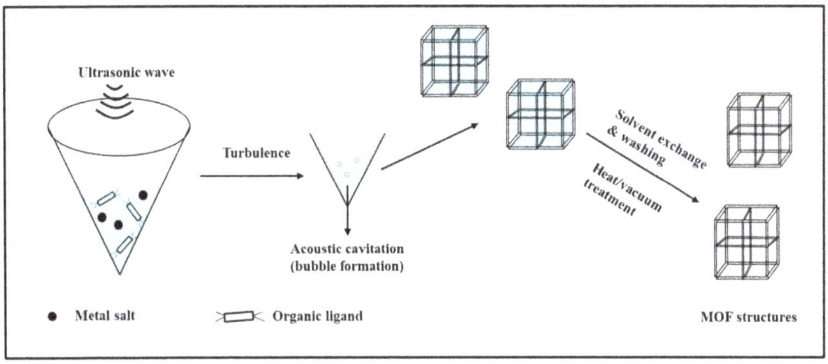

Figure 5.7. MOF materials synthesized through the sonochemical method.

5.2.6 Sonochemical method

The sonochemical method is used to synthesize a desired chemical compound by applying a high energy ultrasound wave to the reaction mixture. It is a cyclic mechanical vibration and it has a higher upper limit than human hearing. It is considered as a homogeneous process with more rapid nucleation and production of small sized particles than the conventional solvothermal synthesis. The details of the sonochemical synthesis process are shown in figure 5.7, where the pyrox reactor combined with the sonicator bar and the power adjustments along with the cooling rates are observed and monitored throughout the process. The desired crystals are observed at the end of the extreme heating and cooling rates. Jung developed high quality MOF-177 crystals using simple sonochemical synthesis [31]. The synthesized MOF-177 showed better crystallinity with a greater BET surface area of 4898 m^2 g^{-1} which was superior to the conventional (4833 m^2 g^{-1}) and microwave route of MOF-177 synthesis (4197 m^2 g^{-1}), and also the product showed the high yield of 95.6%. Wiwasuku developed Zn-MOF through dual Lewis basic amino and carbonyl groups by using the sonochemical method [32]. The resultant Zn-MOF delivered highly sensitive and selective detection of the Al^{3+} ions with a very low detection limit of 30 nM.

5.3 MOF based materials as electrocatalysts for OER

MOF based materials are known for their intrinsic properties such as porosity, surface area, flexibility towards structural addition, and, more importantly, susceptibility to various metal centers. According to their utilization, MOF based materials are categorized into two major categories, direct MOF based OER catalysts and MOF derived OER catalysts [33]. The way to identify both types of MOF structures is given in figure 5.4. Normally, MOFs deliver low conductivity and decreased activity. To overcome these issues, the construction of MOFs is tuned with exposure of active sites, structural modification, and enhanced conductivity via some familiar engineering strategies, as described in figure 5.8. Both direct MOF based materials and MOF derived materials for utilization as electrocatalysts for the OER reaction are discussed and elaborated below.

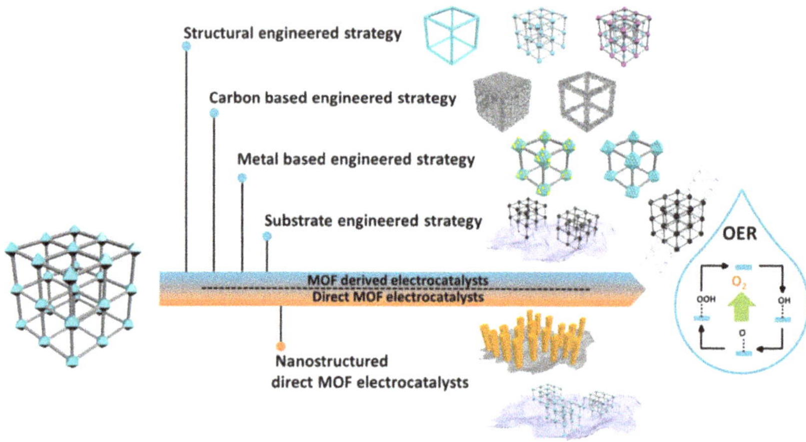

Figure 5.8. MOF based electrocatalysts for electrochemical OER. (Reproduced with permission from [33]. Copyright 2018 Wiley.)

5.3.1 Direct MOF based OER catalyst

Mostly, MOF based materials present insulating properties. Added to that, their poor conductivity and the blockage of active sites by the organic ligands lead to major restrictions on their feasible utilization in electrocatalytic applications. In recent times a new class of MOF structures has emerged which shows high conductivity, high porosity, and excellent charge mobility [19].

This development in a new class of nanostructured MOFs has been considered as a promising finding to improve the electrocatalytic performance for OER applications. Normally, two main strategies have been used for designing the conductive MOF materials. One is the high charge transfer in MOFs observed from the covalent linkages of the coordination polymers. For example, to improve the electrical conductivity of the MOF material, Takaishi *et al* developed an interest in MOFs through the utilization of Cu[Cu(pdt)2] (pdt—2,3-pyrazinedithiolate) [34]. The observed MOF materials could achieve an electrical conductivity value of 6×10^{-4} S cm^{-1} at a temperature of 300 K with the better thermal activation value of 0.193 eV. Further, Sun and co-workers have reviewed many MOF type materials and their relative improved conductivity values [19].

The second approach is designing conductive MOFs through their non-covalent interaction nature. For example, introducing π–π stacking on the organic ligands improves their charge transfer process. Through the utilization of this strategy, tetrathiafulvalene (TTF) based materials were reported. Based on these strategies, many conductive MOFs have been synthesized and utilized for electrocatalytic OER processes. The fabrication process has also been improved, with synthesis control over porosity and surface area which provide more possibilities to improve the electrocatalytic OER performance [33].

Yaqoob *et al* synthesized Co BTC MOF with various concentrations of rGO composites as Co BTC, Co BTC-1 wt% rGO, Co BTC-2 wt% rGO, Co BTC-5 wt%

rGO, and Co BTC-8 wt% rGO. Among all the catalysts, Co BTC-5 wt% rGO has shown a lower overpotential value of 0.29 V versus RHE. Thus the conductivity of the Co BTC-5 wt% rGO was improved significantly which reflected in the electrocatalytic OER activity [35]. The related plots were given in figure 5.9. Ma *et al* synthesized Fe-MOF nanosheet arrays on nickel foam with rare Earth erbium doping via hydrothermal synthesis. The Er0.4 Fe-MOF/NF catalyst provided a lower overpotential value of 248 mV at a fixed current density of 100 mA cm^{-2}. It also delivered a long-term stability of 100 h. The related OER plots are given in figure 5.10. This enhancement is mainly attributed to the synergistic effects of the Fe and Er sites [36].

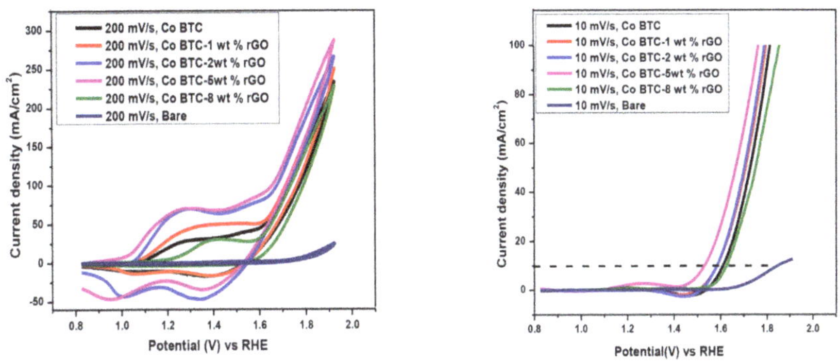

Figure 5.9. (a) The CV curves of Co BTC MOF and 1, 2, 5, 8 wt % rGO composites at 200 mV s^{-1}. (a) The LSV curves observed at a 10 mV s^{-1} scan rate. (Reproduced with permission from [35]. Copyright 2020 Elsevier.)

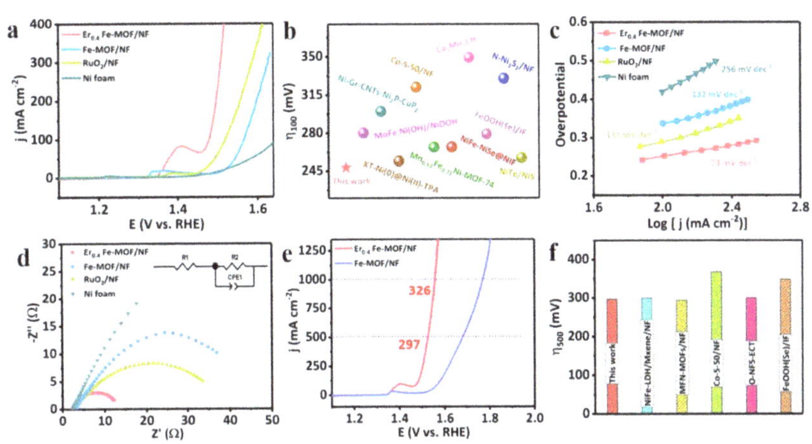

Figure 5.10. (a) LSV curves observed in 1.0 M KOH. (b) Overpotential value of Er0.4 Fe-MOF/NF at 100 mA cm^{-2} compared with other values in the literature. (c) Tafel slopes and (d) EIS of Ni foam, Fe-MOF/NF, commercial RuO_2, and $Er_{0.4}$ Fe-MOF/NF. (e) LSV curves and (f) OER performance comparison histogram. (Adapted from [36]. CC BY 4.0.)

Recently, 1D MOF based fibrous materials were utilized for an electrocatalytic OER reaction. Their notable properties are uni-directional flow of electrons, a highly porous nature, and structural uniformity. Electrospinning is a highly recommendable technique for the synthesis of fibrous materials [37, 38]. Kundu *et al* reported the utilization of bi- and tri-metallic combinations of ZIF based fibrous materials in electrocatalytic OER process [39–41]. For example, CoNi-ZIF fibrous materials are synthesized via a wet chemical process followed by an ES process and delivered a lower overpotential of 332 mV to achieve a 10 mA cm^{-2} current density in the electrocatalytic OER process [42]. In another work, tri-metallic (Co, Ni, Fe) fibers were reported for their exceptional OER performance in 1 M KOH electrolyte solution [43]. Further, Co-incorporated MOF-5 microfibers have been synthesized and tested for electrocatalytic OER in 1 M KOH where it delivered the lower overpotential of 240 mV at a 10 mA cm^{-2} current density [39].

Two-dimensional MOF nanosheets are being paid great attention because of their fascinating properties, exposed active sites, porous structure, and feasible electronic transport, which enhance the OER activity and also shift the value to the lower overpotential region. Zha *et al* have developed Co-based MOF-on-MOF heterojunction nanostructures which can be subjected to the electrocatalytic OER in 1 M KOH. They delivered an overpotential of 335 mV at 10 mA cm^{-2} with excellent stability. Here, the reason behind the excellent OER performance is mainly attributed to the heterojunction in the catalyst, the exposure of many active sites, and the expanded surface area [44]. Two-dimensional MOF materials have shown better electrochemical OER activity along with their structure [45, 46].

5.3.2 MOF derived OER catalyst

The direct utilization of MOF based materials has been restricted by their poor conductive nature. Interestingly, some additional aspects such as tailoring the surface morphology, tuning the pore size, and compatibility for introduced metal ions were achieved through MOF derived catalysts through the use of the MOF as a sacrificial template. Structural engineering is one of the notable ways to achieve nanostructured architecture, hollow structures, and multishell arrangements, etc [47]. Many techniques have been applied to achieve the desired structural aspects, notably, thermal annealing, etching, secondary growth methods, etc. Due to this the number of active sites was improved and the desired surface structure was also obtained. Moreover, the introduction of various active metal centers leads to changes in physicochemical properties and also enhancing the electrochemical activity. Etching is one of the finest ways to produce hollow structures [48, 49]. For example, Han *et al* utilized Ni–Co Prussian blue analogue cubes as a sacrificial precursor material to form a cage-like structure for ammonia etching treatment. In this etching process, they observed that etching occurred on the corners of the cubes and faster toward the body diagonal direction [50]. Considering the poor electrocatalytic performance of the Ni–Co PBA nanocage, it was subjected to a pyrolysis process to achieve a Ni–Co mixed oxide on the nanocage, which showed better

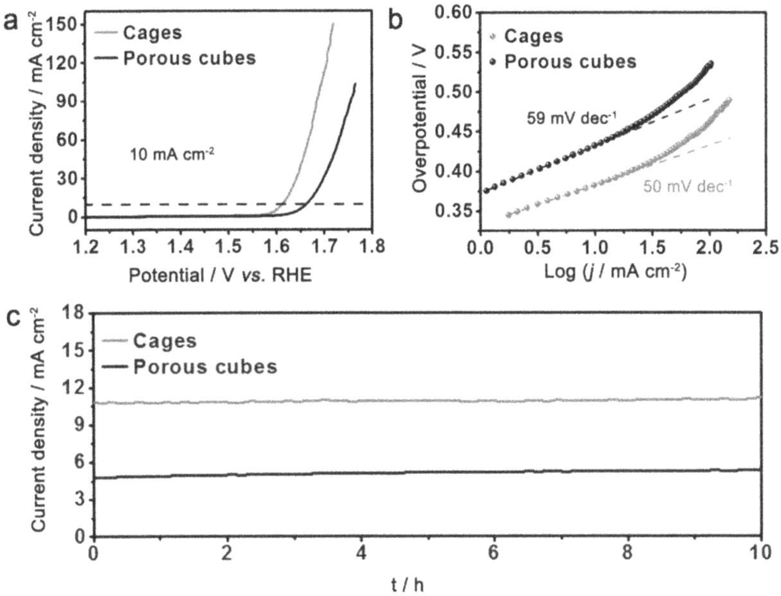

Figure 5.11. (a) Polarization curves, (b) Tafel plots, and (c) chronoamperometry curves of the Ni–Co mixed oxide cages. (Reproduced with permission from [50]. Copyright 2016 Wiley.)

electrocatalytic OER performance than the pristine material and also showed excellent stability (figure 5.11).

Another surface engineering technique is the thermal treatment process. The main aspects of the thermal process are the applied temperature and atmospheric conditions. During an annealing process, the MOF can be converted easily into a carbon matrix or metal oxide or corresponding carbides. These carbon-based materials achieved from the annealing process have shown some unique properties, such as high conductivity, larger surface area, structural flexibility, and thermal and chemical stability. Qian *et al* synthesized heterometallic oxides and/or carbides derived from MOF precursors and the synthesized sample was named InOF-16-Ar/O_2-T ($T = 550$ °C, 700 °C, and 1000 °C), among all the catalysts, InOF-16-Ar-1000 delivered better OER performance than the others [16].

In order to form the multishell on the observed material, a secondary growth method was proposed. Through this process new reactions have been carried out on the active chemical reaction sites of the current MOF. Due to that, many combinations such as MOF/MOF, MOF/LDH, and MOF/oxide have been achieved. The synergism between the modified structures ensured the electron transfer process and the formation of the porous architecture which is essential for the enhancement of electrocatalytic OER activity. For example, Meng *et al* [51] developed an MOF derived binary metal center catalyst by controlling the iron species. The synthesis process is given in figure 5.12. The synthesis of Fe-SBU into ZIF-67 can induce the distinct confinement effect which can help to improve the

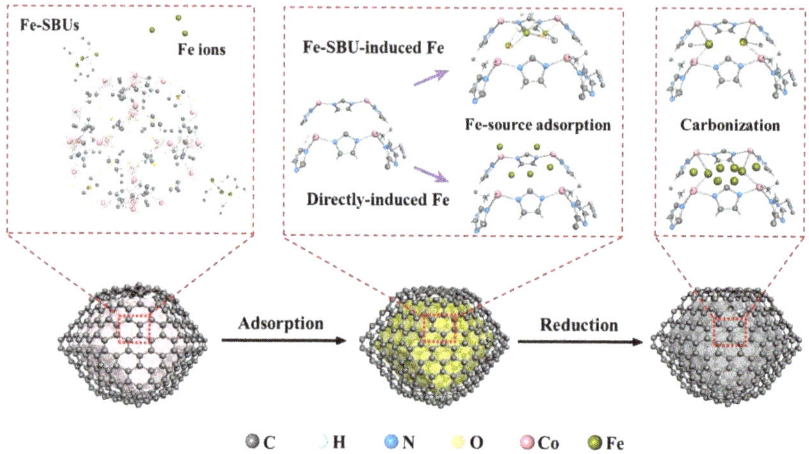

Figure 5.12. Schematic illustration of binary metal center electrocatalyst synthesis. (Reproduced with permission from [51]. Copyright 2021 Royal Society of Chemistry.)

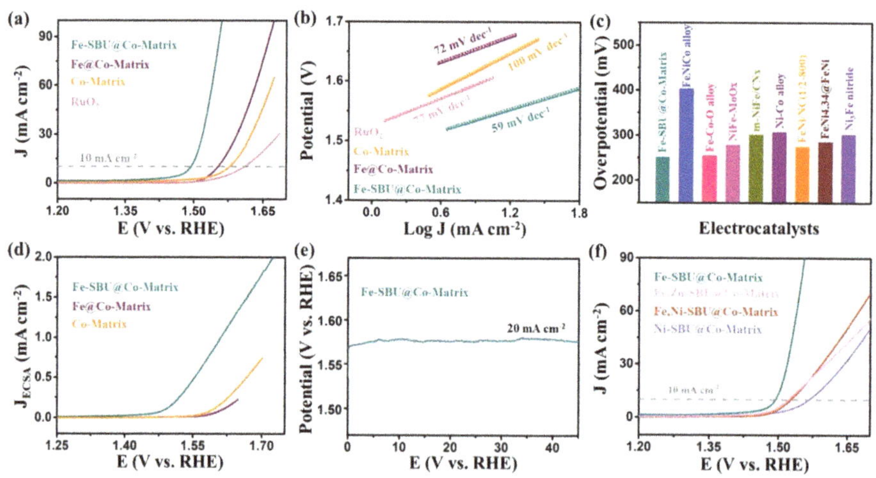

Figure 5.13. (a) Polarization curves at a scan rate of 5 mV s^{-1}. (b) Tafel slopes resultant from polarization data. (c) Comparison of the overpotential values for different electrocatalysts to reach 10 mA cm^{-2}. (d) Polarization curves of MOF derived catalysts based on ECSA. (e) Electrochemical stability of the Fe-SBU@Co-matrix at $J = 50$ mA cm^{-2}. (f) Polarization curves of different metal-containing catalysts at a scan rate of 5 mV s^{-1}. (Reproduced with permission from [51]. Copyright 2021 Royal Society of Chemistry.)

electrocatalytic OER activity of the catalyst in alkaline conditions where it showed the lower overpotential of 249 mV @ 10 mA cm^{-2} (figure 5.13).

Engineering of carbon-based materials onto an MOF matrix is the one of the ways being considered to improve the electrocatalytic OER performance of the catalyst. Normally, MOFs are constructed using metal ions and organic motifs. Thus it is easy to achieve carbon formation along with metal nodes. Carbon could allow better coordination toward the metal ions whereby it can reduce the corrosive behavior from the external environment, which can limit active sites, and also show improved stability. Other than that, it can also establish better modification toward heteroatoms (nitrogen, boron, sulfur, phosphorous) through suitable chemical treatment. Zhang *et al* synthesized MOF derived multi-metal oxides $MoO_2/Co_2Mo_3O_8/Fe_2Mo_3O_8$ (MCF) anchored on a N,P-doped carbon matrix and utilized the same for the oxygen evolution reaction in an alkaline medium. Here, MCF/NPCCNT-40 hybrid catalyst delivered a lower overpotential value of 292 mV @ 10 mA cm^{-2} current density. The better results in electrocatalytic OER activity were mainly observed from heteroatom dopants under a pyrolysis process [52].

Metal-based engineering towards an MOF derived catalyst is another good approach to enhance the electrocatalytic OER activity. The synergism between the metal ions could provide superior electrocatalytic performance. Due to that, researchers are trying to achieve better electrocatalytic performance from non-noble metals rather than noble metals. Dong *et al* developed an MOF derived Zn-doped $CoSe_2$ catalyst and subjected it to an electrocatalytic OER process under 1 M KOH conditions. Due to the synergism between the bimetal ions, it provided better electrocatalytic OER performance than the commercial noble metal catalysts. Similarly, many bimetallic and multi-metallic combinations are utilized to alter the MOF topology and properties which can deliver supremacy for the electrocatalytic OER process [53].

5.4 Summary and outlook

The positive outcomes observed in recent years have been evidence for the development of MOF based electrocatalyst in the field of the OER process. In this chapter, the various synthesis protocols related to the MOF and MOF derived electrocatalysts, the properties associated with the synthesis process, the impact of morphological outcomes from the synthesis, and the pictorial representations of each synthesis process have been discussed thoroughly. Two major categories of MOF based materials in the OER process were summarized. In the case of direct MOF OER based OER catalysts, various dimensional aspects with the morphological outcomes and ability to enhance the electrocatalytic activity, and the exposure of active sites have also been elaborated. In the case of MOF derived OER catalysts, we focused on the four important engineering strategies, namely structural aspects, carbon-based, metal-based, and substrate utilization, to improve the OER performance. Some suggestions for better research outcomes and possible directions are as follows:

- The in depth exploration of MOF based OER electrocatalysts is highly recommended. In the case of direct MOF based electrocatalysts, although they possess low conductivity, they have shown a very good electrochemical response. The creation of an active phase on the structure and the formation of suitable oxides leads to better electrochemical performance for the OER reaction.
- *In situ* measurements in MOFs are highly anticipated to find the real active sites over the catalytic surface. Surface sensitive characterization such as XAS

and XPS can help to identify the real active phase responsible for the electrocatalytic OER reaction. Added to that, computer modeling is also considered as a vital tool to find the exact mechanism of OER, which also help us to develop new MOF based materials towards the OER process.

- In the future, the development of a new MOF based catalyst is highly crucial. Currently, transition metal (in particular Co, Ni, Fe and Mo) based materials are commonly utilized to catalyze the OER in alkaline media rather than in acidic media because of stability issues. Mostly Ir and Ru based derivatives are used as better catalysts for acidic OER. Thus, combining these aspects with MOF based materials is meaningful, specifically, the introduction of Ir and Ru based metals into the transition metal MOF based materials.
- Practically speaking, designing a multifunctional catalyst will be highly desirable. Since MOF based materials have wider properties such as the manipulation of compositions with good compatibility, tuning structural aspects, and the creation of more metal centers make MOFs excellent candidates from the multifunctional catalyst perspective. MOFs with different functionalities are still less explored to date. With better coordination of knowledge, it could be a promising field for other energy related applications as well.

Acknowledgments

The authors thank CSIR-CECRI, Karaikudi, India, for constant support. SSS wishes to acknowledge UGC, New Delhi, India, for the SRF award.

References

[1] Anantharaj S, Ede S R, Sakthikumar K, Karthick K, Mishra S and Kundu S 2016 Recent trends and perspectives in electrochemical water splitting with an emphasis to sulphide, selenide and phosphide catalysts of Fe, Co and Ni: a review *ACS Catal.* **6** 8069–97

[2] Edwards P P, Kuznetsov V L, David W I F and Brandon N P 2008 Hydrogen and fuel cells: towards a sustainable energy future *Energy Policy* **36** 4356–62

[3] Arulraj A, Thachnatharen N, Rajkumar C and Mangalaraja R V 2022 Engineering 2D chalcogenides for energy and environmental remediation *Nanostructured Materials for Sustainable Energy and Environmental Remediation* (Bristol: IOP Publishing) pp 2-1–28

[4] Arulraj A, Murugesan P K, Zamorano R C A T and Mangalaraja R V 2023 Nanoarchitectonics of layered metal chalcogenides-based ternary electrocatalyst for water splitting *Energies* **16** 1669

[5] Anantharaj S, Ede S R, Karthick K, Sam Sankar S, Sangeetha K, Karthik P E and Kundu S 2018 Precision and correctness in the evaluation of electrocatalytic water splitting: revisiting activity parameters with a critical assessment *Energy Environ. Sci.* **11** 744–71

[6] Praveen Kumar M, Sasikumar M, Arulraj A, Rajasudha V, Murugadoss G, Kumar M R, Gouse Peera S and Mangalaraja R V 2022 NiFe layered double hydroxide electrocatalyst prepared via an electrochemical deposition method for the oxygen evolution reaction *Catalysts* **12** 1470

[7] Chen Z, Qing H, Zhou K, Sun D and Wu R 2020 Metal–organic framework-derived nanocomposites for electrocatalytic hydrogen evolution reaction *Prog. Mater Sci.* **108** 100618

[8] Reier T, Oezaslan M and Strasser P 2012 Electrocatalytic oxygen evolution reaction (OER) on Ru, Ir, and Pt catalysts: a comparative study of nanoparticles and bulk materials *ACS Catal.* **2** 1765–72

[9] Duan Y *et al* 2019 Scaled-up synthesis of amorphous NiFeMo oxides and their rapid surface reconstruction for superior oxygen evolution catalysis *Angew. Chem. Int. Ed.* **58** 15772–7

[10] Park K R, Jeon J E, Ali G, Ko Y H, Lee J, Han H and Mhin S 2019 Oxygen evolution reaction of Co–Mn–O electrocatalyst prepared by solution combustion synthesis *Catalysts* **9** 564

[11] Wu W, Liu J, Chen G, Chen Y and Xu C 2019 Thin porous nanosheets of NiFe layered-double hydroxides toward a highly efficient electrocatalyst for water oxidation *Int. J. Hydrogen Energy* **45** 1948–58

[12] Fan K *et al* 2016 Nickel-vanadium monolayer double hydroxide for efficient electrochemical water oxidation *Nat. Commun.* **7** 1–9

[13] Chen P, Xu K, Tong Y, Li X, Tao S, Fang Z, Chu W, Wu X and Wu C 2016 Cobalt nitrides as a class of metallic electrocatalysts for the oxygen evolution reaction *Inorg. Chem. Front.* **3** 236–42

[14] Walter C, Menezes P W, Orthmann S, Schuch J, Connor P, Kaiser B, Lerch M and Driess M 2018 A molecular approach to manganese nitride acting as a high performance electrocatalyst in the oxygen evolution reaction *Angew. Chem. Int. Ed.* **57** 698–702

[15] Jiang J, Liu Q, Zeng C and Ai L 2017 Cobalt/molybdenum carbide@N-doped carbon as a bifunctional electrocatalyst for hydrogen and oxygen evolution reactions *J. Mater. Chem.* A **5** 16929–35

[16] Qian J, Li T T, Hu Y and Huang S 2017 A bimetallic carbide derived from a MOF precursor for increasing electrocatalytic oxygen evolution activity *Chem. Commun.* **53** 13027–30

[17] Anjum M A R, Okyay M S, Kim M, Lee M H, Park N and Lee J S 2018 Bifunctional sulfur-doped cobalt phosphide electrocatalyst outperforms all-noble-metal electrocatalysts in alkaline electrolyzer for overall water splitting *Nano Energy* **53** 286–95

[18] Xiao X, He C T, Zhao S, Li J, Lin W, Yuan Z, Zhang Q, Wang S, Dai L and Yu D 2017 A general approach to cobalt-based homobimetallic phosphide ultrathin nanosheets for highly efficient oxygen evolution in alkaline media *Energy Environ. Sci.* **10** 893–9

[19] Sun L, Campbell M G and Dincə M 2016 Electrically conductive porous metal-organic frameworks *Angew. Chem. Int. Ed.* **55** 3566–79

[20] Dong R, Zheng Z, Tranca D C, Zhang J, Chandrasekhar N, Liu S, Zhuang X, Seifert G and Feng X 2017 Immobilizing molecular metal dithiolene-diamine complexes on 2D metal-organic frameworks for electrocatalytic H_2 production *Chem. Eur. J.* **23** 2234–4

[21] Yaqoob L, Noor T, Iqbal N, Nasir H and Mumtaz A 2021 Electrocatalytic performance of $NiNH_2BDC$ MOF based composites with rGO for methanol oxidation reaction *Sci. Rep.* **11** 1–18

[22] Arul P, Huang S T, Gowthaman N S K, Govindasamy M and Jeromiyas N 2020 Surfactant-free solvothermal synthesis of Cu-MOF via protonation–deprotonation approach: a morphological dependent electrocatalytic activity for therapeutic drugs *Microchim. Acta* **187** 1–13

[23] Xu Y, Li B, Zheng S, Wu P, Zhan J, Xue H, Xu Q and Pang H 2018 Ultrathin two-dimensional cobalt-organic framework nanosheets for high-performance electrocatalytic oxygen evolution *J. Mater. Chem.* A **6** 22070–6
[24] Noor T, Pervaiz S, Iqbal N, Nasir H, Zaman N, Sharif M and Pervaiz E 2020 Nanocomposites of NiO/CuO based MOF with rGO: an efficient and robust electrocatalyst for methanol oxidation reaction in DMFC *Nanomaterials* **10** 1–18
[25] Lee Y R, Kim J and Ahn W S 2013 Synthesis of metal–organic frameworks: a mini review *Korean J. Chem. Eng.* **30** 1667–80
[26] Wu X, Bao Z, Yuan B, Wang J, Sun Y, Luo H and Deng S 2013 Microwave synthesis and characterization of MOF-74 (M = Ni, Mg) for gas separation *Microporous Mesoporous Mater.* **180** 114–22
[27] Cho H Y, Yang D A, Kim J, Jeong S Y and Ahn W S 2012 CO_2 adsorption and catalytic application of Co-MOF-74 synthesized by microwave heating *Catal. Today* **185** 35–40
[28] Al-Ghoul M, Issa R and Hmadeh M 2017 Synthesis, size and structural evolution of metal–organic framework-199 via a reaction-diffusion process at room temperature *Cryst. Eng. Comm.* **19** 608–12
[29] Liu Y, Wei Y, Liu M, Bai Y, Wang X, Shang S, Chen J and Liu Y 2021 Electrochemical synthesis of large area two-dimensional metal–organic framework films on copper anodes *Angew. Chem. Int. Ed.* **60** 2887–91
[30] Wang Z, Li Z, Ng M and Milner P J 2020 Rapid mechanochemical synthesis of metal–organic frameworks using exogenous organic base *Dalton Trans.* **49** 16238–44
[31] Jung D W, Yang D A, Kim J, Kim J and Ahn W S 2010 Facile synthesis of MOF-177 by a sonochemical method using 1-methyl-2-pyrrolidinone as a solvent *Dalton Trans.* **39** 2883–7
[32] Wiwasuku T, Othong J, Boonmak J, Ervithayasuporn V and Youngme S 2020 Sonochemical synthesis of microscale Zn(II)-MOF with dual Lewis basic sites for fluorescent turn-on detection of Al^{3+} and methanol with low detection limits *Dalton Trans.* **49** 10240–9
[33] Shao Q, Yang J and Huang X 2018 The design of water oxidation electrocatalysts from nanoscale metal–organic frameworks *Chem. Eur. J.* **24** 15143–55
[34] Takaishi S, Hosoda M, Kajiwara T, Miyasaka H, Yamashita M, Nakanishi Y, Kitagawa Y, Yamaguchi K, Kobayashi A and Kitagawa H 2009 Electroconductive porous coordination polymer $Cu[Cu(pdt)_2]$ composed of donor and acceptor building units *Inorg. Chem.* **48** 9048–50
[35] Yaqoob L, Noor T, Iqbal N, Nasir H, Sohail M, Zaman N and Usman M 2020 Nanocomposites of cobalt benzene tricarboxylic acid MOF with rGO: an efficient and robust electrocatalyst for oxygen evolution reaction (OER) *Renew. Energy.* **156** 1040–54
[36] Ma Y, Miao Y, Mu G, Lin D, Xu C, Zeng W and Xie F 2021 Highly enhanced OER performance by Er-doped Fe-MOF nanoarray at large current densities *Nanomaterials* **11** 1–11
[37] Sankar S S, Karthick K, Sangeetha K, Karmakar A and Kundu S 2020 Transition-metal-based zeolite imidazolate framework nanofibers via an electrospinning approach: a review *ACS Omega* **5** 57–67
[38] Sankar S S, Karthick K, Sangeetha K, Karmakar A, Madhu R and Kundu S 2021 Current perspectives on 3D ZIFs incorporated with 1D carbon matrices as fibers via electrospinning processes towards electrocatalytic water splitting: a review *J. Mater. Chem.* A **9** 11961–2002
[39] Madhu R, Sankar S S, Karthick K, Karmakar A, Kumaravel S and Kundu S 2021 Electrospun cobalt-incorporated MOF-5 microfibers as a promising electrocatalyst for OER in alkaline media *Inorg. Chem.* **60** 9899–911

[40] Sankar S S, Karthick K, Sangeetha K and Kundu S 2019 In situ modified nitrogen-enriched ZIF-67 incorporated ZIF-7 nanofiber: an unusual electrocatalyst for water oxidation *Inorg. Chem.* **58** 13826–35

[41] Sankar S S, Ede S R, Anantharaj S, Karthick K, Sangeetha K and Kundu S 2019 Electrospun cobalt-ZIF micro-fibers for efficient water oxidation under unique pH conditions *Catal. Sci. Technol.* **9** 1847–56

[42] Sankar S S, Karthick K, Sangeetha K, Karmakar A and Kundu S 2020 Polymeric nanofibers containing CoNi-based zeolitic imidazolate framework nanoparticles for electrocatalytic water oxidation *ACS Appl. Nano Mater.* **3** 4274–82

[43] Sankar S S, Manjula K, Keerthana G, Ramesh Babu B and Kundu S 2021 Highly stable trimetallic (Co, Ni, and Fe) zeolite imidazolate framework microfibers: an excellent electrocatalyst for water oxidation *Cryst. Growth Des.* 1800–9

[44] Zha Q, Yuan F, Qin G and Ni Y 2020 Cobalt-based MOF-on-MOF two-dimensional heterojunction nanostructures for enhanced oxygen evolution reaction electrocatalytic activity *Inorg. Chem.* **59** 1295–305

[45] Sankar S S, Karthick K, Kumaravel S, Karmakar A, Ragunath M and Kundu S 2021 Temperature-controlled structural variations of meticulous fibrous networks of NiFe-polymeric zeolite imidazolate frameworks for enhanced performance in electrocatalytic water-splitting reactions *Inorg. Chem.* **60** 12467–80

[46] Sankar S S, Karthick K, Sangeetha K, Gill R S and Kundu S 2020 Annexation of nickel vanadate ($Ni_3V_2O_8$) nanocubes on nanofibers: an excellent electrocatalyst for water oxidation *ACS Sustain. Chem. Eng.* **8** 4572–9

[47] Ali M, Pervaiz E, Noor T, Rabi O, Zahra R and Yang M 2021 Recent advancements in MOF-based catalysts for applications in electrochemical and photoelectrochemical water splitting: a review *Int. J. Energy Res.* **45** 1190–226

[48] Yang D, Chen Y, Su Z, Zhang X, Zhang W and Srinivas K 2021 Organic carboxylate-based MOFs and derivatives for electrocatalytic water oxidation *Coord. Chem. Rev.* **428** 213619

[49] Safaei M, Foroughi M M, Ebrahimpoor N, Jahani S, Omidi A and Khatami M 2019 A review on metal–organic frameworks: synthesis and applications *TrAC Trends Anal. Chem.* **118** 401–25

[50] Han L, Yu X Y and (David) Lou X W 2016 Formation of Prussian-blue-analog nanocages via a direct etching method and their conversion into Ni–Co-mixed oxide for enhanced oxygen evolution *Adv. Mater.* **28** 4601–5

[51] Meng L, Zhang L, Zhu Y, Jiang H, Kaneti Y V, Na J, Yamauchi Y, Golberg D, Jiang H and Li C 2021 Highly dispersed secondary building unit-stabilized binary metal center on a hierarchical porous carbon matrix for enhanced oxygen evolution reaction *Nanoscale* **13** 1213–9

[52] Zhang X, Chen Y, Chen M, Yu B, Wang B, Wang X, Zhang W and Yang D 2021 MOF derived multi-metal oxides anchored N,P-doped carbon matrix as efficient and durable electrocatalyst for oxygen evolution reaction *J. Colloid Interface Sci.* **581** 608–18

[53] Dong Q, Wang Q, Dai Z, Qiu H and Dong X 2016 MOF-derived Zn-doped $CoSe_2$ as an efficient and stable free-standing catalyst for oxygen evolution reaction *ACS Appl. Mater. Interfaces* **8** 26902–7

IOP Publishing

Materials Technology for the Energy and Environmental Nexus, Volume 1

R V Mangalaraja, Arunachalam Arulraj, Nalandhiran Pugazhenthiran and Chan Siew Hwa

Chapter 6

Hierarchical nanostructured functional materials for flexible supercapacitors

A Gowrisankar, K Sureka and T Selvaraju

Rapid global economic growth is leading to the consumption of more fossil fuels, and the issue of global warming has arisen, which has led to the search for new renewable energy sources. The electricity generated by such renewable energy sources must be efficiently stored in a new eco-friendly energy storage devices that can be integrated into existing energy storage technologies. In general, supercapacitors (SCs) fill the energy gap between conventional capacitors and batteries. Until now, many active electrode materials, such as transition metal oxides (TMOs), conducting polymers, and carbonaceous materials, have been studied for the electrochemical performance of SCs. All the electrode materials have their own merits and each TMO differs in its crystal structure and redox behaviour.

In recent years, the hierarchical nanostructure-based functional materials (HNFMs) have received significant attention in a variety of energy storage and conversion devices, including fuel cells, solar cells, light-emitting diodes, sensors, batteries, supercapacitors, thermoelectric devices, and memory devices. Due to their ordered low dimensional building blocks, hierarchical nanostructures could offer advantages such as more active sites and synergistic properties. Although there are many studies about HNFMs as energy storage materials, to the best of our knowledge, there are few reviews focusing on the development of HNFMs as capacitor electrodes. In this chapter, we focus specifically on the recent advances in the fabrication of flexible supercapacitor devices and their electrochemical performance as supercapacitor materials. The synthesis approaches for HNFMs are categorized, and the electrochemical capacitive performances of various HNFMs are discussed. Finally, the challenges and the promising future of this particular material are discussed.

6.1 Introduction

Energy is at the heart of the climate challenge. Because the world's population is growing, our over-reliance on fossil fuels such as coal, oil, and gas has increased greenhouse gas emissions in our atmosphere. There is an urgent need to develop alternative renewable energy generation and storage systems to overcome today's energy crisis. Solar power, wind power, and hydropower can be substituted for the traditional energy sources. However, these new energy sources need efficient energy storage and conversion devices [1–3]. Among the various energy storage devices, supercapacitors have become attractive energy storage devices that are being investigated widely to replace traditional capacitors and batteries. Because they convey extraordinary advantages, such as a longer cycle life (>10 000 cycles), fast dynamics of charge propagation, better environmental friendliness, and high power density, they are ideal candidates for numerous portable and miniaturized electronic devices [4, 5]. The global supercapacitor market was valued at $5 billion in 2021 and is expected to reach $25 billion by 2030, growing at a compound annual growth rate of 23.5% between 2022 and 2030 [6] (figure 6.1).

6.1.1 Supercapacitors—an overview

According to the energy storage mechanism, there are three types of SCs: electrical double layer capacitors (EDLCs), pseudocapacitors (PCs), and hybrid capacitors. The EDLC stores energy on the electrode surface via fast charge adsorption/desorption, while the PCs store energy through surface faradaic reactions between the electrolyte ions and electrode materials. Carbon materials, including activated carbon, carbon nanotubes (CNTs), and graphene, have been used extensively in the fabrication of EDLC electrodes. Figures 6.2(a)–(c) show schematic representations of the charge storage mechanism of EDLC-type SCs. The promising

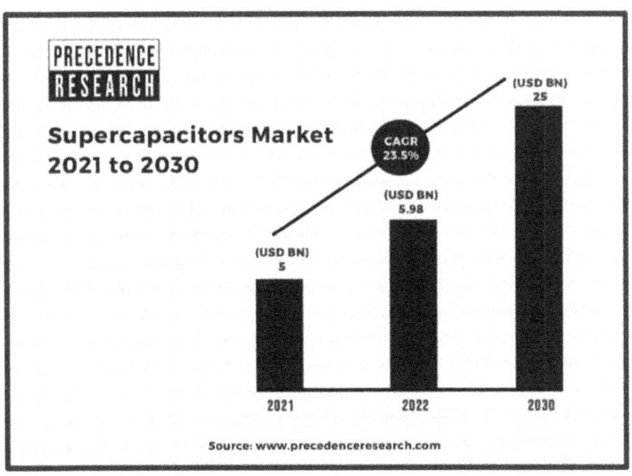

Figure 6.1. The global supercapacitor market size and corresponding growth rate.

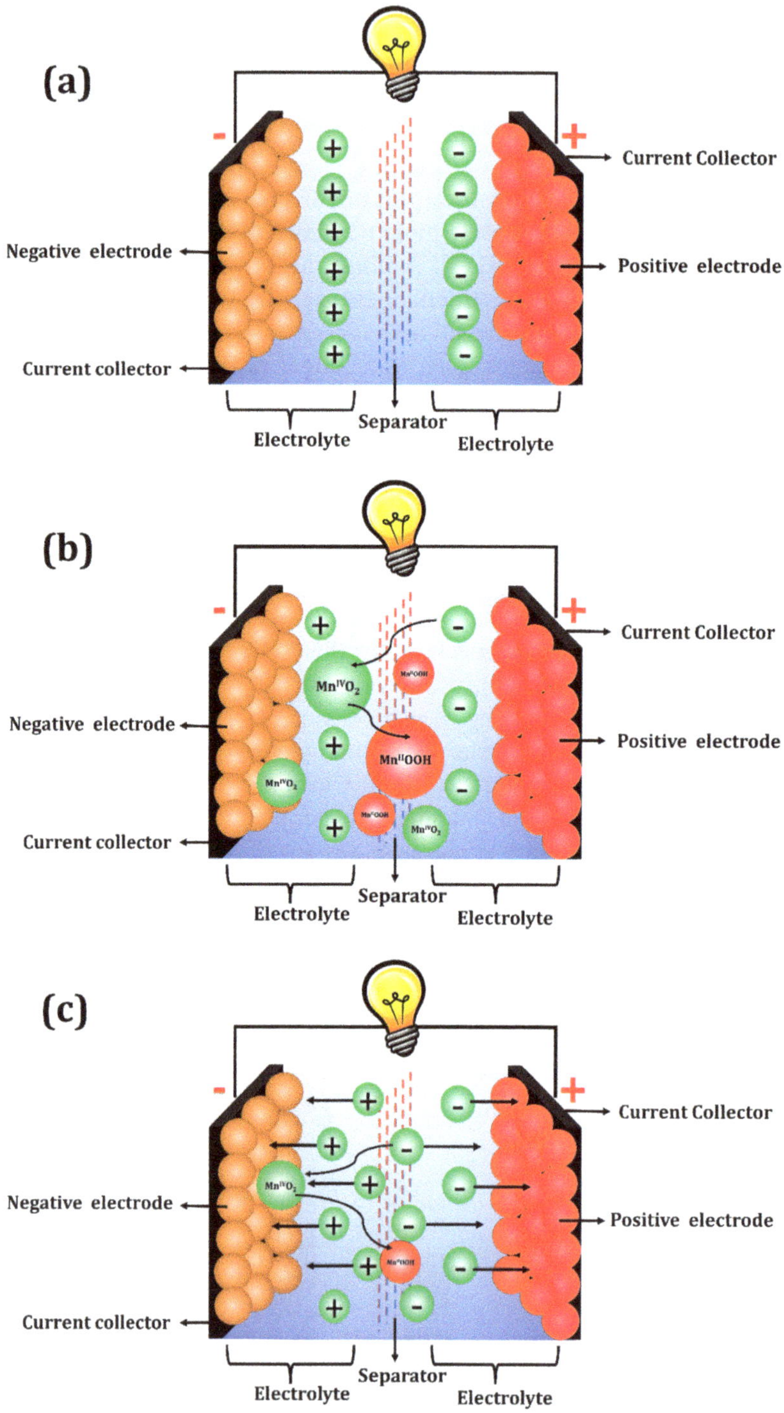

Figure 6.2. Schematic representations of the three types of supercapacitors based on their energy storage mechanism: (a) EDLC, (b) pseudocapacitor, and (c) hybrid supercapacitor.

pseudocapacitive materials include transition metal oxides, chalcogenides, hydroxides, conductive polymers, and so on. In a comparison, PCs are superior to EDLCs because of their high faradaic constant. On the other hand, fewer life cycles are detected for PCs. They have been surpassed by the hybrid capacitor, which is a combination of an EDLC and PC, where metal oxide and carbon are combined or a lithium-based metal oxide is used. The hybrid capacitor has overcome the limitations of EDLCs and PCs in terms of improving the number of life cycles and power storage. Most devices in today's electronic age are simplified based on the requirements. The fundamental aspects in the simplification of device materials are long life, good mechanical strength, portability, and compactness [7–11].

6.1.2 Fabrication of different types of supercapacitor devices

Figure 6.3 presents the fabrication steps for SC electrode devices [12]. First, the electrode material needs to be coated on current collectors such as graphene sheets, stainless steel, nickel foam, and carbon fibre paper. Second, the polymer gel electrolytes, such as PVA/H_2SO_4, PVA/H_3PO_4, and PVA/KOH, are dropped on top of the electrode material surface or applied by dipping the electrode material-coated substrate into the polymer gel electrolyte. These polymer gel electrolytes have high ionic conductivity, a wide operating voltage range, and superior flexibility, and they function as an ion-permeable membrane and absorbent electrolyte to maintain wettability. Furthermore, solid gel electrolytes provide a supporting structure to allow distortion testing without suffering an undesirable short circuit between the electrodes [13]. Additionally, the semi-solid gel electrolyte offers easy ion transport with a minimal distance between the electrode/electrolyte/electrode interfaces to enhance the charge storage properties. As a result, the structural architecture of the as-prepared SC is composed of lightweight devices known as 'movable electronics' [14, 15].

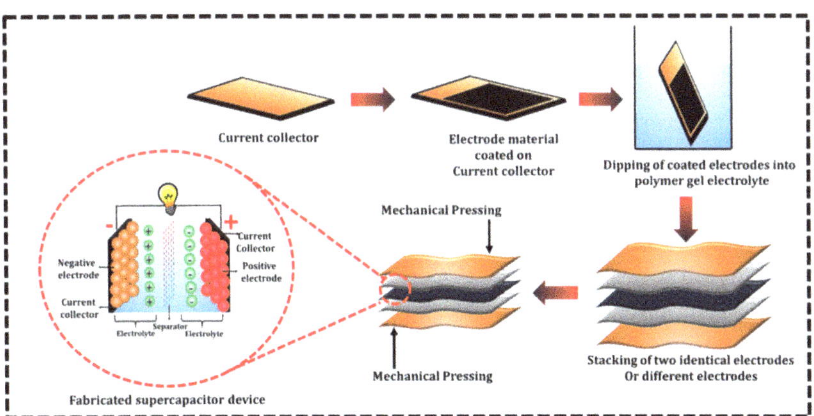

Figure 6.3. Schematic representation of the fabrication of a supercapacitor device.

6.1.3 Electrochemical cell set-up

Two different electrode systems are used to measure the electrochemical performance of any supercapacitor device. Generally, an electrochemical cell set-up is composed of either two or three electrodes contained in a small vial. A typical three-electrode system includes a working, reference, and counter electrode, all linked to an electrochemical workstation. Here, the potential difference is measured between the working and reference electrodes. Likewise, a two-electrode system consists of a working electrode and a counter electrode. At this juncture, the potential difference between the working and counter electrodes is measured [16]. A schematic representation of the three- and two-electrode systems is given in figures 6.4(a) and (b).

6.1.4 The importance of the electrode materials

The electrochemical performance of any supercapacitor device is primarily dependent on the characteristics of the respective materials used for electrode fabrication [17]. These electrode materials should have a high specific surface area (SSA), high electrical conductivity, corrosion resistance, and thermal and chemical stability [18]. A wide range of electrode materials have been successfully utilized for SC applications [19]. This includes carbonaceous materials [20, 21], conducting polymers (CPs) [22, 23], transition metal oxides (TMOs) [24, 25], nitrides [26], carbides [27], metal sulphides [28], chalcogenides [29], and their nanocomposites, along with some novel materials such as layered double hydroxides (LDHs) [30], MXenes [31], metal organic frameworks (MOFs) [32], etc. The most commonly used electrode materials in many supercapacitors are metal oxides and CPs, and their capacitance arises from redox reactions at the electrode/electrolyte interface. Recently, improving the power density and specific capacitance of the electroactive materials used in the SCs has become a major focus of today's researchers. In addition, so far, more substantial developments on hierarchical nanostructures have been made to enhance the electrochemical performance, but it is still a challenge to achieve a more flexible, inexpensive, lightweight, and environmentally friendly SC device [15]. Most electronic devices in the current microelectronic era are simplified based on the

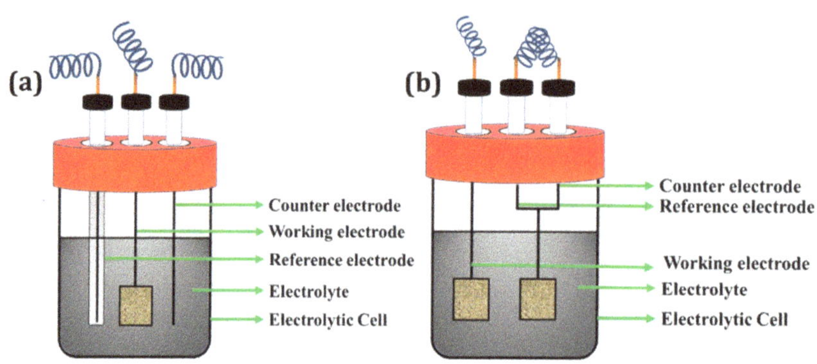

Figure 6.4. A schematic representation of the (a) three- and (b) two-electrode system.

requirements. The fundamental aspects in the simplification of device materials are long life, good mechanical strength, lightness, and compactness. These are the major unified standards in the design of flexible SCs. Finally, this chapter describes the various hierarchical nanostructured based electroactive materials and their formation mechanisms for the potential real-time application of SCs [11, 33].

6.2 Hierarchical nanostructured based functional materials (HNFMs)

With the rapid advancements in flexible energy storage devices, the use of portable and wearable optoelectronic devices such as supercapacitors has gained much attention in day-to-day life [34]. Free standing or self-standing flexible electrode materials are required to achieve higher power and energy density with flexible SCs [35]. Generally, the functional properties of any nanomaterial depends on its unique electronic structure, size, and morphology. Making nanomaterials into hierarchical and systematized structures is a coherent way of preparing cutting-edge functional materials. The HNFMs are self-assembling systems that are developed from small nano-units such as nanoparticles, nanorods, nanowires, nanotubes, nanosheets, etc [36]. This concept was initially developed by Masakazu Aono, who called it 'nano-architectonics' [37]. Moreover, these HNMF based electrode materials could offer a more specific surface area, abundant and accessible electroactive sites, and a short ion diffusion pathway during the rapid charge–discharge process. Also, TMO based HNFMs exhibit higher intrinsic properties, larger specific capacitances, high-rate capability, and excellent cycle stability compared to usual electrode materials [38]. The following sections describe the recent advancements made in HNFMs as electrode materials for SC applications.

6.2.1 MXene based HNFMs

MXenes are a new class of advanced two-dimensional (2D) functional materials that consist of nitrides, carbonitrides, and carbide materials. Recently, the MXene based HNFMs have been studied widely for various applications such as energy storage devices, including flexible supercapacitors. Owing to the unique physiochemical and electrical properties of MXenes, the transportation of bulk electroactive ions is significantly enhanced during the charging and discharging processes [31]. Moreover, the MXene based HNFMs provide more redox-active metal sites on the electrode material's surface [39].

Meanwhile, Chandra Sekhar Rout *et al* have recently developed a solid-state flexible asymmetric supercapacitor (ASC) [40] device using titanium carbide ($Ti_3C_2T_x$) as the negative electrode and composites of nickel cobalt sulphide ($NiCo_2S_4$) incorporated with rGO nanosheets, multiwall carbon nanotubes (MWCNTs), and $Ti_3C_2T_x$ as the positive electrode, with both electrodes separated by PVA-KOH solid gel film. Likewise, the group has developed more than five ASC devices using $NiCo_2S_4$, $Ti_3C_2T_x$, and carbon-based nanostructures (table 6.1). Among them, the $NiCo_2S_4@Ti_3C_2T_x//Ti_3C_2T_x$ system exhibits a higher areal capacitance of 57.47 mF cm^{-2} with a maximum areal energy density of 11.5 μWh

Table 6.1. A comparison of as-fabricated asymmetric devices. (Reproduced with permission from [40]. Copyright 2022 from Springer Nature.)

Electrode materials	Areal capacitance (mF cm^{-2})	Areal energy density μWh cm^{-2}	Areal power density (mW cm^{-2})
$NiCo_2S_4//Ti_3C_2T_x$	16.95	2.35	0.62
$NiCo_2S_4$@rGO//$Ti_3C_2T_x$	28.22	4.75	1.17
$NiCo_2S_4$@MWCNT//$Ti_3C_2T_x$	23.10	3.88	1.37
$NiCo_2S_4$@$Ti_3C_2T_x//Ti_3C_2T_x$	57.47	11.5	1.24
$NiCo_2S_4$@rGo@MWCNT//$Ti_3C_2T_x$	37.73	6.34	0.35

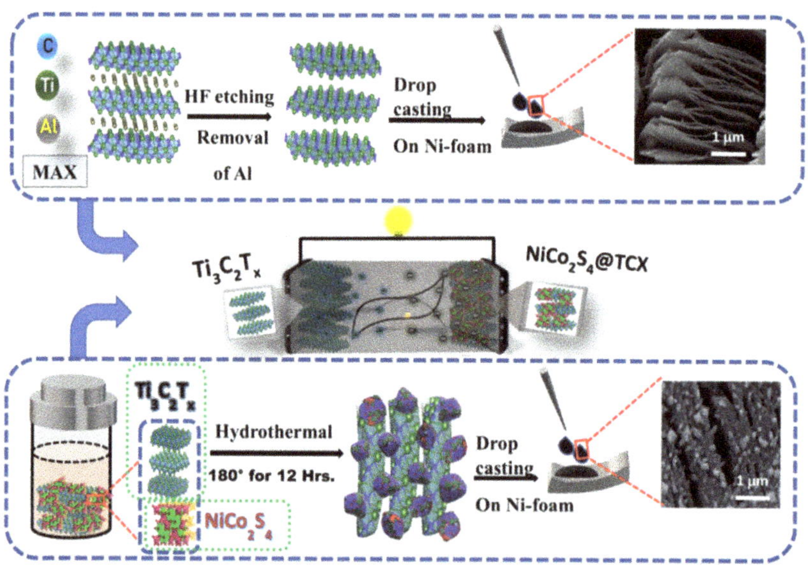

Scheme 6.1. Schematic representation of the fabrication of a $NiCo_2S_4$@$Ti_3C_2T_x//Ti_3C_2T_x$ asymmetric supercapacitor. The upper panel shows HF-etched MXene preparation followed by drop casting on Ni foam as the negative electrode. The lower panel represents the hydrothermal synthesis of $NiCo_2S_4$@$Ti_3C_2T_x$ composite for the positive electrode. The middle panel illustrates the assembly of a $NiCo_2S_4$@$Ti_3C_2T_x//Ti_3C_2T_x$ ASC device. (Reproduced with permission from [40]. Copyright 2022 Springer Nature.)

cm^{-2} at the power density of 1.24 mW cm^{-2}. A detailed schematic representation of the flexible $NiCo_2S_4$@$Ti_3C_2T_x//Ti_3C_2T_x$ ASC device is shown in scheme 6.1. The upper panel shows the preparation of $Ti_3C_2T_x$ by the removal of Al atoms from the MAX phase, and the field emission scanning electron microscopy (FESEM) image shows the HF-etched $Ti_3C_2T_x$ with well-arranged multi-layered sheets. In addition, the lower panel shows the hydrothermal synthesis of $NiCo_2S_4$@$Ti_3C_2T_x$ composite material and its respective FESEM image.

Bending tests were also performed to determine the mechanical flexibility of the fabricated devices (SCs). Figures 6.5(a)–(d) show digital photographs of the $NiCo_2S_4$@$Ti_3C_2T_x/Ti_3C_2T_x$ SCs at various bending angles, such as 0°, 60°, 90°,

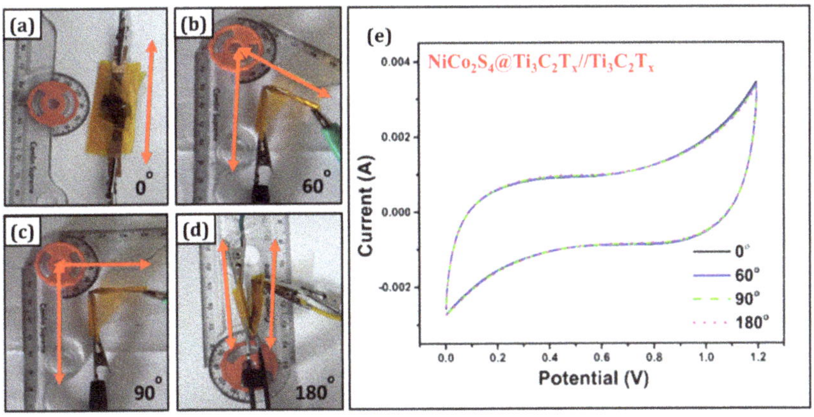

Figure 6.5. (a)–(d) Digital photographs of the $NiCo_2S_4@Ti_3C_2T_x//Ti_3C_2T_x$ device under different bending angles and (e) the respective CV curves recorded at different bending angles at a 100 mV s^{-1} scan rate. (Reproduced with permission from [40]. Copyright 2022 Springer Nature.)

and 180°. Further, the flexibility of the SCs was monitored by recording cyclic voltammetry (CV) curves at different angles and the results are displayed in figure 6.5(e). Surprisingly, the CV curves show no noticeable variation at different bending angles compared to the CV curves without being bent. These results confirm that the $NiCo_2S_4@Ti_3C_2T_x//Ti_3C_2T_x$ device has remarkable flexibility while maintaining crystalline stability. The following features contributed to the excellent electrochemical performance of the $NiCo_2S_4@Ti_3C_2T_x/Ti_3C_2T_x$ device: (i) the synergistic effect of $NiCo_2S_4$ nanobeads and $Ti_3C_2T_x$ sheets, (ii) prevention of volume expansion during rapid charge/discharge cycles, (iii) the unique layered structure of $Ti_3C_2T_x$ sheets improving the bulk ion diffusion process, and (iv) high flexibility and mechanical strength.

Yang and Ji *et al* have developed a high energy density flexible asymmetric supercapacitor [41] with a negative electrode of $Ti_3C_2T_x$-MXene composited with MoO_3 nanobelts and a positive electrode of acidified carbon nanotubes (A-CNTs) coupled with potassium-doped MnO_2 nanowires. As shown in scheme 6.2(a), the van der Waals interaction between $Ti_3C_2T_x$ nanosheets and MoO_3 nanobelts makes a stable and paper-like composite material. Materials such as $Ti_3C_2T_x/MoO_3$ provide a short ion diffusion path during the charging and discharging process. Meanwhile, the free standing A-CNTs/K_xMnO_2 composite material was obtained by treating the MnO_2 nanowires with an alkali metal ion such as K^+, followed by incorporation with acidified carbon nanotubes (scheme 6.2(b)). Here, the introduction of A-CNTs provides strong hybridized interlayer areas where rapid transportation of electrolyte ions takes place during the intercalation/deintercalation processes.

To investigate the practical applications of as-synthesized composite material, the ASC device was fabricated using A-CNTs/K_xMnO_2 as the positive electrode and $Ti_3C_2T_x/MoO_3$ as the negative electrode. Here, PVA-Na_2SO_4 solid gel film acted as a separator. Figure 6.6(a) shows a comparative CV curve of A-CNTs/K_xMnO_2 (0–1 V) and $Ti_3C_2T_x/MoO_3$ (from 0 to −1 V) recorded in a three-electrode system. The

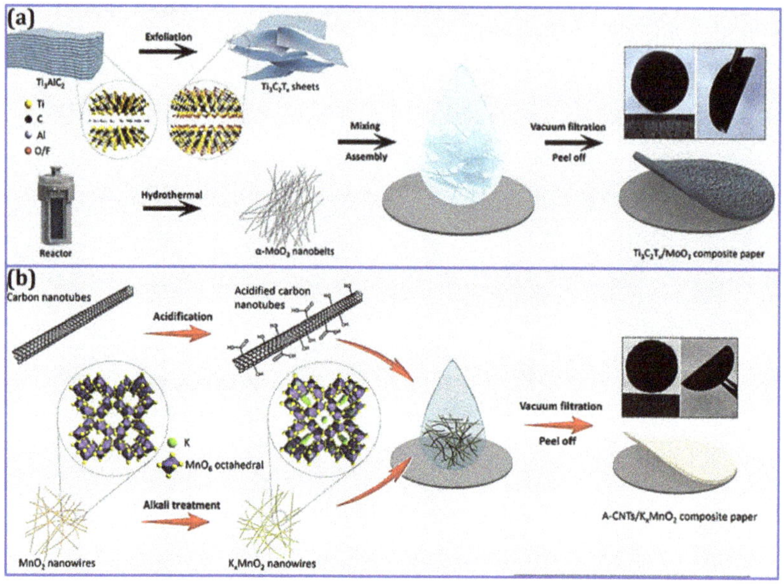

Scheme 6.2. Illustration of the synthesis of paper-like (a) $Ti_3C_2T_x/MoO_3$ and (b) A-CNTs/K_xMnO_2 composites. (Reproduced with permission from [41]. Copyright 2021 Elsevier.)

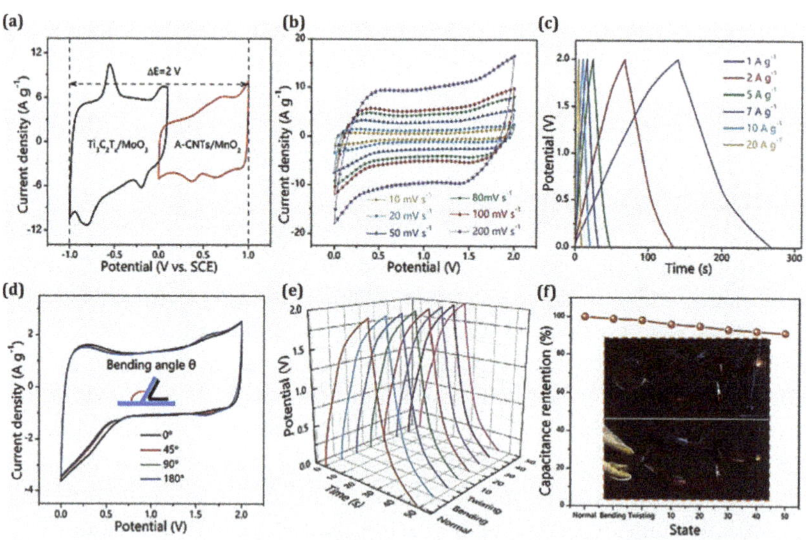

Figure 6.6. Electrochemical performance of the assembled A-CNTs/$K_xMnO_2/Ti_3C_2T_x/MoO_3$ ASC device. (a) Comparative CV curves of $Ti_3C_2T_x/MoO_3$ (negative electrode) and A-CNTs/K_xMnO_2 (positive electrode) at a scan rate of 20 mV s^{-1} in a three-electrode system. (b) CV and (c) GCD curves of assembled ASC devices at various scan rates and current densities, respectively, at a fixed operating potential from 0 to 2 V. (d) CV curves of the flexible ASC device under different bending angles at 20 mV s^{-1}. (e) GCD curves at 5 A g^{-1} current density and (f) specific capacitance retention ratio of a flexible ASC device in normal, bending, and twisting states, and after being bent repeatedly. The inset figure shows a picture of a typical ASC device lighting up a 1.8 V LED. (Reproduced with permission from [41]. Copyright 2021 Elsevier.)

operating potential window was optimized and fixed from 0 to 2 V as the most suitable potential window where the outcome of the CV curves is a good shape without any overvoltage. Figure 6.6(b) displays the CV profile of the ASC device at different scan rates (10–200 mV s^{-1}), suggesting that the assembled A-CNTs/K_xMnO_2//$Ti_3C_2T_x$/MoO_3 ASC device can work stably up to 2 V without any obvious polarization. Interestingly, the obtained quasi-rectangular shape of CV curves is reminiscent of the respective galvanostatic charge discharge (GCD) curves measured at various current densities from 1 to 20 A g^{-1} (figure 6.6(c)). Accordingly, the result suggests that the electrochemical charging and discharging mechanism of the ASC device obeys the principles as typical Faraday pseudocapacitors. The maximum C_{sp} of 65.6 F g^{-1} was observed at 1 A g^{-1} current density for an ASC device with a high energy density of 36.4 Wh kg^{-1} at a power density of 863.5 W kg^{-1}.

Furthermore, flexibility and mechanical stability are two very important parameters in practical applications of supercapacitors. No noticeable changes were observed in all the CV and GCD curves recorded at different bending statuses from 0° to 180° which illustrates the good flexibility of the ASC device (figure 6.6(d) and (e)). The bending durability of the A-CNTs/K_xMnO_2//$Ti_3C_2T_x$/MoO_3 ASC device was further tested for 50 bending and twisting cycles. As depicted in figure 6.6(f), the capacitance retention of 91.5% was preserved even after multiple twisting and bending angles, indicating that the ASC device maintains stable capacitance behaviour when folded. For a practical demonstration, the ASC device was successfully used to glow a 1.8 V LED bulb at both normal and twisted angles, as shown in the inset of figure 6.6(f). The key results of this work are mainly due to the rapid self-assembly and ion permeability of the as-prepared electrode materials, which offer great potential for the future of portable and wearable supercapacitors.

6.2.2 Metal organic framework (MOF) based HNFMs

In recent years, metal organic frameworks (MOFs) have received much attention on account of their applicability in diverse fields, especially materials chemistry [42]. MOFs give rise to a variety of different dimensional architectures with high porosity fabricated using small organic linkers and metal clusters. Due to their inherent advantages such as mechanical strength, pore structure, morphology, and easily tunable metal precursors, the fabrication of rationally designed MOFs for energy storage and conversion applications is possible [43].

Kader Dagsi Kiransan *et al* have developed a lightweight and portable flexible symmetric supercapacitor [44] using bimetallic MOF doped graphene sponges (GS) as an advanced electrode material. Schemes 6.3(a) and (b) depict a representative illustration of the synthesis of bimetallic MOFs such as ZnCo-MOF doped GS (ZnCo-MOF/GS). Primarily, the obtained purple coloured ZnCo-MOF was ultrasonically combined with graphene oxide (GO), followed by annealing at 350 °C to obtain a porous ZnCo-MOF/GS electrode material. As presented in the digital photographs of figure 6.7(a), the ZnCo-MOF/GS sponge material is able to stand on a delicate flower because of its flexible and lightweight nature. Furthermore, even after 20 times of squeezing, the sponge nature of ZnCo-MOF/GS was preserved. The

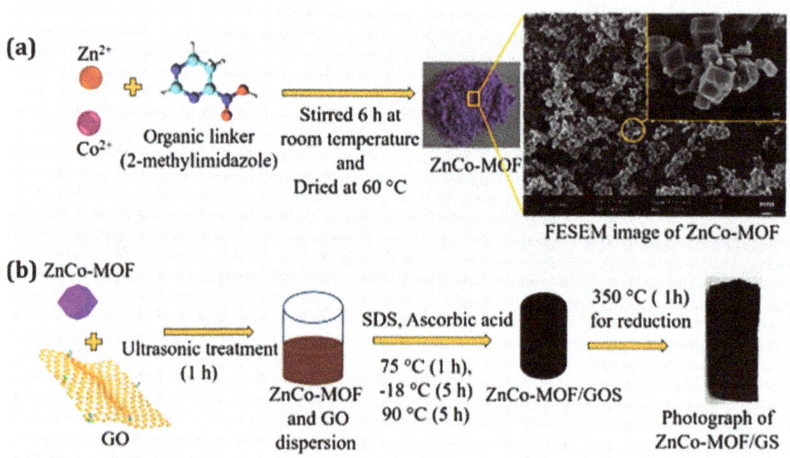

Scheme 6.3. Schematic representation of the synthesis of (a) ZnCo-MOF and (b) ZnCo-MOF/GS.

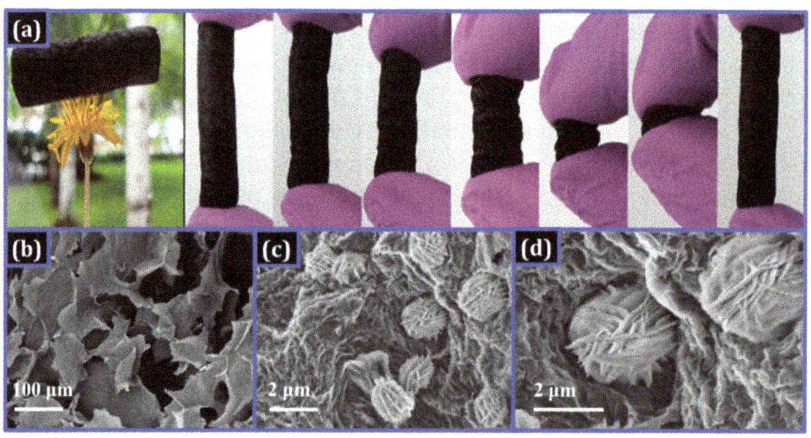

Figure 6.7. (a) Digital photographs of ultralight weight and flexible ZnCo-MOF/GS, (b) SEM images of GS, and (c and d) ZnCo-MOF/GS at different magnifications. (Reproduced from [44]. CC BY 4.0.)

morphological studies also show that ZnCo-MOF has Brussels sprout-like structures with a sharp-edged surface embedded between the pores of the three-dimensional (3D) layered GS (figures 6.7(b)–(d)). Therefore, the close contact between ZnCo-MOF and porous GS improves the conductivity of an electrode material such as ZnCo-MOF/GS.

Figure 6.8(a) shows a digital photograph of a developed ZnCo-MOF/GS/ZnCo-MOF/GS symmetric SC with a separator of PVA-KOH. As can be seen from the CV curves of the symmetric supercapacitor device shown in figure 6.8(b), the current density has increased with respect to the increase in scan rates. Moreover, the specific capacitance (C_{sp}) of the symmetric SC device was calculated using GCD curves gathered at different current densities (figure 6.8(c)). The symmetric SC device displayed a high C_{sp} of 302 F g^{-1} at 1 A g^{-1} current density, and it showed

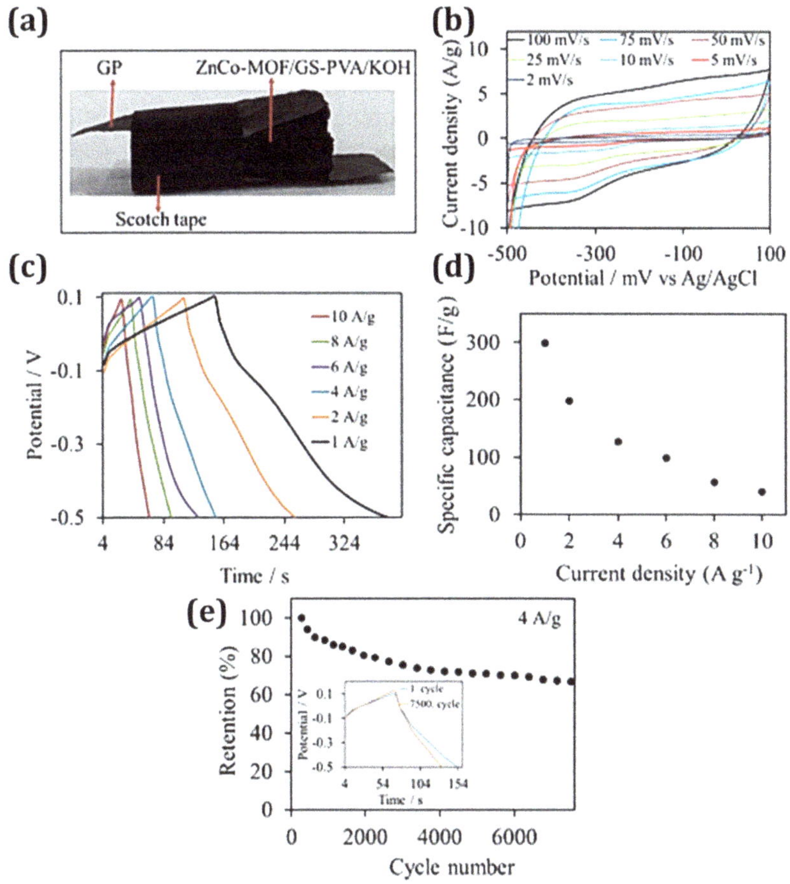

Figure 6.8. Electrochemical performance of a ZnCo-MOF/GS//ZnCo-MOF/GS symmetric supercapacitor. (a) Digital photograph of the assembled symmetric supercapacitor device, (b) CV, and (c) GCD at various scan rates (2–100 mV s^{-1}) and current densities (1–10 A g^{-1}), (d) the relationship between specific capacitances (F g^{-1}) and current densities (A g^{-1}), and (e) long-term cyclic stability study at 4 A g^{-1}. The inset figure shows the GCD curves measured before and after 7500 cycles.

52 F g^{-1} even at 10 A g^{-1} current density (figure 6.8(d)). The long-term cyclic stability test was conducted for a symmetric SC device using continuous charging and discharging at 4 A g^{-1} current density for 7500 cycles. Notably, the device exhibited 67.5% of its initial capacitance after 7500 cycles, suggesting good rate capability. Such retention behaviour indicates that the pores of the electrode material are fully accessible at initial GCD cycles, and further decrement of capacitance might be due to the detachment of ZnCo-MOF/GS after many swelling and shrinking cycles. Additionally, the GCD curves before and after 7500 cycles were recorded and are shown in the inset of (figure 6.8(e)). This result confirms that a meagre change in the shape of the GCD curves was observed after 7500 cycles, which is due to the minimal weight loss of the electrode material.

To further demonstrate the practicability of the ZnCo-MOF/GS//ZnCo-MOF/GS symmetric SC device, tests on the cyclic stability in various bending states was carried out, and the results are shown in figure 6.9. A digital photograph of a bent type symmetric SC device is shown in figure 6.9(a). Because of its high durability and mechanical stability, the ZnCo-MOF/GS//ZnCo-MOF/GS symmetric SC device exhibits the same C_{sp} values even after bending at different angles, such as 90° and 180°, 1500 times (figure 6.9(b)). In addition, the GCD results were used to calculate the energy and power densities of the symmetric SC device, which delivers a reliable and maximum energy density of 108 Wh kg^{-1} at a power density of 5037 W kg^{-1} (figure 6.9(c)). Moreover, the obtained results were compared with previously reported literature, which clearly indicates their superiority over existing ones. Furthermore, the developed symmetric SC device was used to successfully light a green LED light (figure 6.9(d)). These results benefit from the excellent pseudocapacitance of ZnCo-MOF and the dielectric capacitance behaviour of GS, as well as the synergistic effect between the two. Overall, this study revealed that materials modified with bimetallic MOFs can be considered as ideal electrode materials for flexible energy storage devices.

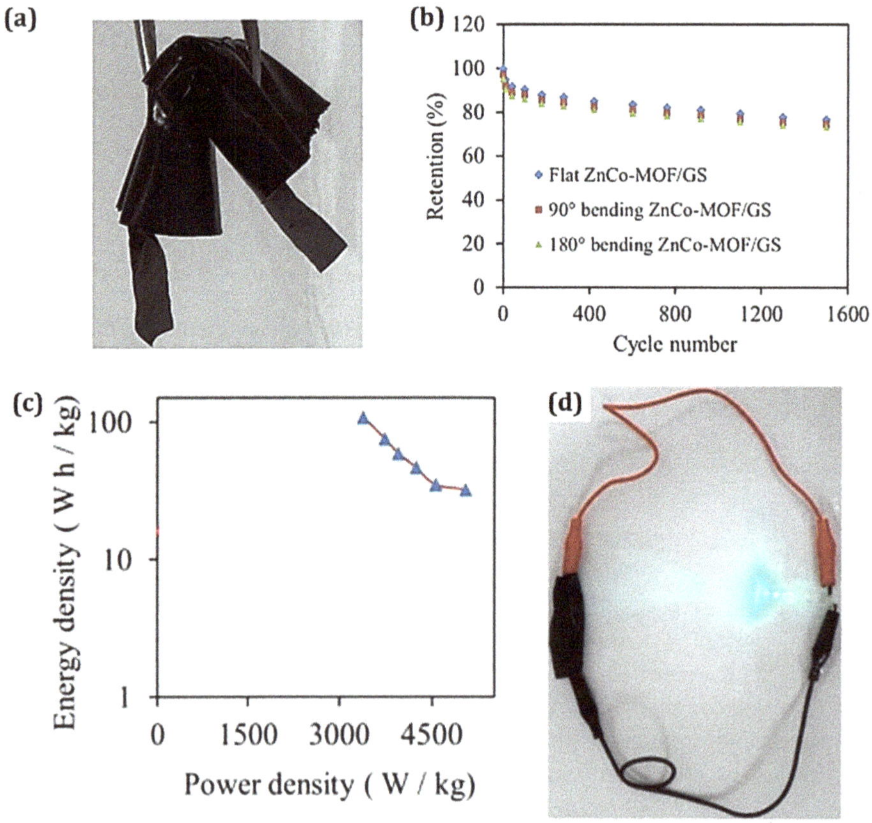

Figure 6.9. Electrochemical performance of an MOF/GS//ZnCo-MOF/GS symmetric SC device. (a) Digital photograph of the bent ZnCo-MOF/GS//ZnCo-MOF/GS symmetric SC device, (b) the capacitance retention behaviour after bending inward 90° (red) and 180° (green) and without bending (blue) for 1500 cycles at 4 A g^{-1} current density, (c) Ragone plot, and (d) its ability to light up an LED bulb.

Yan *et al* have demonstrated the synthesis of a novel core–shell heterostructure [45] based on Mn_2O_3 nanosheets and Cu_xO nanowires attached to Cu foam for flexible ASC applications. First, the Cu foam (CF) was pre-treated through a chemical surface oxidation process, followed by Cu_xO nanowires grown on Cu foam by the calcination method (Cu_xO/CF). Second, manganese precursor was added to the Cu_xO/CF to form Mn_2O_3/Cu_xO/CF, where the core is Cu_xO nanowires and the shell is Mn_2O_3 nanosheets. Similarly, the core–shell heterostructures were further prepared by changing the reaction time (6, 9, and 12 h) of hydrothermal treatment under identical conditions. The resulting samples were labelled as Mn_2O_3/Cu_xO/CF-6, Mn_2O_3/Cu_xO/CF-9, and Mn_2O_3/Cu_xO/CF-12. SEM images show that the morphology of all three samples was similar. However, it can be seen that the non-uniform distribution of Mn_2O_3/Cu_xO/CF-6 shows partially exposed Cu_xO nanowires in figure 6.10(a). However, after 9 h of reaction time, uniform growth of Mn_2O_3 nanosheets on CuxO/CF was observed (figure 6.10(b)). As can be seen in figure 6.10(c), a certain amount of agglomeration was found in the Mn_2O_3/Cu_xO/CF-12 heterostructure, which led to a slow ion diffusion process during the redox reactions. Furthermore, TEM images confirm that Mn_2O_3 nanosheets were attached uniformly on the surface of Cu_xO nanowires, which is consistent with the SEM results (figures 6.10(d)–(i)). These findings support the formation of Mn_2O_3/Cu_xO core–shell heterostructures on Cu foam (scheme 6.4).

In order to investigate the practical application of Mn_2O_3/Cu_xO/CF-9 in flexible energy storage systems, the ASC device was fabricated using the obtained samples as the positive electrode and activated carbon as the negative electrode on Ni foam as

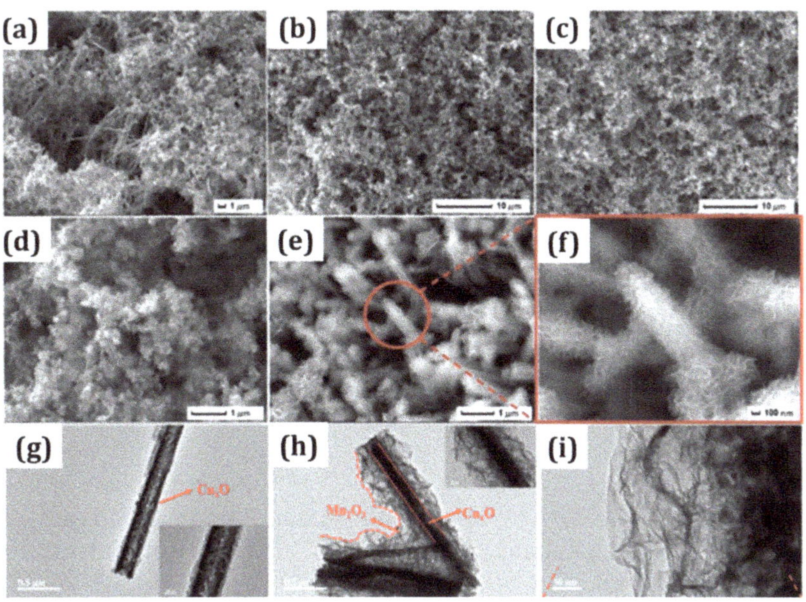

Figure 6.10. SEM images of (a) Mn_2O_3/Cu_xO/CF-6, (b) Mn_2O_3/Cu_xO/CF-9, (c) Mn_2O_3/Cu_xO/CF-12, (d) Mn_2O_3 nanosheets, and (e and f) core–shell structure of Mn_2O_3 nanosheets grown on Cu_xO nanowires. TEM image of (g and h) Cu_xO/CF and (i) Mn_2O_3/Cu_xO/CF. (Reproduced with permission from [45]. Copyright 2022 Elsevier.)

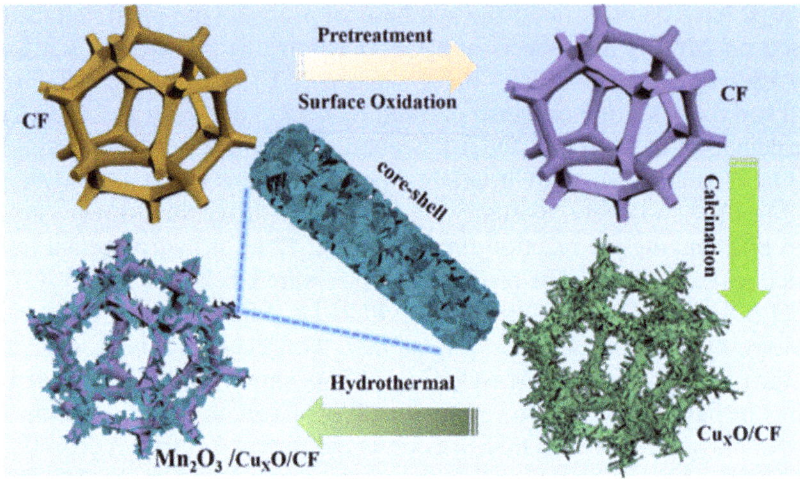

Scheme 6.4. Schematic representation of the synthesis process of $Mn_2O_3/Cu_xO/CF$ electrode materials. (Reproduced with permission from [45]. Copyright 2022 Elsevier.)

the current collector. Here, PVA-KOH solid gel electrolyte film acted as a separator. Figures 6.11(a) and (b) show the CV profile at a scan rate of 100 mV s^{-1} and the GCD profile at a current density of 2 mA cm^{-2} with potential windows ranging from 0 to 1 and 0 to 1.3 V. Obviously, all the GCD curves show linear and symmetrical shapes, and they are in good agreement with the respective CV curves. Subsequently, the potential window was optimized and fixed from 0 to 1.3 V as the most suitable and stable operating potential window. The C_{sp} values were calculated from the obtained GCD curves recorded at different current densities, such as 2, 3, 5, 8, and 10 mA cm^{-2} at the optimized potential window. The assembled $Mn_2O_3/Cu_xO/CF$-9/AC/NF ASC device has a maximum C_{sp} of 512.7 mF cm^{-2} at 2 mA cm^{-2} current density, as shown in figure 6.11(c). The specific capacitance value of the ASC device is gradually decreased with respect to the increase in current density. This is due to the decrease in the mass transportation of ions during the charge transfer process. Figure 6.11(d) shows the long-term cycling stability of the $Mn_2O_3/Cu_xO/CF$-9//AC/NF ASC device, with 76.1% capacitance retention achieved after 10 000 cycles at 5 mA cm^{-2} current density. This long-term electrochemical stability of the $Mn_2O_3/Cu_xO/CF$-9//AC/NF ASC device is mainly due to the rough surface of the Cu_xO nanowires, which can provide more active sites to grow conductive Mn_2O_3 nanosheets. Thereby, the bulk diffusion of electrolyte ions would take place during the continuous charging and discharging cycles, leading to a rapid redox reaction.

The schematic representation of the assembled $Mn_2O_3/Cu_xO/CF$-9//AC/NF ASC device shown in figure 6.12(a). In addition, the flexibility and mechanical stability of the as-developed $Mn_2O_3/Cu_xO/CF$-9//AC/NF ASC device was further tested at different bending angles such as 0°, 90°, and 180° (figure 6.12(b)). As shown in figure 6.12(c), the charging time is constant for different bending angles, but the discharging time is slightly shorter, dropping by 9.8% and 12.7% at 90° and 180°, respectively. The above results show that the superior electrochemical properties of the as-prepared $Mn_2O_3/Cu_xO/CF$ are mainly attributed to the distinctive core–shell

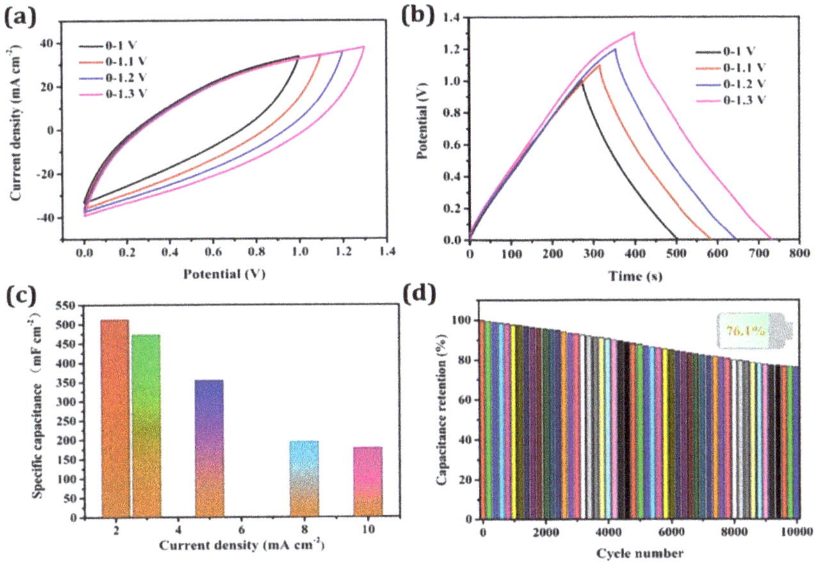

Figure 6.11. (a) CV recorded at scan rate 100 mV s^{-1} and (b) GCD curves at different voltage windows at 2 mA cm^{-2}, (c) plot of specific capacitances at different current densities, and (d) cycling stability after 10 000 cycles under 5 mA cm^{-2} and of a Mn_2O_3/Cu_xO/CF-9//AC/NF ASC device. (Reproduced with permission from [45]. Copyright 2022 Elsevier.)

architecture, which reduces agglomeration and increases the active site for electron and ion transport. Thus, the Mn_2O_3/Cu_xO/CF core–shell material could be a suitable candidate for future flexible energy storage systems.

6.2.3 Layered double hydroxide based HNFMs

The transition metal hydroxides as layered double hydroxides (LDHs) have been investigated extensively as ideal electrode materials for many applications in comparison to the traditional carbonaceous materials and conducting polymers, owing to their interesting properties such as: (i) very low environmental toxicity; (ii) low cost precursor materials; (iii) easy synthesis; (iv) long life span; (v) excellent ion exchange capability; (vi) high specific capacitance; and (vii) high redox activity and easy functionalization. Any LDHs with the general formula $[M_{(1-x)}{}^{2}{+}M_{x}{}^{3+}(OH)_2]_{layer}$ $[A_{x/n}{}^{n-}.zH_2O]_{interlayer}$, where M^{2+} and M^{3+} are divalent and trivalent metal cations occupying the octahedral interstices of the layers, and A^{n-} represents organic or inorganic anions entrenched in the lamellar structure's interlayers. However, combining LDH with other transition oxides is a feasible technology for forming hybrids for energy storage applications. It would be a most promising research project to explore the performances of LDH-containing electrode materials prepared via constructing various heterostructures with well-defined structures, compositions, crystal phases, energy band gaps, and interfaces [46–48].

Han *et al* have demonstrated wire shaped supercapacitors [49] with high flexibility and power density for flexible and wearable energy storage devices. A flexible hybrid

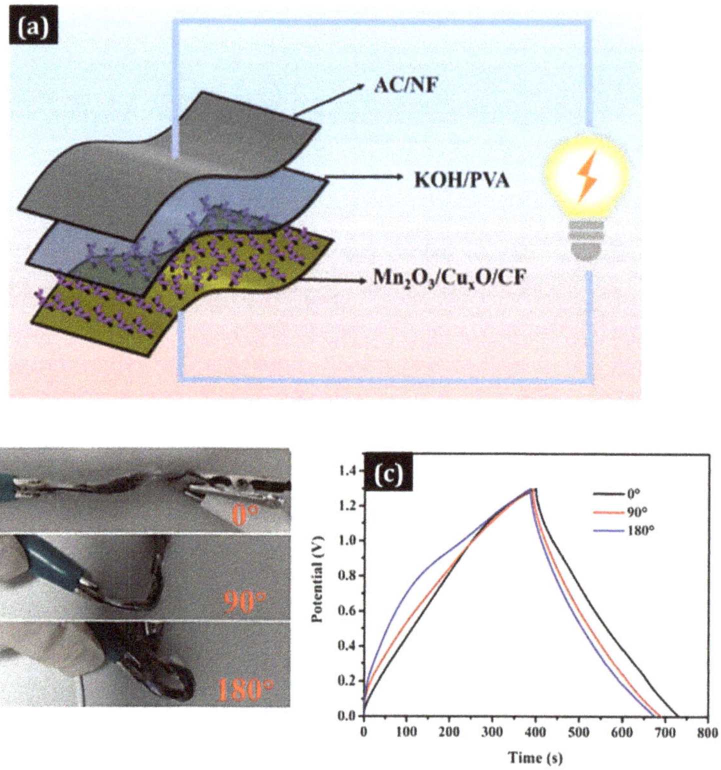

Figure 6.12. (a) Schematic illustration of the assembled Mn_2O_3/Cu_xO/CF-9//AC/NF ASC device, (b) digital photographs of the flexible solid-state Mn_2O_3/Cu_xO/CF-9//AC/NF ASC device, and (c) their GCD curves under various bending angles. (Reproduced with permission from [45]. Copyright 2022 Elsevier.)

electrode consists of NiO nanobelts as the core and CoNi-LDH nanoplates as the shell (NiO@CoNi-LDH) on Ni wire. Figure 6.13(a) depicts the fabrication steps for NiO@CoNi-LDH core–shell nanoarrays. First, NiO nanobelt arrays (NBAs) were grown on Ni foam using $Ni(SO_4)_{0.3}(OH)_{1.4}$ as the precursor material, followed by calcination at 600 °C for 2 h in a nitrogen atmosphere. Following that, the developed Ni/NiO NBAs were treated with $CoCl_2{\cdot}6H_2O$ and $Ni(NO_3)_2{\cdot}6H_2O$ in deionized water using the electrodeposition method for 200 s at a deposition potential of −1.0 V. As a result, Ni/NiO@CoNi-LDH nanoarrays were synthesized and used as positive electrodes for the ASC device. The morphological studies of the as-prepared composite materials were analysed by SEM, and the obtained images are shown in figures 6.13(b)–(g). The Ni foam has a smooth surface (figure 6.13(a)) on which the vertically aligned $Ni/Ni(SO_4)_{0.3}(OH)_{1.4}$ NBAs with lengths of several m and widths of 0.6–1μm were grown, as shown in figures 6.13(b)–(d). When Ni/Ni $(SO_4)_{0.3}(OH)_{1.4}$ NBAs were post-heated at 600 °C, they were transformed into well-organized NiO NBAs with the same morphology as Ni/Ni $(SO_4)_{0.3}(OH)_{1.4}$ NBAs (figure 6.13(e) and (f)). Finally, CoNi-LDH nanoplates were deposited on the NiO NBAs with a lateral size of 80–100 nm and a thickness of 6–10 nm, which are

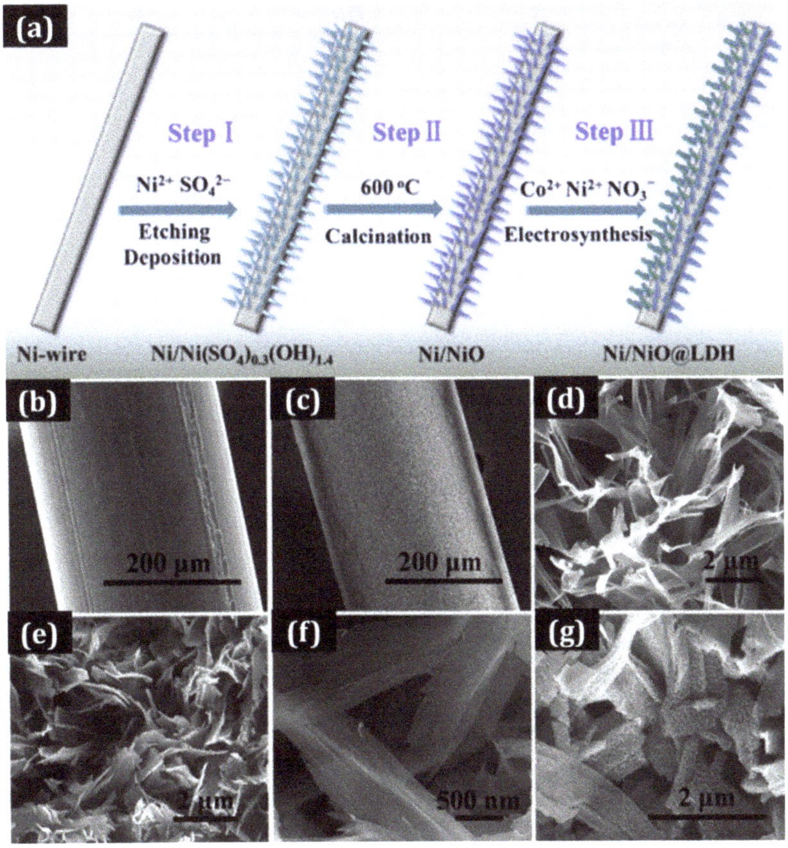

Figure 6.13. (a) Schematic illustration for the preparation of Ni/NiO@CoNi-LDH wire electrodes, (b) SEM images of bare Ni wire, (c and d) $Ni/Ni(SO_4)_{0.3}(OH)_{1.4}$ NBAs, (e and f) Ni/NiO NBAs, and (g) Ni/NiO@LDH NAs. (Reproduced with permission from [49]. Copyright 2020 Elsevier.)

perpendicularly anchored on the NiO nanobelts cores to form hierarchical core–shell architecture (figure 6.13(g)).

Further, a solid-state wire shaped ASC device has been fabricated by employing pseudo-type Ni/NiO@LDH and EDLC-type Ni@AC as the positive and negative electrodes, respectively. Figure 6.14(a) depicts a schematic representation of a wire shaped Ni/NiO@LDH//Ni@AC ASC device. Moreover, the obtained nonrectangular CV curves showed no obvious changes even at a higher scan rate, suggesting the occurrence of an excellent charge transfer process at the electrode and electrolyte interface (figure 6.14(b)). Figure 6.14(c) showed the nonlinear GCD curves at different current densities, consistent with the respective CV curves. From the CV and GCD curves, it was understood that the pseudocapacitive property was mainly responsible for the charge storage mechanism in the wire shaped ASC device. The specific capacitance of the wire shaped ASC device was calculated and plotted as a function of current densities (figure 6.15(d)). At the current density of 0.25 A cm^{-3}, ASC device exhibited the maximum specific capacitance of 34.9 F cm^{-3} and

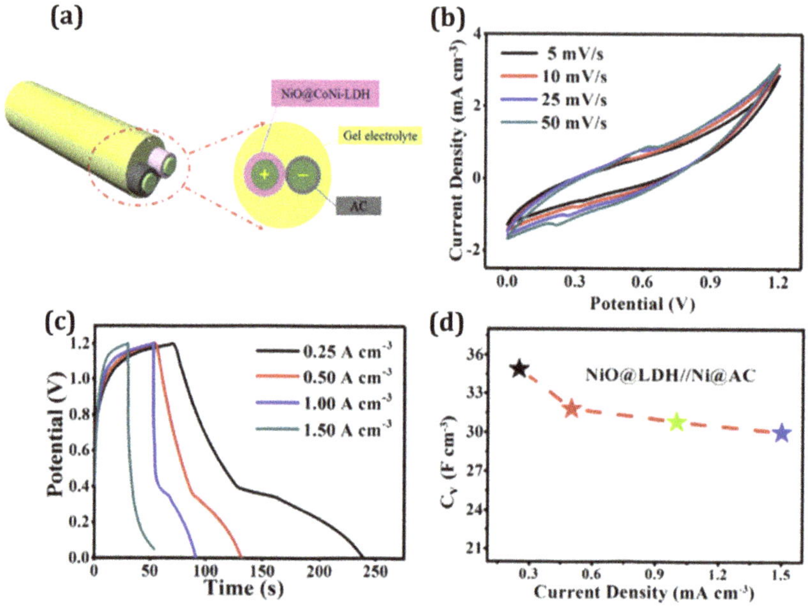

Figure 6.14. (a) Schematic illustration of the wire shaped supercapacitor, (b) CV, and (c) GCD profile at different scan rates and current densities at fixed potential voltage window from 0 to 1.2 V, and (d) the plot between C_{sp} and current densities. (Reproduced with permission from [49]. Copyright 2020 Elsevier.)

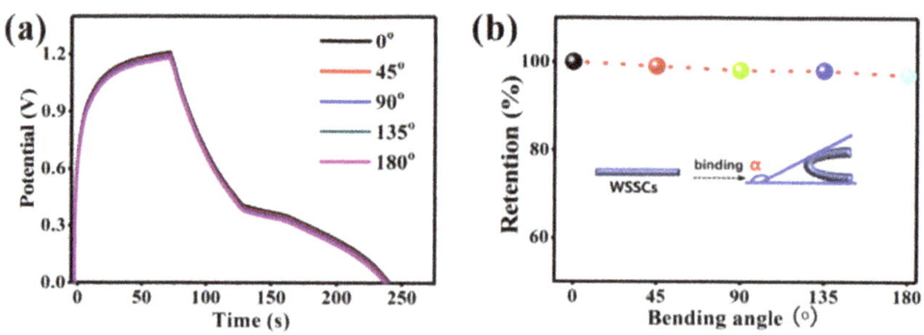

Figure 6.15. (a) GCD profiles and (b) capacitance retention at different bending angles of the Ni/NiO@LDH//Ni@AC wire shaped ASC device. (Reproduced with permission from [49]. Copyright 2020 Elsevier.)

maintained the specific capacitance of 30.0 F cm^{-3} by the increased current density to 1.5 A cm^{-3}, indicating the superior rate capability of the ASC device.

The flexibility properties of the ASC device were tested by bending it at 0°, 45°, 90°, 135°, and 180°, as shown in figure 6.15(a). When the device is bent to 180°, there is no discernible difference in the discharge curves, and approximately 98% of the capacitance remains (figure 6.15(a) and (b)). A wire shaped Ni/NiO@LDH//Ni@AC ASC device with excellent mechanical flexibility and outstanding stability can serve as a promising flexible power source in the fields of portable and miniaturized electronics. Table 6.2 summarizes the comparison of various HNFMs based flexible electrode materials.

Table 6.2. Summary of the reported HNFMs based supercapacitor devices.

S. No.	Electrode materials	Method	Device type	Capacitance	Electrolyte	Current collector	Reference
1	D-Ti_3C_2/MoO_3@IL	Solvothermal	SSC	1680 F g^{-1} @ 1 A g^{-1}	2 M KOH	Ni foam	[50]
2	Ti_3C_2 MXene@graphene	Hydrothermal		352.4 mAh g^{-1} @ 1 A g^{-1}	1 M KOH	Ni foam	[31]
3	$Ti_3C_2T_x$/CMC-PANI	*In situ* chemical oxidation polymerization	ASC	1161.4 mF cm^{-2} @ 1 mA cm^{-2}	1 M H_2SO_4	Film	[39]
4	NiCo-LDH/$(NiCo)Se_2$	Ion exchange reaction method	ASC	142.3 F g^{-1} @ 1 A g^{-1}	6 M KOH	Ni foam	[51]
5	$(NiCo)Se_2$@ZIF-L	Hydrothermal	Hybrid	188.8 mAh g^{-1} @ 1 A g^{-1}	6 M KOH	Ni foam	[52]
6	CoAl-LDH@NiCo-LDH	Hydrothermal	Flexible ASC	2633.6 F g^{-1} @ 1 A g^{-1}	6 M KOH	CC	[53]
7	$NiCo_2O_4$@NiMn-LDH	Hydrothermal	Flexible ASC	4010.4 mF cm^{-2} @ 1 mA cm^{-2}	3 M KOH	CC	[54]
8	$Co_3S_4/Ti_3C_2T_x$	Hydrothermal	ASC	602 F g^{-1} @ 1 A g^{-1}	3 M KOH	Ni foam	[55]
9	NCNRS@NCNSs	Ion exchange co-precipitation method	ASC	2640.2 F g^{-1} @ 1 A g^{-1}	6 M KOH	Ni foam	[38]
10	NPCM-900/Zn	Ion exchange method	Hybrid	708.2 mAh g^{-1} @ 1 A g^{-1}	6 M KOH	—	[56]
11	ZnCo-LDH/ZnCo-NA	Hydrothermal	ASC	1576 F g^{-1} @ 1 A g^{-1}	6 M KOH	Direct electrode	[57]
12	V_2CT_x@NiCoMnOH	Electrostatic adsorption process	ASC	827.45 F g^{-1} @ 1 A g^{-1}	3 M KOH	Ni foam	[58]
13	CCA@MOF	Solvothermal	ASC	1877.45 F g^{-1} @ 1 A g^{-1}	6 M KOH	—	[59]
14	$NiCo_2O_4$@NiMn-MOF	Solvothermal	Hybrid	243.0 mAh g^{-1} @ 1 A g^{-1}	2 M KOH	CC	[60]
15	$NiCo_2S_4$@NiV-LDH	Hydrothermal		1778.8 C g^{-1} @ 1 A g^{-1}	6 M KOH	Ni foam	[61]
16	MXene@$CoSe_2/Ni_3Se_4$	Ion exchange method	ASC	283 mAh g^{-1} @ 1 A g^{-1}	3 M KOH	Ni foam	[62]
17	Co@NiMn-LDH/V_2CT_x	Hydrothermal	ASC	1005 F g^{-1} @ 1 A g^{-1}	6 M KOH	Ni foam	[63]
18	CCO-ZIF-67/Co-NiMn-LDH	Solvothermal	ASC	2955.5 F g^{-1} @ 1 A g^{-1}	2 M KOH	Ni foam	[64]

19	$Ti_3C_2T_x$@LDH-Ag	Hydrothermal	Inplane hybrid SCs	592 C g^{-1} @ 1 A g^{-1}	5 M KOH	Ni foam	[65]
20	$Ti_3C_2T_x$/g-C_3N_4	Thermal polymerization	SSC	2955.5 F g^{-1} @ 1 A g^{-1}	1 M H_2SO_4	Film	[26]
21	1D@3D V_2O_5@Co_3S_4	Sulfidation	ASC	1493.6 F g^{-1} @ 1 A g^{-1}	3 M KOH	Ni foam	[66]
22	B-NiS/$Ti_3C_2T_x$ @2D/2D hydrogel	Hydrothermal	Flexible SCs	10.92 F g^{-1} @ 1 A g^{-1}	1 M KOH	—	[27]
23	Ti_3C_2-DA-$NiMoS_4$	Hydrothermal	ASC	1288 F g^{-1} @ 1 A g^{-1}	—	—	[67]
24	$NiCo_2S_4$@NiMn-LDH	Hydrothermal	ASC	822.64 C g^{-1} @ 1 A g^{-1}	2 M KOH	Cellulose paper	[68]
25	$Ti_3C_2T_x$@NiO-rGO	Chemical bath deposition	SSC	966 F g^{-1} @ 1 A g^{-1}	1 M KOH	Ni foam	[69]
26	P-MXene @ Ni_3S_4/CuS	Hydrothermal	ASC	1917 F g^{-1} @ 1 A g^{-1}	6 M KOH	Ni foam	[70]
27	$MnCo_2O_4$@NiCo-LDH	Hydrothermal, electrodeposition	ASC	4550.0 F g^{-1} @ 1 A g^{-1}	6 M KOH	Ni foam	[71]
28	NiCo-S/$Ti_3C_2T_x$/CC	Hydrothermal	Flexible ASC	1690 F g^{-1} @ 1 A g^{-1}	1 M KOH	CC	[72]
29	Mo-CoP@NiFe-LDH	Hydrothermal	Hybrid	2796 F g^{-1} @ 1 A g^{-1}	6 M KOH	Ni foam	[73]
30	NiCo-LDH/TiO_2/C NFAs	Electrodeposition	ASC	1717.8 F g^{-1} @ 1 A g^{-1}	1 M KOH	CC	[74]
31	$NiCo_2O_4$@NiCo-LDH	Hydrothermal, electrodeposition	ASC	157 F g^{-1} @ 1 A g^{-1}	6 M KOH	CC	[75]
32	PPy@NiCo- LDH@SSy	Hydrothermal, electrodeposition	YSC	1196 F g^{-1} @ 1 A g^{-1}	2 M KOH	SS	[76]
33	V_2CT_x@C	Solvothermal	SSC	551 F g^{-1} @ 1 A g^{-1}	1 M KOH	Graphite sheet	[77]
34	$Co(OH)_2Ni(OH)_2$	Solvothermal	ASC	16.3 mF cm^{-2} @ 2 mA cm^{-2}	2 M KOH	Ni foam	[78]
35	ZMA-LDH@Fe_2O_3/ 3DHPCNF	Hydrothermal	Solid-state SSC	3437 mF cm^{-2} @ 1 mA cm^{-2}	3 M KOH	Film	[79]
36	CoFe-LDH/$Ti_3C_2T_x$	Co-precipitation	Hybrid	149.2 mg g^{-1} @ 50 mA g^{-1}	1 M NaCl	Graphite sheet	[80]
37	Ni/Co-MOF@TCT-NH_2	Polymerization	ASC	1924 F g^{-1} @ 1 A g^{-1}	3 M KOH	Film	[81]

(Continued)

Table 6.2. (*Continued*)

S. No.	Electrode materials	Method	Device type	Capacitance	Electrolyte	Current collector	Reference
38	Carbonized wood@ZIFs	Hydrothermal	SSC	5155.3 mF cm^{-2} @ 1 mA cm^{-2}	3 M KOH	CC	[82]
39	NiCoAl-LDH@$NiCo_2S_4$	Hydrothermal	ASC	1775 F g^{-1} @ 1 A g^{-1}	6 M KOH	CC	[83]
40	s-Ti_3C_2/$NiCo_2O_4$	Facile method	SSC	73.8 F g^{-1} @ 1 A g^{-1}	3 M KOH	Ni foam	[84]
41	NiCo-LDH-210	Solvothermal	Hybrid	2203.6 F g^{-1} @ 1 A g^{-1}	3 M KOH	CNF	[85]
42	q-2D-MOF/rGO	Acoustic micro-fluidic platform	SSC	292 F g^{-1} @ 1 A g^{-1}	1 M H_2SO_4	GO sheet	[42]
43	NiCo-LDH@Ni-NTNW	Electrodeposition	Hybrid	76.7 mF cm^{-3} @1 mA cm^{-2}	1 M KOH	AC sheet	[86]
44	P-Co_3O_4@NC@Ov-NiMn-LDH	Hydrothermal	ASC	1.33 mAh g^{-1} @ 1 A g^{-1}	6 M KOH	Ni foam	[87]
45	BNPC-2	Carbonization process		433.4 F g^{-1} @ 1 A g^{-1}	6 M KOH	CC	[88]
46	Ni_3S_2@NMVL/rGO-15	Hydrothermal	Hybrid	1412.0 C g^{-1}@ 1 A g^{-1}	2 M KOH	Ni foam	[89]
47	CeO_2@$(NiCo)_3S_4$	Hydrothermal	ASC	1319 F g^{-1} @ 1 A g^{-1}	2 M KOH	—	[90]
48	$CoNiS_4$/MXene	Hydrothermal	ASC	751 C g^{-1} @ 1 A g^{-1}	3 M KOH	—	[91]
49	$NiCo_2S_4$@$Ti_3C_2T_x$	Hydrothermal	ASC	1076 F g^{-1} @ 1 A g^{-1}	3 M KOH	Ni foam	[40]
50	Co_3S_4/CoS/MoS_2	Hydrothermal		805.7 F g^{-1} @ 1 A g^{-1}	3 M KOH	CC	[92]

6.3 Concluding remarks and future perspectives

In this chapter, we have described the recent progress made in hierarchical functional nanomaterials for flexible energy storage systems. In particular, the preparation, properties, and practical application of HNFM based SC devices were elaborated. Despite the impressive progress made in HNFMs, some challenges are yet to be addressed in terms of enhancing the energy density and miniaturization of the HNFM based SC devices. Eventually, the efficiency of every SC device can be tuned via the elemental composition of the electrode material. As discussed in this chapter, by developing 2D and 3D hierarchical architectures with the use of small nanostructures, the inherent properties of SC devices are improved. In fact, the future technologies are mainly dependent on how HNFMs can be manipulated at the nanoscale level for miniaturized electronic devices. Many steps have been taken to increase the energy and power density of portable miniaturized electronic devices. Unfortunately, while upscaling the electrode material at an industrial level, certain nanostructures, such as sheets, tubes, wires, fibres, and spheres, etc, might collapse, and their nanostructures could not be reproduced at bulk level. Hence, it is mandatory to design a new bulk production method where the specific morphologies of the nanostructures can be reproduced.

In this electronic era, many researchers are aiming to improve the mechanical flexibility of the HNFM with or without a flexible substrate. As a result, free standing electrode films with good mechanical flexibility as well as portability are urgently needed to overcome the barriers in the fabrication of HNFM based SC devices. Therefore, further studies toward a better understanding of the preparation of free standing, cost-effective, and binder free HNFMs based electrode materials are highly desirable. Undoubtedly, the future of advanced nanotechnology will depend mainly on the advancements made in portable, flexible, and wearable electronic devices. Henceforth, HNFMs would have a wider perspective on the commercialization of SC devices and have many paths for job opportunities in several technological fields.

Abbreviations

ASC	asymmetric supercapacitor
SSC	symmetric supercapacitor
CC	carbon cloth
SS	stainless steel
IL	ionic liquid
YSC	yarn shaped supercapacitor
MOF	metal organic framework
CMC-PANI	carboxymethylcellulose-polyaniline
LDH	layered double hydroxide
NCNRs	nickel cobalt nanorods
NCNNs	nickel cobalt nanosheets
NPCM	N-doped porous carbon monoliths
Ni-NTNW	nickel nanotube network
NMV-L	nickel manganese vanadium doping layered double hydroxide
BNPC	boron (B) and nitrogen (N) co-doped 3D hierarchical porous carbon

CCA	carbonized chitosan aerogels
CCO	cotton carbon
1D	one-dimensional
2D	two-dimensional
3D	three-dimensional
DA	dopamine
rGO	reduced graphene oxide
CNF	cellulose nanofiber
BNNS	boron nitride nanosheets
ZIF	zeolitic imidazolate framework
g-C_3N_4	graphitic carbon nitride
CV	Cyclic Voltammetry
GCD	Galvanostatic Charge Discharge
μm	micrometre
nm	nanometre

References

[1] Khaligh A and Onar O C 2018 *Energy Sources* (Amsterdam: Elsevier)

[2] Owusu P A and Asumadu-Sarkodie S 2016 A review of renewable energy sources, sustainability issues and climate change mitigation *Cogent Eng.* **3** 1–14

[3] Dincer I and Acar C 2015 A review on clean energy solutions for better sustainability *Int. J. Energy Res.* **39** 585–606

[4] Jayalakshmi M and Balasubramanian K 2008 Simple capacitors to supercapacitors—an overview *Int. J. Electrochem. Sci.* **3** 1196–217

[5] Arulraj A, Ilayaraja N, Rajeshkumar V and Ramesh M 2019 Direct synthesis of cubic shaped Ag_2S on Ni mesh as binder-free electrodes for energy storage applications *Sci. Rep.* **9** 10108

[6] Inayat A and Raza M 2019 District cooling system via renewable energy sources: A review *Renewable and Sustainable Energy Reviews* **107** 360–373

[7] Choi H and Yoon H 2015 Nanostructured electrode materials for electrochemical capacitor applications *Nanomaterials* **5** 906–36

[8] Liu B C, Li F, Ma L and Cheng H 2010 Advanced materials for energy storage *Adv. Mater.* **22** 28–62

[9] Wang Y, Guo J, Wang T, Shao J, Wang D and Yang Y 2015 Mesoporous transition metal oxides for supercapacitors *Nanomaterials* **5** 1667–89

[10] Wen Y, Zhao Y, Lu L, Zhang S J and Xu B 2017 Electrochemical performance of Zn–Al double layered hydroxide for supercapacitor application *IOP Conf. Ser.: Mater. Sci. Eng.* **250** 012055

[11] Palchoudhury S, Ramasamy K, Gupta R K and Gupta A 2019 Flexible supercapacitors: a materials perspective *Front. Mater.* **5** 1–9

[12] Chee W K, Lim H N, Zainal Z, Huang N M, Harrison I and Andou Y 2016 Flexible graphene-based supercapacitors: a review *J. Phys. Chem.* C **120** 4153–72

[13] Zhang Q, Liu K, Ding F and Liu X 2017 Recent advances in solid polymer electrolytes for lithium batteries *Nano Res.* **10** 4139–74

[14] Liu G, Lu S, Xu W, He G, Zheng Y and Cheng Y 2018 Fabrication of graphene/copper-nickel foam composite for high performance supercapacitors *New J. Chem.* **42** 9455–62

[15] Meng F, Li Q and Zheng L 2017 Flexible fiber-shaped supercapacitors: design, fabrication, and multi-functionalities *Energy Storage Mater.* **8** 85–109

[16] Arulraj A and Nishanthi S T 2020 *2D Materials–Based Flexible Supercapacitors for High Energy Storage Devices* (Amsterdam: Elsevier) pp 417–36

[17] Iro Z S, Subramani C and Dash S S 2016 A brief review on electrode materials for supercapacitor *Int. J. Electrochem. Sci.* **11** 10628–43

[18] Abdel Maksoud M I A *et al* 2021 *Advanced Materials and Technologies for Supercapacitors Used in Energy Conversion and Storage: A Review* (Berlin: Springer International)

[19] Zhang Y, Li L, Su H, Huang W and Dong X 2015 Binary metal oxide: advanced energy storage materials in supercapacitors *J. Mater. Chem.* A **3** 43–59

[20] Krishnan S G, Arulraj A, Jagadish P, Khalid M, Nasrollahzadeh M, Fen R, Yang C-C and Hegde G 2023 Pore size matters!—a critical review on the supercapacitive charge storage enhancement of biocarbonaceous materials *Crit. Rev. Solid State Mater. Sci.* **48** 1–56

[21] Gowrisankar A, Sherryn A L and Selvaraju T 2020 Impact of morphological variation by phase-oriented MnO_2-based hierarchical ternary composites for the fabrication of solid-state symmetric supercapacitor *Ionics (Kiel)* **26** 2563–79

[22] Duan X, Deng J, Wang X and Liu P 2017 Preparation of rGO/G/PANI ternary nano-composites as high performance electrode materials for supercapacitors with spent battery powder as raw material *Mater. Des.* **129** 135–42

[23] Asen P and Shahrokhian S 2017 A high performance supercapacitor based on graphene/polypyrrole/Cu_2O–$Cu(OH)_2$ ternary nanocomposite coated on nickel foam *J. Phys. Chem.* C **121** 6508–19

[24] Liu Y, Cai X, Luo B, Yan M, Jiang J and Shi W 2016 MnO_2 decorated on carbon sphere intercalated graphene film for high-performance supercapacitor electrodes *Carbon NY* **107** 426–32

[25] Krishnan S G, Arulraj A, Khalid M, Reddy M V and Jose R 2021 Energy storage in metal cobaltite electrodes: opportunities and challenges in magnesium cobalt oxide *Renew. Sustain. Energy Rev.* **141** 110798

[26] Vivek E, Arulraj A, Krishnan S G, Khalid M and Potheher I V 2021 Novel nanostructured $Nd(OH)_3$/g-C_3N_4 nanocomposites (nanorolls anchored on nanosheets) as reliable electrode material for supercapacitors *Energy Fuels* **35** 15205–12

[27] Yu T, Lei X, Chen H, Fan K, Wang D and Gao Y 2022 Conductive hydrogels with 2D/2D β-NiS/$Ti_3C_2T_x$ heterostructure for high-performance supercapacitor electrode materials *Ceram. Int.* **48** 1382–93

[28] Lin T W, Sadhasivam T, Wang A Y, Chen T Y, Lin J Y and Shao L D 2018 Ternary composite nanosheets with MoS_2/WS_2/graphene heterostructures as high-performance cathode materials for supercapacitors *Chem. Electro. Chem.* **5** 1024–31

[29] Chen X, Liu G, Zheng W, Feng W, Cao W, Hu W and Hu P A 2016 Vertical 2D MoO_2/$MoSe_2$ core–shell nanosheet arrays as high-performance electrocatalysts for hydrogen evolution reaction *Adv. Funct. Mater.* **26** 8537–44

[30] Li S, Cheng P, Luo J, Zhou D, Xu W, Li J, Li R and Yuan D 2017 High-performance flexible asymmetric supercapacitor based on CoAl-LDH and rGO electrodes *Nano-Micro Lett.* **9** 1–10

[31] Arulraj A, Mangalaraja R V and Khalid M 2022 MXene: Pioneering 2D Materials. Fundamental Aspects and Perspectives of MXenes ed M Khalid, A N Grace, A Arulraj and A Numan (Berlin: Springer) pp 1–16

[32] Gao S, Sui Y, Wei F, Qi J, Meng Q and He Y 2018 Facile synthesis of cuboid Ni-MOF for high-performance supercapacitors *J. Mater. Sci.* **53** 6807–18
[33] Gowrisankar A, Saravanakumar T and Selvaraju T 2020 *Graphene-Based Composite Materials for Flexible Supercapacitors* ed A Pandikumar and P Rameshkumar (Amsterdam: Elsevier)
[34] Ghosh D and Kim S O 2015 Chemically modified graphene based supercapacitors for flexible and miniature devices *Electron. Mater. Lett.* **11** 719–34
[35] Chen T and Dai L 2014 Flexible supercapacitors based on carbon nanomaterials *J. Mater. Chem.* A **2** 10756–75
[36] Shen J, Wang Y, Gu K, Zhu Y, Hu Y and Li C 2020 Construction of hierarchical functional nanomaterials for energy conversion and storage *Particuology* **48** 34–47
[37] Shrestha R G, Maji S, Shrestha L K and Ariga K 2020 Nanoarchitectonics of nanoporous carbon materials in supercapacitors applications *Nanomaterials* **10** 1–27
[38] Zheng K, Liao L, Zhang Y, Tan H, Liu J, Li C and Jia D 2022 Hierarchical NiCo-LDH core/shell homostructural electrodes with MOF-derived shell for electrochemical energy storage *J. Colloid Interface Sci.* **619** 75–83
[39] Xu H, Cui L, Lei Z, Xu M and Jin X 2022 MXene/carboxymethylcellulose-polyaniline ($Ti_3C_2T_x$/CMC-PANI) film as flexible electrode for high-performance asymmetric supercapacitors *Electrochim. Acta* **436** 141408
[40] Pathak M and Rout C S 2022 Hierarchical $NiCo_2S_4$ nanostructures anchored on nanocarbons and $Ti_3C_2T_x$ MXene for high-performance flexible solid-state asymmetric supercapacitors *Adv. Compos. Hybrid Mater* **5** 1404–22
[41] Pan Z, Yang C, Li Y, Hu X and Ji X 2022 Rational design of A-CNTs/K_xMnO_2 and $Ti_3C_2T_x$/MoO_3 free-standing hybrid films for flexible asymmetric supercapacitor *Chem. Eng. J.* **428** 131138
[42] Ehrnst Y, Ahmed H, Komljenovic R, Massahud E, Shepelin N A, Sherrell P C, Ellis A V, Rezk A R and Yeo L Y 2022 Acoustotemplating: rapid synthesis of freestanding quasi-2D MOF/graphene oxide heterostructures for supercapacitor applications *J. Mater. Chem.* A **10** 7058–72
[43] Wang Z, Huang J, Mao J, Guo Q, Chen Z and Lai Y 2020 Metal–organic frameworks and their derivatives with graphene composites: preparation and applications in electrocatalysis and photocatalysis: a review *J. Mater. Chem.* A **8** 2934–61
[44] Erçarlkcl E, Dağcl Klranşan K and Topçu E 2022 Three-dimensional ZnCo-MOF modified graphene sponge: flexible electrode material for symmetric supercapacitor *Energy Fuels* **36** 1735–45
[45] Li Y, Jiang H, Yan X, Zhang W, Zhang M, Zhu W, Pan J, Sufyan Javed M, Cheng W and Guan Y 2021 Rationally designed Mn_2O_3/Cu_xO core–shell heterostructure generated on copper foam as binder-free electrode for flexible asymmetric supercapacitor *Appl. Surf. Sci.* **566** 150715
[46] Zhang Y, Xu H and Lu S 2021 Preparation and application of layered double hydroxide nanosheets *RSC Adv.* **11** 24254–81
[47] Mishra G, Dash B and Pandey S 2018 Layered double hydroxides: a brief review from fundamentals to application as evolving biomaterials *Appl. Clay Sci.* **153** 172–86
[48] Yin J, Zhou G, Gao X, Chen J, Zhang L, Xu J, Zhao P and Gao F 2019 α-and β-Phase Ni–Mg hydroxide for high performance hybrid supercapacitors *Nanomaterials* **9** 1–14

[49] Yin Q, Li D, Zhang J, Zhao Y, Luo J, Shao M and Han J 2020 An all-solid-state fiber-type supercapacitor based on hierarchical Ni/NiO@CoNi-layered double hydroxide core–shell nanoarrays *J. Alloys Compd.* **813** 152187

[50] Ashraf I, Ahmad S, Dastan D, Wang C, Garmestani H and Iqbal M 2022 Fabrication of ionic liquid based D-Ti_3C_2/MoO_3 hybrid electrode system for efficient energy storage applications *Electrochim. Acta* **429** 141036

[51] Jia H, Zhu X, Song T, Pan J, Peng F, Li L and Liu Y 2022 Hierarchical nanocomposite of carbon-fiber-supported NiCo-based layered double-hydroxide nanosheets decorated with $(NiCo)Se_2$ nanoparticles for high performance energy storage *J. Colloid Interface Sci.* **608** 175–85

[52] Yang Q, Liu Y, Deng C, Sun L and Shi W 2022 In-situ construction of heterostructure (Ni, Co)Se_2 nanoarrays derived from cone-like ZIF-L for high-performance hybrid supercapacitors *J. Colloid Interface Sci.* **608** 3049–58

[53] Wang Y, Yang H, Lv H, Zhou Z, Zhao Y, Wei H and Chen Z 2022 High performance flexible asymmetric supercapacitor constructed by cobalt aluminum layered double hydroxide @ nickel cobalt layered double hydroxide heterostructure grown *in situ* on carbon cloth *J. Colloid Interface Sci.* **610** 35–48

[54] Xiao H, Ma Y, Xu M, Liu R, Li X, Wang X, Wang Y, Liu Y and Yuan G 2022 Constructing nickel cobaltate @nickel-manganese layered double hydroxide hybrid composite on carbon cloth for high-performance flexible supercapacitors *J. Colloid Interface Sci.* **611** 149–60

[55] Luo L, Zhou Y, Yan W, Du G, Fan M and Zhao W 2022 Construction of advanced zeolitic imidazolate framework derived cobalt sulfide/MXene composites as high-performance electrodes for supercapacitors *J. Colloid Interface Sci.* **615** 282–92

[56] Song P, Chen C, Shen X, Zeng S, Premlatha S, Ji Z, Zhai L, Yuan A and Liu Q 2022 Metal–organic frameworks-derived carbon modified wood carbon monoliths as three-dimensional self-supported electrodes with boosted electrochemical energy storage performance *J. Colloid Interface Sci.* **620** 376–87

[57] Liao L, Zheng K, Zhang Y, Li X, Jiang D and Liu J 2022 Self-templated pseudomorphic transformation of ZIF into layered double hydroxides for improved supercapacitive performance *J. Colloid Interface Sci.* **622** 309–18

[58] Yu T, Li S, Zhang L, Li F, Wang J, Pan H and Zhang D 2023 *In situ* growth of ZIF-67-derived nickel-cobalt-manganese hydroxides on 2D V_2CT_x MXene for dual-functional orientation as high-performance asymmetric supercapacitor and electrochemical hydroquinone sensor *J. Colloid Interface Sci.* **629** 546–58

[59] Zhao K, Sun X, Fu H, Guo H, Wang L, Li D and Liu J 2022 *In situ* construction of metal–organic frameworks on chitosan-derived nitrogen self-doped porous carbon for high-performance supercapacitors *J. Colloid Interface Sci.* **632** 249–59

[60] Zhang J, Huang J, Wang L, Sun P, Wang P, Yao Z and Yang Y 2022 Coupling bimetallic NiMn-MOF nanosheets on $NiCo_2O_4$ nanowire arrays with boosted electrochemical performance for hybrid supercapacitor *Mater. Res. Bull.* **149** 111707

[61] Pan J, Li S, Li F, Zhang W, Guo D, Zhang L, Zhang D, Pan H, Zhang Y and Ruan Y 2022 Design and construction of core–shell heterostructure of Ni–V layered double hydroxide composite electrode materials for high-performance hybrid supercapacitor and L-tryptophan sensor *J. Alloys Compd.* **890** 161781

[62] Yang Y, Huang X, Sheng C, Pan Y, Huang Y and Wang X 2022 *In-situ* formation of MOFs derivatives $CoSe_2$/Ni_3Se_4 nanosheets on MXene nanosheets for hybrid supercapacitor with enhanced electrochemical performance *J. Alloys Compd.* **920** 165908

[63] Zhang Y, Cao J, Li J, Yuan Z, Li D, Wang L and Han W 2022 Self-assembled cobalt-doped NiMn-layered double hydroxide (LDH)/V_2CT_x MXene hybrids for advanced aqueous electrochemical energy storage properties *Chem. Eng. J.* **430** 132992
[64] Hao C, Wang X, Wu X, Guo Y, Zhu L and Wang X 2022 Composite material CCO/Co–Ni–Mn LDH made from sacrifice template CCO/ZIF-67 for high-performance supercapacitor *Appl. Surf. Sci.* **572** 151373
[65] Zhang C, Guo R, Wang H, Xie X and Du C 2022 Composite electrodes with NiCoAl-LDH coated $Ti_3C_2T_x$ MXene and incorporated Ag nanowires for screen-printable in-plane hybrid supercapacitors on textiles *Appl. Surf. Sci.* **598** 153796
[66] Xue Z, Tao K and Han L 2022 Stringing metal–organic framework-derived hollow Co_3S_4 nanopolyhedra on V_2O_5 nanowires for high-performance supercapacitors *Appl. Surf. Sci.* **600** 154076
[67] Xu J, Yang X, Zou Y, Zhu L, Xu F, Sun L, Xiang C and Zhang J 2022 High density anchoring of $NiMoS_4$ on ultrathin Ti_3C_2 MXene assisted by dopamine for supercapacitor electrode materials *J. Alloys Compd.* **891** 161945
[68] Xue Y, Liu X, Han L, Xie Z, Liu L, Li Y, Hua Y, Wang C, Zhao X and Liu X 2022 Fabrication of hierarchical $NiCo_2S_4$ nanotubes@NiMn-LDH nanosheets core–shell hybrid arrays on Ni foam for high-performance asymmetric supercapacitors *J. Alloys Compd.* **900** 163495
[69] Chen W, Peng Y, Qiu Z, Zhang X and Xu H 2022 3D hierarchical $Ti_3C_2T_X$ @NiO-reduced graphene oxide heterostructure hydrogel as free-standing electrodes for high performance supercapacitor *J. Alloys Compd.* **901** 163614
[70] Guo H, Zhang J, Yang F, Wang M, Zhang T, Hao Y and Yang W 2022 Sandwich-like porous MXene/Ni_3S_4/CuS derived from MOFs as superior supercapacitor electrode *J. Alloys Compd.* **906** 163863
[71] Wang Y, Wang Z, Zheng X, Teng X, Xu L, Yuan Y, Liu X, Fu A, Li Y and Li H 2022 Core–sheath heterostructure of $MnCo_2O_4$ nanowires wrapped by NiCo-layered double hydroxide as cathode material for high-performance quasi-solid-state asymmetric supercapacitors *J. Alloys Compd.* **904** 164047
[72] He L, Wu F, Bai M, Ruan H, Fang L, Hu J, Luo H, Zhang S, Hu B and Zhou M 2022 NiCo–S nanosheets grown on $Ti_3C_2T_x$-wrapped carbon cloth for high-performance flexible supercapacitor *J. Alloys Compd.* **908** 164493
[73] Chen X, Li S, Liu Y, Xie K and Wang Y 2022 MOF-derived Mo-CoP@NiFe LDH hierarchical nanosheets for high-performance hybrid supercapacitors *J. Alloys Compd.* **919** 165842
[74] Wang J, Mao J, Ma F, Qi K, Liu Y, Cheng L and Chen R 2022 Nickel–cobalt layered double hydroxide fabricated on TiO_2/C nanofiber arrays as free standing electrode for high performance supercapacitors *J. Alloys Compd.* **920** 165909
[75] Li S, Luo Y, Wang C, Wu M, Xue Y, Yang J and Li L 2022 A novel hierarchical core–shell structure of $NiCo_2O_4$@NiCo-LDH nanoarrays for higher-performance flexible all-solid-state supercapacitor electrode materials *J. Alloys Compd.* **920** 165986
[76] Wang S, Wang J, Liao S, Chen J and Wei Q 2022 Hierarchical core–shell polypyrrole@NiCo layered double hydroxide arrays grown on stainless steel yarn with high flexibility for 1D symmetric yarn-shaped supercapacitors *J. Alloys Compd.* **926** 166811
[77] Yadav A, Singal S, Soni P, Singh G and Sharma R K 2023 Structural engineering and carbon enrichment in V_2CT_x MXene: an approach for enhanced supercapacitive charge storage *J. Alloys Compd.* **934** 167859

[78] Zhou J J *et al* 2022 Controllable transformation of CoNi-MOF-74 on Ni foam into hierarchical-porous $Co(OH)_2/Ni(OH)_2$ micro-rods with ultra-high specific surface area for energy storage *Chem. Eng. J.* **428** 1–9

[79] Poudel M B and Kim H J 2022 Confinement of Zn–Mg–Al-layered double hydroxide and α-Fe_2O_3 nanorods on hollow porous carbon nanofibers: a free-standing electrode for solid-state symmetric supercapacitors *Chem. Eng. J.* **429** 132345

[80] Lei J, Xiong Y, Yu F and Ma J 2022 Flexible self-supporting CoFe-LDH/MXene film as a chloride ions storage electrode in capacitive deionization *Chem. Eng. J.* **437** 135381

[81] Yue L, Chen L, Wang X, Lu D, Zhou W, Shen D, Yang Q, Xiao S and Li Y 2023 Ni/Co-MOF@aminated MXene hierarchical electrodes for high-stability supercapacitors *Chem. Eng. J.* **451** 138687

[82] Zhang Z, Deng S, Wang D, Qing Y, Yan G, Li L and Wu Y 2023 Low-tortuosity carbon electrode derived from Wood@ZIF-67 for supercapacitor applications *Chem. Eng. J.* **454** 140410

[83] Li Y, Yan X, Zhang W, Zhou W, Zhu Y, Zhang M, Zhu W and Cheng X 2022 Hierarchical micro-nano structure based NiCoAl-LDH nanosheets reinforced by $NiCo_2S_4$ on carbon cloth for asymmetric supercapacitor *J. Electroanal. Chem.* **905** 115982

[84] Wang L, Cao J, Zhou Y H and Liu X 2022 Design and characterization of monolayer Ti_3C_2 MXene/$NiCo_2O_4$ nanocones hybrid architecture for asymmetric supercapacitors *J. Electroanal. Chem.* **923** 116787

[85] Zhao X, Li H, Zhang M, Pan W, Luo Z and Sun X 2022 Hierarchical nanocages assembled by NiCo-layered double hydroxide nanosheets for a high-performance hybrid supercapacitor *ACS Appl. Mater. Interfaces* **14** 34781–92

[86] Amin K M, Krois K, Muench F, Etzold B J M and Ensinger W 2022 Hierarchical pipe cactus-like Ni/NiCo-LDH core–shell nanotube networks as a self-supported battery-type electrode for supercapacitors with high volumetric energy density *J. Mater. Chem.* A **10** 12473–88

[87] Zhu F, Sun L, Liu Y and Shi W 2022 Dual-defect site regulation on MOF-derived P-Co_3O_4@NC@Ov-NiMnLDH carbon arrays for high-performance supercapacitors *J. Mater. Chem.* A **10** 21021–30

[88] Wang L, Wang C, Wang H, Jiao X, Ouyang Y, Xia X, Lei W and Hao Q 2018 ZIF-8 nanocrystals derived N-doped carbon decorated graphene sheets for symmetric supercapacitors *Electrochim. Acta* **289** 494–502

[89] Zhang G, Yang J, Xuan H, Yang J, Liang X, Li Y and Han P 2022 Design and construction of hierarchical Ni_3S_2@V-doped NiMn-LDH heterostructure on rGO/Ni foam as an advanced electrode for battery-supercapacitor hybrid devices *J. Alloys Compd.* **896** 163125

[90] Xue Z, Yang C, Tao K and Han L 2022 Heterostructure of metal–organic framework-derived straw-bundle-like CeO_2 decorated with $(Ni,Co)_3S_4$ nanosheets for high-performance supercapacitor *Appl. Surf. Sci.* **592** 153231

[91] Qi J Q, Huang M Y, Ruan C Y, Zhu D D, Zhu L, Wei F X, Sui Y W and Meng Q K 2022 Construction of $CoNi_2S_4$ nanocubes interlinked by few-layer $Ti_3C_2T_x$ MXene with high performance for asymmetric supercapacitors *Rare Met.* **41** 4116–26

[92] Qiao F, Liu W, Wang S, Lin F, Chen Y, Yuan M, Weng Z, Wang S, Zheng J and Zhao Y 2021 Hierarchical Co_3S_4/CoS/MoS_2 leaf-like nanoflakes array derived from Co-ZIF-L as an advanced anode for flexible supercapacitor *J. Alloys Compd.* **870** 159393

IOP Publishing

Materials Technology for the Energy and Environmental Nexus, Volume 1

R V Mangalaraja, Arunachalam Arulraj, Nalandhiran Pugazhenthiran and Chan Siew Hwa

Chapter 7

Sulfide-based 2D nanomaterials for photocatalytic activity

Mehdi Abedi[1], Zahra Rahiminezhad[1], Marzieh Mehrzadeh, Vahid Alimardani, Younes Ghasemi, Divya Chauhan, Mohammad Ashfaq, R V Mangalaraja and Neetu Talreja

Recently, wastewater contamination has become one of the biggest global issues as the scarcity of drinking water is increasing every day. In this context, several advanced technologies have been adopted to treat different contaminants, including chemical and biological contaminants. Among them, the advanced oxidation process (AOP) has proved its efficiency in removing chemical contaminants. Numerous semiconductor materials such as metals, carbon-based materials, and two-dimensional (2D) materials have been used to treat chemical and biological pollutants. Recently, 2D nanomaterials (NMs), in particular sulfide-based 2D NMs such as WS_2 and MoS_2, have gained significant interest due to their exceptional characteristics, in particular tunable bandgaps and surface functional groups. Here we discuss the design of photocatalyst materials, in particular WS_2 and MoS_2, and their efficiency towards the degradation of contamination in water systems using a natural energy source. We also discuss the mechanism of the AOP, the remaining challenges, and their advancement for the photodegradation of contamination in the water systems.

7.1 Introduction

Two-dimensional nanomaterials (2D NMs), especially graphene (G), have attracted much attention recently due to their surprising and unique mechanical, optical, and electrical properties. In the wake of the discovery of graphene's extraordinary

[1] These authors have equally contributed to the work.

doi:10.1088/978-0-7503-5724-1ch7

properties, scientists have been motivated to carry out ongoing research into 2D NMs such as transition metal dichalcogenides (TMDCs) due to their appropriate properties and the technological applications in a variety of industries, including lubricants, rechargeable batteries, semiconductors, supercapacitors, photocatalysts, photoactive devices, and hydrodesulfurization catalysts (Radisavljevic *et al* 2011, Wu *et al* 2014, Arulraj *et al* 2022). Usually, chalcogenides have the formula MX_2, where M is a transition metal such as molybdenum (Mo) or tungsten (W), and X is a chalcogen atom including sulfur, selenium, or tellurium. Mo and S atoms occupy alternately the hexagonal corners of the layers of MX_2's hexagonal structure. MX_2 is available in three structural polytypes: rhombohedral, tetragonal, and hexagonal. The material's band structure varies from the 2D TMDCs to the 3D bulk structure. The TMDCs' chemical and physical characteristics depend on how many layers are present (Zhu *et al* 2016). A recent study showed that the single-layer TMDCs exhibited a direct bandgap. In contrast, the bulk materials inherently have an indirect bandgap, thereby 2D MoS_2 and WS_2 have a direct bandgap, unlike the bandgap of graphene (Splendiani *et al* 2010). The bandgap values of the TMDCs would be between 1.5 and 1.8 eV (Yao *et al* 2020, Dong *et al* 2019, Singh *et al* 2021), but for a lot of optoelectronic energy, the bandgap had to be greater than 1.8 eV since this barrier is the boundary of switching between the on-off states in a semiconductor (Lopez-Sanchez *et al* 2013). The 2D MoS_2 and WS_2 have an intrinsic bandgap with an energy greater than 1.8 eV (2.3 eV for WS_2 and 1.8–1.9 eV for MoS_2), which makes them ultimate materials for real-life applications such as supercapacitors, energy storage, optoelectronics, photocatalysis, and biomedicine (Guan *et al* 2021, Faraji *et al* 2022, Ashfaq *et al* 2022). Moreover, scientists have used various synthesis techniques to improve the performance in multiple fields and to add additional elements to the MoS_2 and WS_2. The synthesis of TMDCs and their derivatives and the investigation of their optoelectronic applications have received significant research attention. The heterostructures of 2D NMs can be used to develop devices such as electronic devices, and there is a clear advantage to improving the performance of basic materials (Tedstone *et al* 2016). It has been shown that the edge site of 2D TMDCs exhibit a significant amount of catalytic activity, which can be increased by including the appropriate doping materials, and the number of these active site can be optimized. Doping with different additives provides an effective method for increasing catalytically active sites, despite the basal plane's typically lower catalytic activity than for the edge sites. Some of these techniques for improving MOS_2 and WS_2 performance will be mentioned below. This chapter provides extensive knowledge of the design of photocatalyst material, in particular WS_2 and MoS_2, and their efficiency towards the degradation of contamination in water systems using natural sources of energy. We also discuss the remaining challenges and their advancement for the photodegradation of contamination in water systems.

7.2 AOPs and their mechanisms

The advanced oxidation process (AOP) refers to a process that produces or makes use of a highly reactive, strong transient species (HO•, H_2O_2, O_3, and superoxide anion radicals), most prominently the hydroxyl radical (Dalrymple *et al* 2007). The species can be formed from water using solar, electrical, sound, and other forms of energy. These methods differ in how the hydroxyl radicals are produced, whether generated with or without chemicals such as H_2O_2 or ozone or by using a catalyst (Kurian 2021). Some of additional active oxygen species are also developed, such as the superoxide radical ($O_2\bullet^-$) and its conjugate acid form ($HO_2\bullet$). However, the active oxygen species are not as active as HO•, and their characteristic features are non-selectiveness, strong reactivity, and potent oxidizing ability of HO• radicals (Andreozzi *et al* 1999). Depending on the organic contaminants that are the focus of the approach, HO• radicals have the potential to break down a wide variety of organic pollutants into compounds that are not very hazardous (Domingues *et al* 2021).

Usually, HO• radicals with the contaminants can lead to hydrogen abstraction from the NH, CH, or OH groups, radical–radical interactions, direct electron transfer resulting in oxidized intermediates, or, in the case of complete mineralization, the production of CO_2, H_2O, and inorganic acids (Legrini *et al* 1993, Devi and Kavitha 2013). Despite the HO• radicals' high oxidation potential, the kinetic rates of their interactions with pollutants depend on the compounds' affinities for the oxidant. It is possible to classify the AOPs as either heterogeneous or homogeneous, with the latter group further subdivided by whether or not energy input is required to activate the AOP. The homogeneous technologies are simple, repeatable, and easy to manage. The generation of secondary sludge produced in the neutralization step of the treated water and lower reactivity due to relatively unstable liberating oxygen during the oxidation processes led to the introduction of the heterogeneous AOP technologies. The activation AOPs can be initiated using photochemical, non-photochemical, and hybrid or combined energy inputs, as presented in table 7.1.

The performance of different AOPs depends on innumerable factors, including pH, temperature, the amount of oxidant present, and the catalyst surface size and shape. In the catalytic oxidation, where H_2O_2 is present, the pH affect the degradation rate significantly. The acidic pH can be utilized to promote Fenton-assisted oxidation activity. In contrast, the basic pH can reduce activity by forming an iron complex in the reaction medium. In a low pH reaction medium, H+ ions can form a stable oxonium ion, which makes H_2O_2 more stable and reduces its reactivity toward ferrous ions (Xu *et al* 2009, Bhasarkar and Bal 2021). Therefore, it is critical to have a regulated pH in the solution since this factor significantly affects the catalytic activity and the oxidant employed in the different AOPs. A rise in reaction temperature is generally reported to have a more positive effect on the oxidation efficiency. At higher temperatures, H_2O_2 decomposes into hydroxyl radicals, which may increase the oxidation kinetics and yield (Salem *et al* 2015). During the wastewater treatment, the shape and size of catalysts are the essential parameters that affect the recalcitrant pollutants' degradation. H_2O_2 is unstable, releasing oxygen as HO and HO_2 radicals when it hits active surfaces. The high specific surface area and surface energy of nanomaterials, and particles' shapes, geometries, and surfaces, play a significant role in determining the efficiency of AOPs in the degradation of pollutants. An adsorption process occurs at the catalyst's 'active sites', and the shape of the nanoparticles influences the accessibility of active sites.

Table 7.1. Overview of the different AOPs.

Activation source	Technique	Mechanism	Advantages	Disadvantages	References
Non-photochemical	Fenton.	Generation of HO• radicals in an acidic medium due to the breakdown of H_2O_2 using Fe catalyst.	Does not require sophisticated techniques/equipment and has environmentally viable processes.	Effective under acidic pH.	Deng and Zhao (2015)
	Ozonation	Generation of reactive oxygen species (ROS) using ozone gas.	Degradation of organic pollutants.	Limited shelf life, expensive, and requires high energy.	Gottschalk *et al* (2009)
	Sonolysis	Generation of HO• radicals from water pyrolysis by ultrasound irradiation.	A green approach to destroy toxic water pollutants.	Requirements of high energy and valuable on a lab scale.	Güyer and Ince (2011)
	Electrochemical	Generation of ROS via electricity and without the need for chemicals.	Safer, compatible, and able to control hydroxyl radical generation.	The surface of the electrode is affected during the process.	Garcia-Segura *et al* (2018)
	Radiation	Generation of HO• and hydrated electron generation by gamma irradiation.	Effectively used at different temperatures and requires low energy.	Concern about mineralization and toxicity.	Wang and Wang (2018)

Photochemical	UV oxidation (UV and UV/H_2O_2).	The combination of ultraviolet radiation with H_2O_2, leading to its photolysis, generates two HO• hydroxyl radicals.	UV radiation functions as a disinfectant, physically incapacitating the germs and helping the photolysis of peroxide to produce highly reactive OH radical species.	Photolysis depends on the intensity and time of irradiation.	Kim *et al* (2009)
	Photo-Fenton	The Fenton reaction is based on the use of a mixture of iron salts (Fe^{2+}) and H_2O_2, generating hydroxyl radicals under mild acidic conditions.	The Fenton process is significantly improved upon exposure to UV radiation.	Acidic conditions are required.	Alalm *et al* (2015)
	Photocatalysis	The catalyst absorbs the radiation energy and it is raised from a higher energy level state to an excited state, which leads to the production of highly unstable and reactive radicals.	Low cost.	Slow kinetics and generation of radicals.	Bhasarkar and Bal (2021)
Hybrid	Sono-photolysis, photocatalytic ozonation, photoelectron catalysis, sono-Fenton.	More hydroxyl radical generation.	A higher number of reactive species also results in more significant mineralization.	The experimental portion did not specify or justify the sequence of the paired treatment.	Bhasarkar and Bal (2021)

Nanoparticles may have unique reactivities, selectivities, binding strengths, and configurations depending on their shape. In the next section, we will talk about the AOPs based on nanomaterials that are used to degrade toxins in water that are produced by a variety of industries.

7.3 MoS_2-based 2D NMs

7.3.1 MoS_2 synthesis

Thy key factor for material development is the synthesis procedure. Different synthesis approaches, such as atomic layer deposition, sputtering, chemical vapor deposition (CVP), and physical vapor deposition (PVD), etc, have been investigated for different types of TMDCs (Gupta *et al* 2020).

7.3.1.1 Preparation methods based on a top-down approach

7.3.1.1.1 Mechanical exfoliation method

Novoselov and Geim applied a mechanical exfoliation method that was first used to isolate graphene oxide (GO) (Novoselov *et al* 2004). The mechanical delamination is clean, inexpensive, and scalable, and creates high-quality structures in depositing a layer of MoS_2 on the surface of the substrate (Ottaviano *et al* 2017, Li *et al* 2014). In this method an adhesive tape of MoS_2 was attached to the substrate and removed several times. A layer of MoS_2 was created on the surface due to the van der Waals interaction between the substrate and the stickiness of the MoS_2 tape. This method is efficient for studying the basic properties of MoS_2 dichalcogenide materials and their performance (Gupta *et al* 2020). Jumiati *et al* reported a simple thermal annealing procedure to create single to quadri-layers of mechanically exfoliated MoS_2 nanosheets with a smooth surface in air. Heat treatment converts the MoS_2 nanolayers to MoO_3, inducing hole injection and resulting in p-type channels in the fabricated field-effect transistors (Wu *et al* 2013). Wenli *et al* developed a room temperature NO_2 sensor using mechanical stripping technology. These sensors are more sensitive to NO_2 gas than other MoS_2-based sensors. Due to the thin MoS_2 layer thickness, the adsorption of NO_2 molecules changes the electronic state of the entire material (Li *et al* 2019b). The main challenge of mechanical exfoliation of MoS_2 layers on various materials as the substrate is the relatively small lateral size of the NoS_2 layers. To solve this problem Magda *et al* reported a mechanical exfoliation technique with a chemically enhanced bonding method to increase the thickness of the MoS_2 layer to hundreds of microns. They created a 100 nm thin gold layer on the mica surface to enhance the attachment of sulfur atoms to the gold. This is because gold atoms have a higher affinity to interact with sulfur than silica and silicates (Magda *et al* 2015).

7.3.1.1.2 Exfoliation in the liquid phase

There are two methods for creating thin films of MoS_2 on substrates by liquid-phase exfoliation. In this method, the MoS_2 flakes are generated in the different ways, such as grinding, electrolysis, sonication, stirring, bubbling, and shearing in a liquid.

7.3.1.2 Preparation methods based on a bottom-up approach

7.3.1.2.1 Physical vapor deposition (PVD)

PVD is a widely used technique for creating thin films and surface coatings. PVD is used commonly in industrial environments and combines various techniques to create coatings with excellent properties (Baptista *et al* 2018, Arulraj *et al* 2018). Four steps make up the process: (i) using a high-energy source, such as an electron beam, to vaporize the material to be deposited; (ii) transporting the vaporized material to the substrate surface using a transporting gas; (iii) a reaction between the transporting gas and the vaporized material; and (iv) depositing the vaporized atoms on the target surface. In a report by Zheng *et al*, a highly reproducible PVD process was used as a catalyst to produce an MoS_2–carbon composite. By adopting this technology, the MoS_2 deposited on carbon cloths showed excellent catalytic activity for hydrogen evolution reaction in an acidic solution due to the amorphous nature and numerous exposed edge sites of MoS_2 (Zhang *et al* 2015). However, it is not easy to control the thickness of the deposited layers, necessitating further advancement in the production of 2D NMs (Vishwanath *et al* 2015). Figure 7.1 provides a schematic diagram of the PVD process.

7.3.1.2.2 Sputtering

Sputtering is a thin film technology of the PVD class. In this method, the substrate surface is bombarded by highly energetic ions created by a plasma or ion source, which interact with the surface and cause film formation on the substrate. This method changes the chemical and microstructure of the coated thin film. It is found that the characteristics of a thin film prepared by the sputtering technique is very variable and depends strongly on the sputtering conditions and has a direct effect on the characteristics of the formed film such as atomic composition, crystal phase,

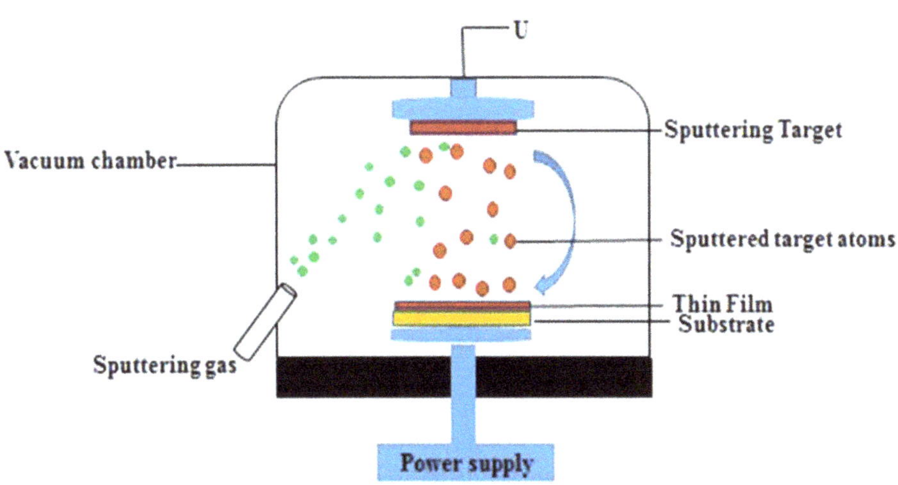

Figure 7.1. Schematic illustration of the PVD process.

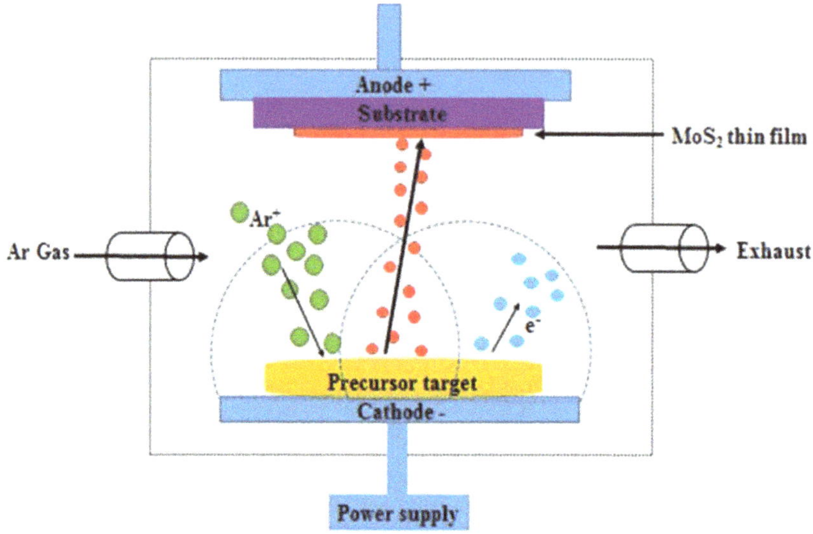

Figure 7.2. Schematic illustration of the magnetron sputtering process.

microstructure, and morphology (Fleischauer and Bauer 1988). A change in the temperature of the sputtering apparatus and substrate can alter the crystallinity of the formed film. Generally, the amorphous crystal structures are generated at low and high temperatures, and crystal phases are generated (Spalvins 1980). Figure 7.2 provides a schematic diagram of the magnetron sputtering process.

7.3.1.2.3 Solvothermal and hydrothermal methods

In the solvothermal (the medium is not water) or hydrothermal (the medium is water) methods, molybdenum atoms are reacted with sulfur or sulfides for several hours in a stainless-steel vessel under controlled high pressure or high temperature. The reaction will be followed by post-annealing. Under these conditions, uniquely sized MoS_2 structures are created to form thin firms. Ammonium tetrathiomolybdate is a precursor that converts to MoS_2 films at high temperatures. Several techniques, such as photo- or electrochemical methods, can be temperature-assisted in this way.

7.3.1.2.4 Chemical vapor deposition (CVD)

A potent technique for creating solid thin films with excellent quality is CVD. Despite being used widely in various industries, it is constantly being improved as it is adjusted for new materials. Polymeric and inorganic thin films can be coated on various substrates and are being manufactured precisely today, pushing CVD synthesis to new heights (Sun *et al* 2021). CVD is the reaction of volatile precursors injected into a chamber (usually under a vacuum). When the chamber is heated to the reaction temperature, the precursor gases react or decompose to form the desired coating and bind to the material's surface (Lau 2016). Figure 7.3 provides a schematic diagram of the CVD process.

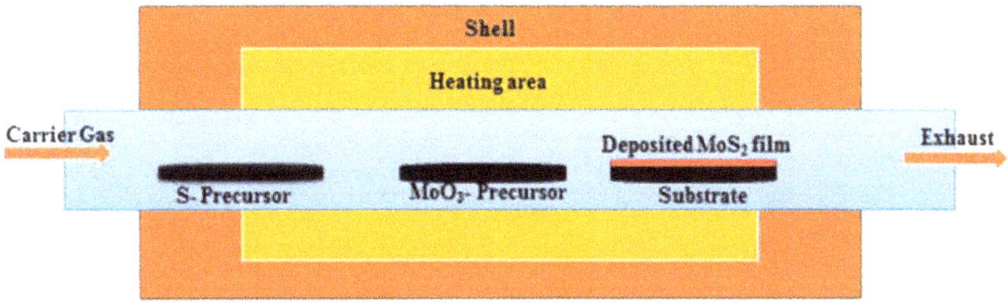

Figure 7.3. Schematic illustration of the CVD process.

7.3.1.2.5 Atomic layer deposition

The atomic layer deposition is a gas-phase chemical process that creates precise thin films on a substrate. In this technique the number of pulses controls the film thickness. Each pulse deposits one layer of atoms on the surface and then the chamber is purged with an inert gas, followed by another pulse for the second layer (Parsons *et al* 2011). This method usually uses hydrogen sulfide and molybdenum (V) chloride as precursors for preparing a thin film on various substrates such as silicon, silicon oxide, quartz, etc (Kalanyan *et al* 2018).

7.3.1.2.6 Thermal decomposition

One of the most popular methods for developing large-scale 2D transition metal dichalcogenides is thermal decomposition (thermolysis). In this method the first substrate is submerged in a solution containing transition metal dichalcogenide precursors such as ammonium thiomolybdates, $(NH_4)_2MoS_4$, or ammonium tetra-thiotungstate,$(NH_4)_2WS_4$, and this is allowed to dry. It is then annealed in at an appropriate temperature and atmosphere. Due to annealing the ammonium thiomolybdates or ammonium tetrathiotungstate coated on a substrate convert to MoS_2 or WS_2. Liu *et al* used a two-step decomposition to prepare 2D MoS_2 on SiO_2/Si wafer. After immersing the SiO_2/Si wafer in ammonium thiomolybdates under a hydrogen atmosphere the temperature was increased to 500 °C for converting $(NH_4)_2MoS_4$ to MoS_2, then under argon and sulfur atmospheres the temperature was increased up to 1000 °C to change the crystalline structure. The two-step thermal decomposition achieved a uniform and homogenized 2D MoS_2 layer on the surface of the SiO_2/Si wafer for a field-effect transistor (FET) application that shows high mobility and current ratio (Liu *et al* 2012). One effective parameter and critical factor during MoS_2 layer synthesis on the substrate by thermal decomposition is H_2 flux. It has been demonstrated that during annealing under a hydrogen atmosphere, some $(NH_4)_2MoS_4$ precursors converted to MoO_3 instead of MoS_2, which decreases the quality of MoS_2 layers. It is reported that in high H_2 flux the amount of MoO_3 increased, and the best H_2 flux for MoO_3 development is 100 sccm (Qi *et al* 2015). Figure 7.4 provides a schematic diagram of the thermal decomposition process.

7.3.1.2.7 Sol–gel methods

The sol–gel surface modification of the substrates is thought to be effective. The main benefit of the sol–gel method is that it produces stable surfaces with a high

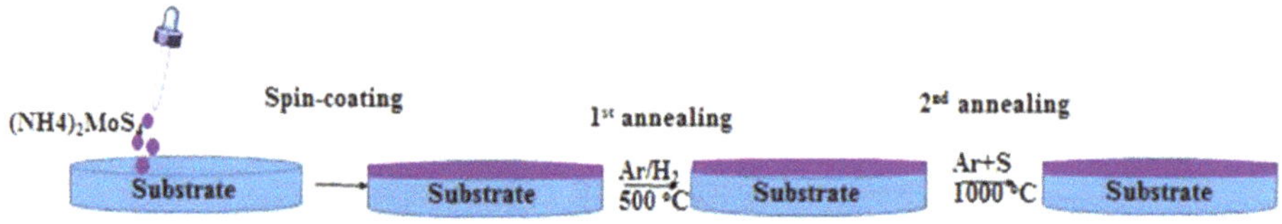

Figure 7.4. Schematic illustration of the thermal decomposition process.

surface area. The experimental set-up that was used had an impact on the material's chemical and physical properties that were obtained using the sol–gel method. The hydrolysis of precursors and polycondensation of hydrolyzed products in acidic or basic media are the two main reactions in sol–gel processes. Phung *et al* prepared ultrathin MoS_2 film on TiO_2–glass by the sol–gel method for methylene blue decomposition. Compared to pure MoS_2 and TiO_2, the thin films of the MoS_2/TiO_2 have more photocatalytic activity and successfully decomposed about 60% methylene blue after 150 min (Phung *et al* 2017).

7.3.2 Properties

7.3.2.1 Optical properties

Generally, there is a significant difference between single-layer and multilayers MoS_2 in three areas: absorption coefficient, extinction coefficient, and refractive index. The adsorption coefficient is defined as the rate at which the intensity of electromagnetic radiation, such as light, decreases. The adsorption coefficients of MoS_2 monolayers and multilayers are 400–500 nm and decrease significantly at wavelengths above 500 nm. The extinction coefficient describes the maximum number of wavelengths of light that can pass through a substance. The extinction coefficients for the single-layer and multilayer MoS_2 are 450 nm and 400 nm, respectively, and both the MoS_2 single-layer and multilayers are transparent at wavelengths longer than 500 nm. Therefore, the multilayer MoS_2 retains light better than the single-layer MoS_2. The refractive index is a parameter used to describe how light waves bend when traveling through various media. The refractive index for wavelengths under 250 nm for single-layer and multilayer MoS_2 are <2 and >2, respectively. A larger refractive index for both single and multilayer MoS_2 is observed at 500 nm (Ullah *et al* 2018).

7.3.2.2 Mechanical properties of MoS_2

For the mechanical properties of the material, the Young's modulus for MoS_2 sheets, bulk MoS_2, and single layers were reported to be about $E = 0.33$, 0.24, and 0.27 TPa, respectively. The increased modulus of monolayer MoS_2 compared to bulk MoS_2 is due to stacking defects in flakes with smaller monolayer thickness (Wang *et al* 2022). The MoS_2 exhibited nano-electromechanical properties due to the induction of mechanical strain on the MoS_2 layer, the electronic properties of MoS_2 can be changed from metallic to semiconductor (Chu *et al* 2009, Lee *et al* 2005). The moderate elasticity, flexibility, and low tension of MoS_2 make the material an excellent candidate for semiconductor and optoelectronic equipment development (Wang *et al* 2022).

7.3.2.3 Photoluminescence spectra

As mentioned above, when the MoS_2 changes from a single layer to a bulk, the bandgap converts from direct to indirect, decreasing the binding energy from 1.1 to 0.1 eV. Therefore, in the single-layer state, the MoS_2 exhibits significant photoluminescence properties that do not exist in the bulk form (Venkata Subbaiah *et al* 2016). It has been reported that realistic photoluminescence spectra can be obtained by doping different components, such as ionic liquid, into the MoS_2 materials, which causes the spectrum to shift to another energy region.

7.3.2.4 Electronic band structure

The MoS_2 monolayers' electric characteristics, such as their band structure and density of states, affect how well they work in optoelectronic and electronic devices. The crystal structure of MoS_2 shows that the indirect bandgap increases as the number of layers decreases. Based on theoretical findings, the bandgap of MoS_2 is approximately 1.9 eV. The GW approximation in the electronic structure theory predicted a larger bandgap and hypothesized that monolayer MoS_2 would have a bandgap in the range of 2.7–2.9 eV because of the different environmental conditions on the electronic state situation of the exciton binding energy (Yue *et al* 2012). The framework describes the need to change the bandgap state of MoS_2 in the outer electron shell, which is strongly influenced by the environment, compared to the graphene-based materials where less loading is required. Therefore, the mechanical deformation reduces the bandgap of monolayer MoS_2, converts the direct bandgap into an indirect bandgap, and transforms the semiconductor characterization into conductive materials. Based on the literature, a single layer of MoS_2 converts about 10% of its semiconductor properties to boost the electrical conductivity (Johari and Shenoy 2012).

7.3.3 Photocatalytic applications

The TMDC materials, such as MoS_2, have received a great attention because they are composed of two safe materials (Venkata Subbaiah *et al* 2016). Strong covalent bonds link Mo and S atoms, but weak van der Waals bonds link these layers in their structure and bulk phase. Compared to graphene (G) and graphene oxide (GO), single-layer MoS_2 has a lower surface interference probability and provides the better chemical stability. MoS_2 has unique properties due to the direct tuning of the bandgap, strong optical absorption, and pronounced photoluminescence with energies in the visible range in a variety of fields, including electronics, solid lubricants, biosensors, gas sensors, catalysts, and supercapacitors (Mouloua *et al* 2021). Additionally, a variety of techniques (including phase conversion, exfoliation, and foreign-atom doping) can be used to alter the properties of MoS_2, giving it significant potential for use in AOPs. Moreover, this material displays unusual photocatalytic and electrochemical activity due to its interstice characteristics

(Zhou *et al* 2022). These characteristics further encourage researchers to examine how different catalytic systems work in conjunction with the standard AOPs connected by MoS_2 (Ashraf *et al* 2020). Mo (IV), which has a low valence state, transfers electrons to a variety of oxidant states and undergoes self-oxidation to reach higher valence states such as Mo (V) and Mo (VI). Additionally, the electron paramagnetic resonant principle predicts that low-valence Mo will interact with dissolved oxygen to produce $O_2\bullet^-$. In the presence of OH^- or produced H_2O_2, $O_2\bullet^-$ converts to ROS (Yang *et al* 2021). However, various studies have reached conflicting conclusions about the predominant ROS. They assume that different contaminants produce ROS with ionized capabilities and electron donating/accepting capacities (Lee *et al* 2020). Compared to the organic compound, MoS_2, it showed less sensitivity to H_2O_2 (Li *et al* 2020). One undesirable phenomenon is the decomposition of H_2O_2 to O_2 rather than converting to $OH^{\bullet-}$, which results in less effective oxidation (Zhou *et al* 2021, Chen *et al* 2022). MoS_2 has an excellent catalytic activity and has been used in numerous studies as an incredible catalyst. However, its catalytic activity decreased after some cycle oxidation, causing a decreasing degradation rate. Thermal decomposition in an inert atmosphere, H_2O_2 treatment, and UV radiation are three successful methods for regaining the MoS_2 catalytic activity as the surface is repaired and oxidized Mo (IV) is reduced by nearby S atoms (Su *et al* 2022, Huang *et al* 2020, Du *et al* 2020). One of the most popular strategies for improving the catalytic activity of MoS_2 is elemental doping with other active elements such as Fe, Cu, etc, atoms. When the right amount of dopant is used (sulfite), the substrate's adsorption energy can be improved significantly, whereas adding more Fe causes the layer structure to break down and the adsorption energy to decrease (Huang *et al* 2020, Liu *et al* 2019).

An alternative method for the modification of MoS_2 to improve charge flow is the combination of MoS_2 with an another 2D material (Li *et al* 2016). Qu *et al* showed the efficient conversion of Fe^{3+} ions to Fe^{2+} with high efficiency in an $FeOCl/MoS_2$ membrane flow-through reactor due to Fe–O–Mo bonding (Qu *et al* 2020). Likewise, Liu *et al* demonstrated that the Mo–O–Cu bond in prepared $CuMgAl/MoS_2$ also acts as an efficient electron-transfer tunnel (Liu *et al* 2019). Due to the low redox potential of MoS_2 compared to the redox potentials of iron, copper, and cobalt (0.77 eV, 0.153 eV, and 1.83 eV, respectively), homogeneous metal ions, including Cu^{2+}, Co^{2+}, or Fe^{2+}, incorporated in MoS_2 has drawn a lot of attention recently. In contrast to heterogeneous reactions, homogeneous reactions between Fe(II), Cu(II), and Co(II) and various oxidants move along quickly. They are less affected by mass-transfer limitations (Zhou *et al* 2022). Singh *et al* prepared 2D MoS_2-PVA semiconductor thin film nanostructures by exfoliation, deposited them on glass substrates by a spin coating technique, and calcinated them at 350 °C. The MoS_2 layer demonstrated increased photocatalytic activity due to its higher mobility and better light harvesting capabilities under the UV–vis light for methylene blue (MB) and malachite green degradation (Singh *et al* 2020). In Chen's research, a hierarchical MoS_2/Bi_2MoO_6 composite was employed as a photocatalyst for rhodamine B (RhB) degradation under visible light, and the outcomes were compared to those of pure Bi_2MoO_6. The coupling of MoS_2 and Bi_2MoO_6 improved charge separation, and the photocatalytic activity was superior (Chen *et al* 2015). Lu's study created a 2D NL $BiOIO_3/MoS_2$ with a large specific surface area using hydrothermal techniques. This nanocomposite had an expanded photocatalytic activity wavelength and superior photocatalytic activity toward RhB due to the improved light harvesting capabilities (Lu *et al* 2020). By using the solvothermal technique, Sabarinathan *et al* created a 2D NL MoS_2/TiO_2 nanocomposite that significantly accelerated the degradation of methylene blue by 99.3% under visible light for 12 min in an aqueous suspension, a value three times faster than for pure TiO_2 or MoS_2 (Sabarinathan *et al* 2017). To degrade RhB in an aqueous solution under the UV light irradiation, Vattikuti *et al* constructed a 2D NL MoS_2 nanosheet decorated with SnO_2 nanoparticles. The prepared nanocomposite showed three times more photocatalytic activity than pure MoS_2 due to the suppression of the electron–hole recombination effect of SnO_2 nanoparticles on the MoS_2 substrate (Vattikuti *et al* 2015).

The hydrothermal technique was used to demonstrate the 2D–2D TiO_2/WS_2 nanocomposite used in Wu's experiment and as a photocatalyst for removing RhB from an aqueous solution. Due to improved electron–hole recombination and light harvesting, the prepared nanocomposite increases RhB degradation yields by 4.6- and 9.8-fold, respectively, compared to pure WS_2 layers and TiO_2 nanosheets (Wu *et al* 2019). A TiO_2/WS_2 2D NL biocompatible composite was made in Er-Chieh's study using a microwave-assisted method, which significantly increased the photocatalytic activity compared to pure TiO_2 nanoparticles. This nanocomposite lowers the electron–hole recombination rate in a solution containing 3% TiO_2/WS_2 2D NL nanocomposite. Excellent biocompatibility was shown by this nanocomposite, which has potential applications in medicine (Cho *et al* 2018). In Ma's research, magnetron-sputtered WS_2 nanofilms were prepared and laser-drilled to create pores, partially oxidize WS_2, and fabricate WS_2/WO_3. The results showed a 31-fold increase in photocurrent density after the laser treatment of WS_2 film (Ma *et al* 2017). Vattikuti's research prepared a mesoporous WS_2 structure with a hexagonal pore structure using a soft-templating approach and CTAB as a cationic template. Compared to porous or nonporous WS_2 with smaller surface areas, this photocatalyst's higher surface area led to noticeably greater photocatalytic activity for RhB degradation. Greater surface area and increased catalytic activity were produced by increasing CTAB concentration (Vattikuti and Byon 2015). In Fu's research, solvothermal techniques were used to demonstrate an ultrathin WS_2/BiOBr heterostructure that resembles a flower. This heterostructure exhibits significant photocatalytic activity against various pollutants, including methyl orange, Lanasol Red 5B, tetracycline, oxytetracycline, phenol, and bisphenol. The results showed that the developed nanocomposites significantly increase the rate of pollution degradation when compared to pure WS_2 or BiOB. Additionally, by reducing the concentration of electrolyte ions or increasing the activity of the photocatalysts for hydrogen ions, the degradation rate of pollutants was increased (Fu *et al* 2019).

In general, the MoS_2-based 2D NMs are used efficiently in photocatalytic applications due to their unique characteristics. The incorporation/doping of metals significantly improved the photocatalytic ability attributed to MoS_2-based hybrid materials that might become next-generation water treatment materials.

7.4 WS_2-based 2D NMs

7.4.1 WS_2 properties

7.4.1.1 Optical properties

Due to their specific bandgaps with visible and infrared wavelength energy, single-layer dichalcogenide materials have attracted much attention in optoelectronics and semiconductors (Cong *et al* 2018). These materials, in particular SeS_2, MoS_2, and WS_2, have been studied extensively as semiconductors. Among all these materials, WS_2 has attracted more interest from scientists due to its unique properties, including its stable structure, high emission quantum efficiency, strong spin–orbit coupling, and high binding energy (Yuan and Huang 2015, Zhu *et al* 2015). This dichalcogenide material exhibits different behavior in the bulk and single-layer states. In single-layer situations they are direct-gap semiconductors, but the bulk materials are indirect-gap semiconductors (Zeng *et al* 2013). Furthermore, the single-layer WS_2 exhibits certain behaviors such as exciton–polaritons (Flatten *et al* 2016), single photon emission (Palacios-Berraquero *et al* 2017), biexcitons (Plechinger *et al* 2015), and stimulated emission (Ye *et al* 2015), which are characteristic of the development of the optoelectronic devices.

7.4.1.2 Photoluminescence spectra

Developing on-chip single photon emitters is currently being pursued using the emerging 2D semiconductors. These devices typically need high photon generation capacity, purity, broadening, indistinguishability, and appropriate performance at room temperature to act as single photon emitters, the next generation of quantum materials (Chakraborty *et al* 2015). As reported, single-layer WS_2 can act as a single photon emitter with high efficiency at room temperature due to appropriate excitation bands and defects in their structure (Koperski *et al* 2015, Srivastava *et al* 2015). The exciton–polaritons were produced in a semiconducting structure with a hole that allowed photons to couple with them. The single-layer WS_2 has a comparable intrinsic exciton binding energy to other typical semiconductors, resulting in a stable exciton–polariton.

7.4.1.3 Electronic properties

Based on optical experiments, a WS_2 monolayer displayed a bandgap of about 2 eV, which is an appropriate bandgap for single-layer semiconductor. In addition, its strong spin–orbit coupling properties have received more attention for different electrical devices (Zhu *et al* 2011), resulting in spin-valley coupling (Xiao *et al* 2012). WS_2, in comparison to MoS_2, exhibited about three-fold larger valence band splitting; thus, finding the valley Hall effect is easier (Xiao *et al* 2012, Mak *et al* 2014b). WS_2 can be engineered with other optoelectronic active materials, such as graphene, to create an excellent material with both specific properties in field-effect transistors (Georgiou *et al* 2013). The ability of charge transport in WS_2 strongly depends on the WS_2 layer temperature (Jo *et al* 2014).

7.4.1.4 Mechanical properties

To study the mechanical properties of WS_2 materials, it is essential to know the role of its fundamental effects in flexible field-effect transistors. The Young's modulus of WS_2 dichalcogenide material is very variable and depends strongly on the number of layers. The Young's modulus of the single-layer WS_2 prepared with the CVD was reported to be about 272 GPa (Liu *et al* 2014), which is equal to Young's modulus reported for single-layer MoS_2 (264 GPa) reported for the same method, but is greater than the Young's modulus reported for 5L WSe_2 (170 GPa) (Zhang *et al* 2016, Liu *et al* 2014). Interestingly, as the number and thickness of the layers increases, the Young's modulus decreases; accordingly, the strength and flexibility of the WS_2 decrease (Falin *et al* 2021).

7.4.2 Photocatalytic applications

The W and Mo are transition metals and when combined with elements from group 6 of the periodic table (S, Se, and Te), they form a fantastic materials, known as dichalcogenides materials, with unique optical, electronic, magnetic, and thermal properties due to their unique structure, and have numerous applications in various industries (López-Urías *et al* 2015, Niknam *et al* 2022). This dichalcogenide materials generally have a stable structure in bulk, making them suitable for the use as dry lubricants. However, in a single layer, due to adsorption, visible light converts to WO_2 rapidly due to a photo-oxidation reaction (Kotsakidis *et al* 2019).

According to experimental results, the bandgap of monolayer WS_2 is reported to be about 2.0 eV (Elías *et al* 2013). The theoretical model suggests that WS_2 is expected to have the highest mobility among the TMDC-based semiconductors because of its lower effective mass (Liu *et al* 2011). The solid spin–orbit pairing that causes valence band splitting in WS_2 is also very intriguing (Zhu *et al* 2011). This leads to spin-valley coupling. The valence band of WS_2 is three times that of MoS_2, making it easier to distinguish the Valley Hall effect (Mak *et al* 2014a, Yao 2012).

A variety of parameters effects such as molecule doping, strain, substrates, defects, etc, have been researched to improve the performance of the 2D layer of WS_2. The components that could be included in the crystalline lattice significantly affect the doping strategy used for the TMDCs (Laskar *et al* 2014). Liu *et al* reported that by randomly doping Cu as a thin layer in field-effect transistors, it was possible to increase the electron mobility of 2D WS_2 by a factor of 5–7 and decrease the contact resistance by a factor of 1–3 (Liu *et al* 2020). The bandgap decreased from 1.98 to 1.83 eV due to Zhang *et al*'s introduction of CH using a plasma-assisted method as a dopant in WS_2 monolayers (Zhang *et al* 2019). Both the experimental and theoretical results show that these carbon doped WS_2 monolayers enhanced the electrical and optical performance. Li *et al* showed that the substrate greatly influenced the optical properties of monolayer WS_2 (Li and Wang 2020). They prepared a thin WS_2 layer on sapphire and SiO_2/Si substrates by CVD methods. As a result, the WS_2/sapphire showed a higher binding energy than WS_2/SiO_2/Si. Given that the associated devices will be used in various environments and with varying levels of excitation power, understanding the nonlinear PL characteristics of the monolayer WS_2 on various substrates offers invaluable guidance for upcoming optoelectronic and optical requests (table 7.2).

Table 7.2. Some examples of MoS_2- and WS_2-based materials with photocatalytic degradation activity and energy storage.

Method	Substrate	Oxidized layer	Application	Results	References
PVD	Carbon cloth	MoS_2	Catalysts	High catalytic activity and increased stability in acidic media.	Zhang *et al* (2015)
PVD	Silicon wafer	MoS_2	Environment purification	Suppresses adhesion of bacteria and inactivates bacteria. Kills both Gram-negative and Gram-positive bacteria.	Lian *et al* (2022)
PVD	Silicon dioxide and graphite	MoS_2	Electronic	Formation scalable and uniform film at low temperature (350 °C). The prepared material displays similar properties to the material prepared by exfoliated methods.	Muratore *et al* (2014)
PVD	Silicon	MoS_2	Electronic	A displayed photo-voltaic effect. Potential gradient properties.	Wang *et al* (2014)
PVD	Pyrolytic graphite	MoS_2	Electronic	Create a monolayer and bilayer (BL) MoS_2 structure with a high density of electronic state.	Bussolotti *et al* (2018)
Thermolysis	Sapphire	MoS_2	Optoelectronics	Preparation of large-area and highly crystalline MoS_2 with a simple and scalable approach. Exhibit n-type behaviors with the on/off current.	Liu *et al* (2012)
Thermolysis	SiO_2/Si	MoS_2	Electronics	Investigation of the growing mechanism of the MoS_2 layer. The purity of MoS_2 depend highly on hydrogen flux; in low hydrogen flux, MoO_3 is created in the MoS_2 film structure, but in high hydrogen flux (when it reaches 100 sccm), pure MoS_2 film is achieved.	Qi *et al* (2015)
Chemical bath deposition	Au/Si	MoS_2	Electrocatalysts	A uniform, scalable, and tunable thin layer of bi-, tri-, and few-layer MoS_2 modified by quantum dots using inexpensive precursors.	Vikraman *et al* (2017)

Ultrasonic assisted strategy	$Bi1_2O_{17}Cl_2$	MoS_2	Photocatalyst	An inexpensive and effective photocatalyst with a degradation rate of 92% in 30 min for rhodamine B degradation.	Ma *et al* (2020)
Hydrothermal	Graphene	MoS_2	Photocatalyst	Graphitic carbon nitride nanosheets/layered MoS_2/graphene ternary nanojunction composites act as an effective photocatalysts for the degradation of rhodamine B with appropriate photoelectrochemical activity and stability. Adsorption bandgap of g-C_3N_4 after loading with MoS_2/graphene decreases from 2.84 eV to 2.43 eV.	Tian *et al* (2018)
Ultrasonic chemical method	CeO_2@MoS_2/g-C_3N_4	MoS_2	Photocatalyst	Decreased H_2O adsorption energy. Abundant active sites for water splitting, fast charge transfer.	Zhu *et al* (2019)
Electrospinning and hydrothermal	TiO_2/WO_3@MoS_2	MoS_2	Photocatalyst	The highest hydrogen production achieved in a composite containing 60% MoS_2. Increase in photocatalytic activity effect related to the synergetic effect of WO_3 and MoS_2 as the hole collector and electron collector.	Zhao *et al* (2018)
Hydrothermal	MoS_2/ZnS	MoS_2	Photocatalyst	The highest photocatalytic activity was achieved for a composite continuing 0.010 M of ZnS for methylene blue (MB) decomposition, with an increase in the photocatalytic activity effect related to synergetic interaction and separation of the hole collector and electron collector.	Harish *et al* (2019)
Hydrothermal	n-$BiVO_4$@p-MoS_2	MoS_2	Photocatalyst	The best performance for crystal violet degradation was achieved at 10wt% catalyst under visible-light irradiation.	Zhao *et al* (2016)
Hydrothermal	Ag-C_3N_4/Ag/MoS_2	MoS_2	Photocatalyst	Photocatalytic degradation of RhB and photocatalytic hydrogen production.	Lu *et al* (2017)

(Continued)

Table 7.2. (*Continued*)

Method	Substrate	Oxidized layer	Application	Results	References
Hydrothermal	TiO_2-MoS_2	MoS_2	Photocatalyst	Stable photocatalyst with a high-density active site and charge transfer mobility, RhB degradation, and hydrogen production.	Bai *et al* (2015)
Solvothermal	MoS_2-g-PMMA/PI	MoS_2	Capacitor	The charge–discharge efficiency is 61.7%, with an increase of 47.9% in charge density, 20% in dielectric constant, and 40% in energy density compared to pure polyimide.	Li *et al* (2021)
Hydrothermal	MoS_2-graphene	MoS_2	Capacitor	Long cycle life hybrid Li-ion capacitor with a specific power density of 40 000 W kg^{-1}.	Wang *et al* (2017)
Chemically exfoliated	Carbon electrode/ MoS_2	MoS_2	Water treatment	Effective at removing ions from saline water.	Xing *et al* (2017)
Hydrothermal	MoS_2@MXene	MoS_2	Water treatment	Excellent charge efficiency (90.2%), good cyclic stability (96% retention rate), and rapid distillation rate for ion removal from saline water.	Cai *et al* (2022)
Sonochemical	BiOCl/WS_2	WS_2	Photocatalyst	With 2% BiOCl/WS_2, about 98.4 of malachite green was degraded at 45 min, with high quantum efficiency in the UV and visible area.	Ashraf *et al* (2020)
Hydrothermal	WS_2/QDs/ BiOBr	WS_2	Photocatalyst	The efficacy for pollutant removal was 92% (ciprofloxacin), 99% (Lanasol Red 5B), 95% (rhodamine B), 85% (metronidazole), 96% (tetracycline), and 41% (bisphenol A).	Fu *et al* (2020)
Hydrothermal	Paper electrodes	WS_2	Photocatalyst	Effective degrading of 90% of NFT in 120 min.	Fatima *et al* (2022)
Solvothermal	MoS_2, WS_2, and WSe_2 as nanosheet	MoS_2, WS_2, and WSe_2	Degradation of organic pollutants	Three nanosheets exhibited piezocatalysis activity for organic pollutants under ultrasonic vibration; the amount of activity for this nanosheet was as $MoS_2 > WS_2 > WSe_2$.	Li *et al* (2019a)

Liquid-phase exfoliation	Exfoliated hexagonal WS_2 platelets	WS_2	Photocatalyst	The photocatalytic activity was enhanced 1.71-fold compared to hexagonal WS_2 platelets synthesized using the traditional route. Effective degradation activity for organic pollutants such as Congo red (CR), methyl orange (MO), phenol red (PR), and rhodamine B (RhB).	Koyyada *et al* (2019)
Mechanochemistry	Graphitic carbon nitride (g-C_3N_4)/WO_3/WS_2)	WS_2	Photocatalyst	8.6-fold enhanced photodegrading activity in comparison to pure g-C_3N_4 in visible light, 7.8-fold increased hydrogen generation.	Song *et al* (2021)
CVD	h-WO_3/WS_2 core–shell nanowire	WS_2	Supercapacitors	h-WO_3 nanowires coated with 2D WS_2 retain capacity over 30 000 charge–discharge cycles.	Choudhary *et al* (2016)
PVD	WO_2/WS_2-rGO	WS_2	Sodium-ion capacitors	Ultrathin WO_2/WS_2 nanosheets wrapped with rGO maintain 90% capacity retention over 1000 cycles.	Xu *et al* (2022)
Liquid-phase exfoliation	WS_2/graphene	WS_2	Supercapacitors	Good rate (79.9%) and excellent cyclic stability (102.5%), excellent pseudocapacitive behavior.	Chen *et al* (2019)
Solvothermal	WS_2 nanosheets @carbon fiber cloth	WS_2	Energy storage	High specific capacity (1130 mAh), excellent rate capability (510 mAh), and a smaller initial discharge capacity loss (12.64%)	Wang *et al* (2018)
Hydrothermal	Carbon fiber	WS_2 nanoflowers	Supercapacitors	100% energy stability after 1000 charge–discharge cycles.	Mohan *et al* (2022)
Solvothermal	Carbon fiber cloth	WS_2 nanoplates	Supercapacitors	99% energy stability after 500 cycle charge–discharged, WS_2/CFC significantly increase specific capacitance compared with bulk WS_2.	Shang *et al* (2017)
Hydrothermal	Ni foam	WS_2/$ZnCo_2O_4$	Supercapacitors	96% energy stability after 4000 charge–discharge cycles.	Venkata Guru Raghavendra *et al* (2021)

7.5 Conclusion

Although monolayer transition metal disulfide materials have been investigated extensively in different fields, including the photocatalysts, supercapacitors, and energy storage applications, there are still many exciting features to discover. Although numerous techniques have been used to create the 2D MoS_2 or WS_2 thus far, finding one that is reliable, effective, and productive has not been easy. Studying the mechanisms underlying growth from the theoretical and experimental standpoint is still crucial and imperative. The appropriate performance of the devices that have been developed based on the 2D MoS_2 or WS_2 materials is expected soon, thanks to relevant synthesis techniques and substrate engineering. The requirements for a scale-up outside of laboratory investigations have yet to be met, so much work is still to be done. Commercializing the 2D MoS_2- and WS_2-based materials will bring a completely different future to the fabrication of electronics and optoelectronics. Additionally, the 2D MX_2 materials can be developed for the multifunctional logic circuit applications because of their inherent semiconducting specificity and layered structure.

References

Alalm M G, Tawfik A and Ookawara S 2015 Degradation of four pharmaceuticals by solar photo-Fenton process: kinetics and costs estimation *J. Environ. Chem. Eng.* **3** 46–51

Andreozzi R, Caprio V, Insola A and Marotta R 1999 Advanced oxidation processes (AOP) for water purification and recovery *Catal. Today* **53** 51–9

Arulraj A, Thachnatharen N, Rajkumar C and Mangalaraja R V 2022 *Engineering 2D chalcogenides for energy and environmental remediation Nanostructured Materials for Sustainable Energy and Environmental Remediation* (Bristol: IOP Publishing) pp 2-1–2-28

Arulraj A, Ramesh M, Subramanian B and Senguttuvan G 2018 In-situ temperature and thickness control grown 2D-MoS2 via pulsed laser ablation for photovoltaic devices *Sol Energy* **174** 286–95

Ashfaq M, Talreja N, Chauhan D and Viswanathan M R 2022 Synthesis of Cu-doped 2D-WS_2 nanosheet-based nano-antibiotic materials for inhibiting *E. coli* and *S. aureus* bacterial strains *New J. Chem.* **46** 5581–7

Ashraf W, Bansal S, Singh V, Barman S and Khanuja M 2020 BiOCl/WS_2 hybrid nanosheet (2D/2D) heterojunctions for visible-light-driven photocatalytic degradation of organic/inorganic water pollutants *RSC Adv.* **10** 25073–88

Bai S, Wang L, Chen X, Du J and Xiong Y 2015 Chemically exfoliated metallic MoS_2 nanosheets: a promising supporting co-catalyst for enhancing the photocatalytic performance of TiO_2 nanocrystals *Nano Res.* **8** 175–83

Baptista A, Silva F, Porteiro J, Míguez J and Pinto G 2018 Sputtering physical vapour deposition (PVD) coatings: a critical review on process improvement and market trend demands *Coatings* **8** 402

Bhasarkar J B and Bal D K 2021 Nanomaterial-based advanced oxidation processes for degradation of waste pollutants *Handbook of Nanomaterials for Wastewater Treatment* (Amsterdam: Elsevier)

Bussolotti F *et al* 2018 Electronic properties of atomically thin MoS_2 layers grown by physical vapour deposition: band structure and energy level alignment at layer/substrate interfaces *RSC Adv.* **8** 7744–52

Cai Y, Wang Y, Zhang L, Fang R and Wang J 2022 3D heterostructure constructed by few-layered MXenes with a MoS_2 layer as the shielding shell for excellent hybrid capacitive deionization and enhanced structural stability *ACS Appl. Mater. Interfaces* **14** 2833–47

Chakraborty C, Kinnischtzke L, Goodfellow K M, Beams R and Vamivakas A N 2015 Voltage-controlled quantum light from an atomically thin semiconductor *Nat. Nanotechnol.* **10** 507–11

Chen W, Yu X, Zhao Z, Ji S and Feng L 2019 Hierarchical architecture of coupling graphene and 2D WS_2 for high-performance supercapacitor *Electrochim. Acta* **298** 313–20

Chen Y, Tian G, Shi Y, Xiao Y and Fu H 2015 Hierarchical MoS_2/Bi_2MoO_6 composites with synergistic effect for enhanced visible photocatalytic activity *Appl. Catal.* B **164** 40–7

Chen Z, Lian C, Huang K, Ji J, Yan Q, Zhang J and Xing M 2022 'Small amount for multiple times' of H_2O_2 feeding way in MoS_2-Fe_x heterogeneous Fenton for enhancing sulfadiazine degradation *Chin. Chem. Lett.* **33** 1365–72

Cho E-C, Chang-Jian C-W, Zheng J-H, Huang J-H, Lee K-C, Ho B-C and Hsiao Y-S 2018 Microwave-assisted synthesis of TiO_2/WS_2 heterojunctions with enhanced photocatalytic activity *J. Taiwan Inst. Chem. Eng.* **91** 489–98

Choudhary N, Li C, Chung H-S, Moore J, Thomas J and Jung Y 2016 High-performance one-body core/shell nanowire supercapacitor enabled by conformal growth of capacitive 2D WS_2 layers *ACS Nano* **10** 10726–35

Chu M, Sun Y, Aghoram U and Thompson S E 2009 Strain: a solution for higher carrier mobility in nanoscale MOSFETs *Annu. Rev. Mater. Res.* **39** 203–29

Cong C, Shang J, Wang Y and Yu T 2018 Optical properties of 2D semiconductor WS_2 *Adv. Opt. Mater.* **6** 1700767

Dalrymple O K, Yeh D H and Trotz M A 2007 Removing pharmaceuticals and endocrine-disrupting compounds from wastewater by photocatalysis *J. Chem. Technol. Biotechnol.* **82** 121–34

Deng Y and Zhao R 2015 Advanced oxidation processes (AOPs) in wastewater treatment *Curr. Pollut. Rep.* **1** 167–76

Devi L G and Kavitha R 2013 A review on non-metal ion doped titania for the photocatalytic degradation of organic pollutants under UV/solar light: role of photogenerated charge carrier dynamics in enhancing the activity *Appl. Catal.* B **140** 559–87

Domingues E, Fernandes E, Gomes J and Martins R C 2021 Advanced oxidation processes perspective regarding swine wastewater treatment *Sci. Total Environ.* **776** 145958

Dong J, Zhao Y and Ouyang G 2019 The effect of alloying on the band engineering of two-dimensional transition metal dichalcogenides *Physica* E **105** 90–6

Du M, Yi Q, Ji J, Zhu Q, Duan H, Xing M and Zhang J 2020 Sustainable activation of peroxymonosulfate by the Mo(IV) in MoS_2 for the remediation of aromatic organic pollutants *Chin. Chem. Lett.* **31** 2803–8

Elías A L, Perea-López N, Castro-Beltrán A, Berkdemir A, Lv R, Feng S, Long A D, Hayashi T, Kim Y A and Endo M 2013 Controlled synthesis and transfer of large-area WS_2 sheets: from single layer to few layers *ACS Nano* **7** 5235–42

Falin A *et al* 2021 Mechanical properties of atomically thin tungsten dichalcogenides: WS_2, WSe_2, and WTe_2 *ACS Nano* **15** 2600–10

Faraji S, Bagheri-Mohagheghi M M and Shirpay A 2022 The effect of tungsten (W) concentration and sulfuration on morphology and optical properties and tuning of the band gap of 2D-MoS_2 thin films *Opt. Quantum Electron.* **54** 703

Fatima T, Husain S, Narang J, Khanuja M, Shetti N P and Reddy K R 2022 Novel tungsten disulfide (WS_2) nanosheets for photocatalytic degradation and electrochemical detection of pharmaceutical pollutants *J. Water Process Eng.* **47** 102717

Flatten L C, He Z, Coles D M, Trichet A A P, Powell A W, Taylor R A, Warner J H and Smith J M 2016 Room-temperature exciton-polaritons with two-dimensional WS_2 *Sci. Rep.* **6** 33134

Fleischauer P D and Bauer R 1988 Chemical and structural effects on the lubrication properties of sputtered MoS_2 films *Tribol. Trans.* **31** 239–50

Fu S, Liu X, Yan Y, Li L, Liu H, Zhao F and Zhou J 2019 Few-layer WS_2 modified BiOBr nanosheets with enhanced broad-spectrum photocatalytic activity towards various pollutants removal *Sci. Total Environ.* **694** 133756

Fu S, Yuan W, Liu X, Yan Y, Liu H, Li L, Zhao F and Zhou J 2020 A novel 0D/2D WS_2/BiOBr heterostructure with rich oxygen vacancies for enhanced broad-spectrum photocatalytic performance *J. Colloid Interface Sci.* **569** 150–63

Garcia-Segura S, Ocon J D and Chong M N 2018 Electrochemical oxidation remediation of real wastewater effluents—a review *Process Saf. Environ. Prot.* **113** 48–67

Georgiou T *et al* 2013 Vertical field-effect transistor based on graphene–WS_2 heterostructures for flexible and transparent electronics *Nat. Nanotechnol.* **8** 100–3

Gottschalk C, Libra J A and Saupe A 2009 *Ozonation of Water and Waste Water: A Practical Guide to Understanding Ozone and Its Applications* (New York: Wiley)

Guan Y, Yao H, Zhan H, Wang H, Zhou Y and Kang J 2021 Optoelectronic properties and strain regulation of the 2D WS_2/ZnO van der Waals heterostructure *RSC Adv.* **11** 14085–92

Gupta D, Chauhan V and Kumar R 2020 A comprehensive review on synthesis and applications of molybdenum disulfide (MoS_2) material: past and recent developments *Inorg. Chem. Commun.* **121** 108200

Güyer G T and Ince N H 2011 Degradation of diclofenac in water by homogeneous and heterogeneous sonolysis *Ultrason. Sonochem.* **18** 114–9

Harish S, Prachi, Archana J, Navaneethan M, Shimomura M, Ikeda H and Hayakawa Y 2019 Synergistic interaction of 2D layered MoS_2/ZnS nanocomposite for highly efficient photocatalytic activity under visible light irradiation *Appl. Surf. Sci.* **488** 36–45

Huang L-Z, Wei X, Gao E, Zhang C, Hu X-M, Chen Y, Liu Z, Finck N, Luetzenkirchen J and Dionysiou D D 2020 Single Fe atoms confined in two-dimensional MoS_2 for sulfite activation: a biomimetic approach towards efficient radical generation *Appl. Catal.* B **268** 118459

Jo S, Ubrig N, Berger H, Kuzmenko A B and Morpurgo A F 2014 Mono- and bilayer WS_2 light-emitting transistors *Nano Lett.* **14** 2019–25

Johari P and Shenoy V B 2012 Tuning the electronic properties of semiconducting transition metal dichalcogenides by applying mechanical strains *ACS Nano* **6** 5449–56

Kalanyan B, Beams R, Katz M B, Davydov A V, Maslar J E and Kanjolia R K 2018 MoS_2 thin films from a $(N\,^tBu)_2(NMe_2)_2Mo$ and 1-propanethiol atomic layer deposition process *J. Vac. Sci. Technol.* A **37** 1–11

Kim I, Yamashita N and Tanaka H 2009 Performance of UV and UV/H_2O_2 processes for the removal of pharmaceuticals detected in secondary effluent of a sewage treatment plant in Japan *J. Hazard. Mater.* **166** 1134–40

Koperski M, Nogajewski K, Arora A, Cherkez V, Mallet P, Veuillen J Y, Marcus J, Kossacki P and Potemski M 2015 Single photon emitters in exfoliated WSe_2 structures *Nat. Nanotechnol.* **10** 503–6

Kotsakidis J C, Zhang Q, Vazquez de Parga A L, Currie M, Helmerson K, Gaskill, D K and Fuhrer M S 2019 Oxidation of monolayer WS_2 in ambient is a photoinduced process *Nano Lett.* **19** 5205–15

Koyyada G P, Vattikuti S V, Shome S, Shim J, Chitturi V and Jung J H 2019 Enhanced solar light-driven photocatalytic degradation of pollutants and hydrogen evolution over exfoliated hexagonal WS_2 platelets *Mater. Res. Bull.* **109** 246–54

Kurian M 2021 Advanced oxidation processes and nanomaterials: a review *Cleaner Eng. Technol.* **2** 100090

Laskar M R, Nath D N, Ma L, Lee E W, Lee C H, Kent T, Yang Z, Mishra R, Roldan M A and Idrobo J-C 2014 p-type doping of MoS_2 thin films using Nb *Appl. Phys. Lett.* **104** 092104

Lau K K S 2016 Chemical vapor deposition *Medical Coatings and Deposition Technologies* Eds: David A. Glocker, Shrirang V. Ranade, (Hoboken, NJ: Wiley); pp. 403-55

Lee J, Von Gunten U and Kim J-H 2020 Persulfate-based advanced oxidation: critical assessment of opportunities and roadblocks *Environ. Sci. Technol.* **54** 3064–81

Lee M L, Fitzgerald E A, Bulsara M T, Currie M T and Lochtefeld A 2005 Strained Si, SiGe, and Ge channels for high-mobility metal-oxide-semiconductor field-effect transistors *J. Appl. Phys.* **97** 011101

Legrini O, Oliveros E and Braun A M 1993 Photochemical processes for water treatment *Chem. Rev.* **93** 671–98

Li H, Wu J, Yin Z and Zhang H 2014 Preparation and applications of mechanically exfoliated single-layer and multilayer MoS_2 and WSe_2 nanosheets *Acc. Chem. Res.* **47** 1067–75

Li J, Zhan G, Yu Y and Zhang L 2016 Superior visible light hydrogen evolution of Janus bilayer junctions via atomic-level charge flow steering *Nat. Commun.* **7** 1–9

Li J, Zhang J, Zhang S and Ren K 2021 2D MoS_2 nanosheet-based polyimide nanocomposite with high energy density for high temperature capacitor applications *Macromol. Mater. Eng.* **306** 2100079

Li K and Wang W 2020 Effects of substrates on the optical properties of monolayer WS_2 *J. Cryst. Growth* **540** 125645

Li Q, Hu B, Yang Q, Cai X, Nie M, Jin Y, Zhou L, Xu Y, Pan Q and Fang L 2020 Interaction mechanism between multi-layered MoS_2 and H_2O_2 for self-generation of reactive oxygen species *Environ. Res.* **191** 110227

Li S, Zhao Z, Yu D, Zhao J-Z, Su Y, Liu Y, Lin Y, Liu W, Xu H and Zhang Z 2019a Few-layer transition metal dichalcogenides (MoS_2, WS_2, and WSe_2) for water splitting and degradation of organic pollutants: understanding the piezocatalytic effect *Nano Energy* **66** 104083

Li W, Zhang Y, Long X, Cao J, Xin X, Guan X, Peng J and Zheng X 2019b Gas sensors based on mechanically exfoliated MoS_2 nanosheets for room-temperature NO_2 detection *Sensors* **19** 2123

Lian Z, Li H, Wu T, Zhao J, Cai S and Yang R 2022 Vapor deposition of MoO_x/MoS_2 films on silicon wafer with visible-light responsive photocatalytic antibacterial properties *Appl. Surf. Sci.* **606** 154874

Liu K-K *et al* 2012 Growth of large-area and highly crystalline MoS_2 thin layers on insulating substrates *Nano Lett.* **12** 1538–44

Liu K *et al* 2014 Elastic properties of chemical-vapor-deposited monolayer MoS_2, WS_2, and their bilayer heterostructures *Nano Lett.* **14** 5097–103

Liu L, Kumar S B, Ouyang Y and Guo J 2011 Performance limits of monolayer transition metal dichalcogenide transistors *IEEE Trans. Electron Devices* **58** 3042–7

Liu M, Wang H, Zhu P, Wang J, Tan H and Zheng Z 2019 Tuning the oxidation state of copper in CuMgAl hydrotalcite using partially oxidized MoS_2 for efficient hydroxylation of phenol *Catal. Commun.* **129** 105752

Liu M *et al* 2020 Enhanced carrier transport by transition metal doping in WS_2 field effect transistors *Nanoscale* **12** 17253–64

Lopez-Sanchez O, Lembke D, Kayci M, Radenovic A and Kis A 2013 Ultrasensitive photodetectors based on monolayer MoS_2 *Nat. Nanotechnol.* **8** 497–501

López-Urías F L, Elías A, Perea-López N, Gutiérrez H R, Terrones M and Terrones H 2015 Electronic, magnetic, optical, and edge-reactivity properties of semiconducting and metallic WS_2 nanoribbons *2D Mater.* **2** 015002

Lu D, Wang H, Zhao X, Kondamareddy K K, Ding J, Li C and Fang P 2017 Highly efficient visible-light-induced photoactivity of Z-scheme g-C_3N_4/Ag/MoS_2 ternary photocatalysts for organic pollutant degradation and production of hydrogen *ACS Sustain. Chem. Eng.* **5** 1436–45

Lu M, Xiao X, Wang Y and Chen J 2020 Construction of novel $BiOIO_3$/MoS_2 2D/2D heterostructures with enhanced photocatalytic activity *J. Alloys Compd.* **831** 154789

Ma J, Shi L, Wang Z, Ren T, Geng Z and Qi W 2020 2D layered MoS_2 loaded on $Bi_{12}O_{17}Cl_2$ nanosheets: an effective visible-light photocatalyst *Ceram. Int.* **46** 7438–45

Ma S *et al* 2017 Enhanced photocatalytic activity of WS_2 film by laser drilling to produce porous WS_2/WO_3 heterostructure *Sci. Rep.* **7** 3125

Magda G Z, Pető J, Dobrik G, Hwang C, Biró L P and Tapasztó L 2015 Exfoliation of large-area transition metal chalcogenide single layers *Sci. Rep.* **5** 14714

Mak K F, McGill K L, Park J and McEuen P L 2014a The valley Hall effect in MoS_2 transistors *Science* **344** 1489–92

Mak K F, McGill K L, Park J and McEuen P L 2014b Valleytronics. The valley Hall effect in MoS_2 transistors *Science* **344** 1489–92

Mohan V V, Manuraj M, Anjana P M and Rakhi R B 2022 WS_2 nanoflowers as efficient electrode materials for supercapacitors *Energy Technol.* **10** 2100976

Mouloua D, Kotbi A, Deokar G, Kaja K, El Marssi M, EL Khakani, M A and Jouiad M 2021 Recent progress in the synthesis of MoS_2 thin films for sensing, photovoltaic and plasmonic applications: a review *Materials* **14** 3283

Muratore C, Hu J J, Wang B, Haque M A, Bultman J E, Jespersen M L, Shamberger P J, McConney M E, Naguy R D and Voevodin A A 2014 Continuous ultra-thin MoS_2 films grown by low-temperature physical vapor deposition *Appl. Phys. Lett.* **104** 261604

Niknam S, Dehdast S A, Pourdakan O, Shabani M and Koohi M K 2022 Tungsten disulfide nanomaterials (WS_2 NM) application in biosensors and nanomedicine: a review *Nanomed. Res. J.* **7** 214–26

Novoselov K S, Geim A K, Morozov S V, Jiang D, Zhang Y, Dubonos S V, Grigorieva I V and Firsov A A 2004 Electric field effect in atomically thin carbon films *Science* **306** 666–9

Ottaviano L *et al* 2017 Mechanical exfoliation and layer number identification of MoS_2 revisited *2D Mater.* **4** 045013

Palacios-Berraquero C, Kara D M, Montblanch A R P, Barbone M, Latawiec P, Yoon D, Ott A K, Loncar M, Ferrari A C and Atatüre M 2017 Large-scale quantum-emitter arrays in atomically thin semiconductors *Nat. Commun.* **8** 15093

Parsons G N, George S M and Knez M 2011 Progress and future directions for atomic layer deposition and ALD-based chemistry *MRS Bull.* **36** 865–71

Phung H N T, Tran V N K, Nguyen L T, Phan L K T, Duong P A and Le H V T 2017 Investigating visible-photocatalytic activity of MoS_2/TiO_2 heterostructure thin films at various MoS_2 deposition times *J. Nanomater.* **2017** 3197540

Plechinger G, Nagler P, Kraus J, Paradiso N, Strunk C, Schüller C and Korn T 2015 Identification of excitons, trions and biexcitons in single-layer WS_2 *Phys. Status Solidi—Rapid Res. Lett.* **9** 457–61

Qi F *et al* 2015 Effect of hydrogen on the growth of MoS_2 thin layers by thermal decomposition method *Vacuum* **119** 204–8

Qu S, Wang W, Pan X and Li C 2020 Improving the Fenton catalytic performance of FeOCl using an electron mediator *J. Hazard. Mater.* **384** 121494

Radisavljevic B, Radenovic A, Brivio J, Giacometti V and Kis A 2011 Single-layer MoS_2 transistors *Nat. Nanotechnol.* **6** 147–50

Sabarinathan M, Harish S, Archana J, Navaneethan M, Ikeda H and Hayakawa Y 2017 Highly efficient visible-light photocatalytic activity of MoS_2–TiO_2 mixtures hybrid photocatalyst and functional properties *RSC Adv.* **7** 24754–63

Salem I A, El-Ghamry H A and El-Ghobashy M A 2015 Application of montmorillonite–Cu (II) ethylenediamine catalyst for the decolorization of chromotrope 2R with H_2O_2 in aqueous solution *Spectrochim. Acta* A **139** 130–7

Shang X, Chi J-Q, Lu S-S, Gou J-X, Dong B, Li X, Liu Y-R, Yan K-L, Chai Y-M and Liu C-G 2017 Carbon fiber cloth supported interwoven WS_2 nanosplates with highly enhanced performances for supercapacitors *Appl. Surf. Sci.* **392** 708–14

Singh A K, Kumbhakar P, Krishnamoorthy A, Nakano A, Sadasivuni K K, Vashishta P, Roy A K, Kochat V and Tiwary C S 2021 Review of strategies toward the development of alloy two-dimensional (2D) transition metal dichalcogenides *iScience* **24** 103532

Singh J, Rishikesh, Kumar S, Verma H K and Soni R K 2020 Cost-effective scalable synthesis of few layers MoS_2 based thin film for sunlight enforced photocatalytic activity *Opt. Mater.* **110** 110506

Song T, Xie C, Matras-Postolek K and Yang P 2021 2D layered g-C_3N_4/WO_3/WS_2 S-scheme heterojunctions with enhanced photochemical performance *J. Phys. Chem.* C **125** 19382–93

Spalvins T 1980 Tribological properties of sputtered MoS_2 films in relation to film morphology *Thin Solid Films* **73** 291–7

Splendiani A, Sun L, Zhang Y, Li T, Kim J, Chim C-Y, Galli G and Wang F 2010 Emerging photoluminescence in monolayer MoS_2 *Nano Lett.* **10** 1271–5

Srivastava A, Sidler M, Allain A V, Lembke D S, Kis A and Imamoğlu A 2015 Optically active quantum dots in monolayer WSe_2 *Nat. Nanotechnol.* **10** 491–6

Su X, Guo Y, Yan L, Wang Q, Zhang W, Li X, Song W, Li Y and Liu G 2022 MoS_2 nanosheets vertically aligned on biochar as a robust peroxymonosulfate activator for removal of tetracycline *Sep. Purif. Technol.* **282** 120118

Sun L, Yuan G, Gao L, Yang J, Chhowalla M, Gharahcheshmeh M H, Gleason K K, Choi Y S, Hong B H and Liu Z 2021 Chemical vapour deposition *Nat. Rev. Methods Primers* **1** 5

Tedstone A A, Lewis D J and O'Brien P 2016 Synthesis, properties, and applications of transition metal-doped layered transition metal dichalcogenides *Chem. Mater.* **28** 1965–74

Tian H, Liu M and Zheng W 2018 Constructing 2D graphitic carbon nitride nanosheets/layered MoS_2/graphene ternary nanojunction with enhanced photocatalytic activity *Appl. Catal.* B **225** 468–76

Ullah M S, Yousuf A H B, Es-Sakhi A D and Chowdhury M H 2018 Analysis of optical and electronic properties of MoS_2 for optoelectronics and FET applications *AIP Conf. Proc.* **1957** 020001

Vattikuti S V P and Byon C 2015 Effect of CTAB surfactant on textural, structural, and photocatalytic properties of mesoporous WS_2 *Sci. Adv. Mater.* **7** 2639–45

Vattikuti S V P, Byon C, Reddy C V and Ravikumar R V S S N 2015 Improved photocatalytic activity of MoS_2 nanosheets decorated with SnO_2 nanoparticles *RSC Adv.* **5** 86675–84

Venkata Guru Raghavendra K, Sreekanth T V M, Kim J and Yoo K 2021 Novel hydrothermal synthesis of jasmine petal-like nanoflower $WS_2/ZnCo_2O_4$ as efficient electrode material for high-performance supercapacitors *Mater. Lett.* **285** 129133

Venkata Subbaiah Y P, Saji K J and Tiwari A 2016 Atomically thin MoS_2: a versatile nongraphene 2D material *Adv. Funct. Mater.* **26** 2046–69

Vikraman D, Akbar K, Hussain S, Yoo G, Jang J-Y, Chun S-H, Jung J and Park H J 2017 Direct synthesis of thickness-tunable MoS_2 quantum dot thin layers: optical, structural and electrical properties and their application to hydrogen evolution *Nano Energy* **35** 101–14

Vishwanath S *et al* 2015 Comprehensive structural and optical characterization of MBE grown $MoSe_2$ on graphite, CaF_2 and graphene *2D Mater.* **2** 024007

Wang B, Muratore C, Voevodin A A and Haque M A 2014 Photo-sensitivity of large area physical vapor deposited mono and bilayer MoS_2 *Nano Convergence* **1** 22

Wang Z M, Salamo G and Bellucci S 2022 *Lecture Notes in Nanoscale Science and Technology* (Frankfrut: Springer)

Wang R, Wang S, Jin D, Zhang Y, Cai Y, Ma J and Zhang L 2017 Engineering layer structure of MoS_2–graphene composites with robust and fast lithium storage for high-performance Li-ion capacitors *Energy Storage Mater.* **9** 195–205

Wang S and Wang J 2018 Degradation of carbamazepine by radiation-induced activation of peroxymonosulfate *Chem. Eng. J.* **336** 595–601

Wang T, Sun C, Yang M, Zhang L, Shao Y, Wu Y and Hao X 2018 Enhanced reversible lithium ion storage in stable 1T@2H WS_2 nanosheet arrays anchored on carbon fiber *Electrochim. Acta* **259** 1–8

Wu J, Li H, Yin Z, Li H, Liu J, Cao X, Zhang Q and Zhang H 2013 Layer thinning and etching of mechanically exfoliated MoS_2 nanosheets by thermal annealing in air *Small* **9** 3314–9

Wu W *et al* 2014 Piezoelectricity of single-atomic-layer MoS_2 for energy conversion and piezotronics *Nature* **514** 470–4

Wu Y, Liu Z, Li Y, Chen J, Zhu X and Na P 2019 Construction of 2D–2D TiO_2 nanosheet/layered WS_2 heterojunctions with enhanced visible-light-responsive photocatalytic activity *Chin. J. Catal.* **40** 60–9

Xiao D, Liu G-B, Feng W, Xu X and Yao W 2012 Coupled spin and valley physics in monolayers of MoS_2 and other group-VI dichalcogenides *Phys. Rev. Lett.* **108** 196802

Xing F, Li T, Li J, Zhu H, Wang N and Cao X 2017 Chemically exfoliated MoS_2 for capacitive deionization of saline water *Nano Energy* **31** 590–5

Xu J, Guo H, Huo K and Li Y 2022 Heterogeneous WO_2/WS_2 microspheres synergized with reduced graphene oxides with high rate capacity for superior sodium-ion capacitors *J. Electroanal. Chem.* **920** 116598

Xu X-R, Li X-Y, Li X-Z and Li H-B 2009 Degradation of melatonin by UV, UV/H_2O_2, Fe^{2+}/H_2O_2 and $UV/Fe^{2+}/H_2O_2$ processes *Sep. Purif. Technol.* **68** 261–6

Yang J-C. E, Zhu M-P, Duan X, Wang S, Yuan B and Fu M-L 2021 The mechanistic difference of 1T-2H MoS_2 homojunctions in persulfates activation: structure-dependent oxidation pathways *Appl. Catal.* B **297** 120460

Yao W 2012 Valley and spin coupled physics in monolayers of MoS_2 and other group VI dichalcogenides *11th Int. Conf. on Condensed Matter Theory and Computational Materials Science* 2012

Yao W, Kang Z, Deng J, Chen Y, Song Q, Ding X L, Lu F and Wang W 2020 Synthesis of 2D $MoS_{2(1-x)}Se_{2x}$ semiconductor alloy by chemical vapor deposition *RSC Adv.* **10** 42172–7

Ye Y, Wong Z J, Lu X, Ni X, Zhu H, Chen X, Wang Y and Zhang X 2015 Monolayer excitonic laser *Nat. Photonics* **9** 733–7

Yuan L and Huang L 2015 Exciton dynamics and annihilation in WS_2 2D semiconductors *Nanoscale* **7** 7402–8

Yue Q, Kang J, Shao Z, Zhang X, Chang S, Wang G, Qin S and Li J 2012 Mechanical and electronic properties of monolayer MoS_2 under elastic strain *Phys. Lett.* A **376** 1166–70

Zeng H *et al* 2013 Optical signature of symmetry variations and spin-valley coupling in atomically thin tungsten dichalcogenides *Sci. Rep.* **3** 1608

Zhang F *et al* 2019 Carbon doping of WS_2 monolayers: bandgap reduction and p-type doping transport *Sci. Adv.* **5** eaav5003

Zhang R, Koutsos V and Cheung R 2016 Elastic properties of suspended multilayer WSe_2 *Appl. Phys. Lett.* **108** 042104

Zhang X, Zhang Y, Yu B-B, Yin X-L, Jiang W-J, Jiang Y, Hu J-S and Wan L-J 2015 Physical vapor deposition of amorphous MoS_2 nanosheet arrays on carbon cloth for highly reproducible large-area electrocatalysts for the hydrogen evolution reaction *J. Mater. Chem.* A **3** 19277–81

Zhao J, Zhang P, Fan J, Hu J and Shao G 2018 Constructing 2D layered MoS_2 nanosheets-modified Z-scheme TiO_2/WO_3 nanofibers ternary nanojunction with enhanced photocatalytic activity *Appl. Surf. Sci.* **430** 466–74

Zhao W, Liu Y, Wei Z, Yang S, He H and Sun C 2016 Fabrication of a novel p–n heterojunction photocatalyst n-$BiVO_4$@p-MoS_2 with core–shell structure and its excellent visible-light photocatalytic reduction and oxidation activities *Appl. Catal.* B **185** 242–52

Zhou H, Xie Z, Liu Y, Lai B, Ong W-J, Wang S and Duan X 2022 Recent advances in molybdenum disulfide-based advanced oxidation processes *Environ. Funct. Mater.* **1** 1–9

Zhou H, Zhang H, He Y, Huang B, Zhou C, Yao G and Lai B 2021 Critical review of reductant-enhanced peroxide activation processes: trade-off between accelerated Fe^{3+}/Fe^{2+} cycle and quenching reactions *Appl. Catal.* B **286** 119900

Zhu B, Chen X and Cui X 2015 Exciton binding energy of monolayer WS_2 *Sci. Rep.* **5** 9218

Zhu C, Wang Y, Jiang Z, Xu F, Xian Q, Sun C, Tong Q, Zou W, Duan X and Wang S 2019 CeO_2 nanocrystal-modified layered $MoS_2/g\text{-}C_3N_4$ as 0D/2D ternary composite for visible-light photocatalytic hydrogen evolution: interfacial consecutive multi-step electron transfer and enhanced H_2O reactant adsorption *Appl. Catal.* B **259** 118072

Zhu J, Wu J, Sun Y, Huang J, Xia Y, Wang H, Wang H, Wang Y, Yi Q and Zou G 2016 Thickness-dependent bandgap tunable molybdenum disulfide films for optoelectronics *RSC Adv.* **6** 110604–9

Zhu Z Y, Cheng Y C and Schwingenschlögl U 2011 Giant spin–orbit-induced spin splitting in two-dimensional transition-metal dichalcogenide semiconductors *Phys. Rev.* B **84** 153402

Chapter 8

Recent progress in transition metal nitrides for electrochemical and photocatalytic applications

Kiruthiga Ramakrishnan, Kumaramangalam Jeyalakshmi, Udayabhaskar Rednam, R V Mangalaraja, Prabhu Saravanan and Padmanaban Annamalai

Worldwide energy crisis in meeting present demands requires high-performance energy storage and conversion devices. In this search, batteries, supercapacitors, and photocatalysts have come to the fore. The performance enhancement of the above devices depends on the tuning of the components in their respective systems, in which respect a popular material that has been under constant examination by researchers in recent years is transition metal nitrides (TMNs). The TMNs are two-dimensional (2D) structures with excellent physicochemical properties suitable for applications in interdisciplinary fields. In this chapter, a brief description of TMN based electrodes for supercapacitor applications is discussed. This chapter also summarizes the recent research advances in TMN based photocatalysts for the most challenging water splitting and CO_2 reduction reactions.

8.1 Introduction

Flexible electronics are a new and promising technology for high-performance portable electronic devices in the near future. They require the development of small and lightweight supercapacitors (SCs), lithium-ion batteries (LIBs), and sodium ion batteries (NIBs) as flexible energy storage devices. The advancement of improved electrode materials with the necessary features is required to meet future demands. The electrode materials applied in energy storage devices are very important, due to the demands for energy by the humans in day-to-day life. Recently, extensive

doi:10.1088/978-0-7503-5724-1ch8

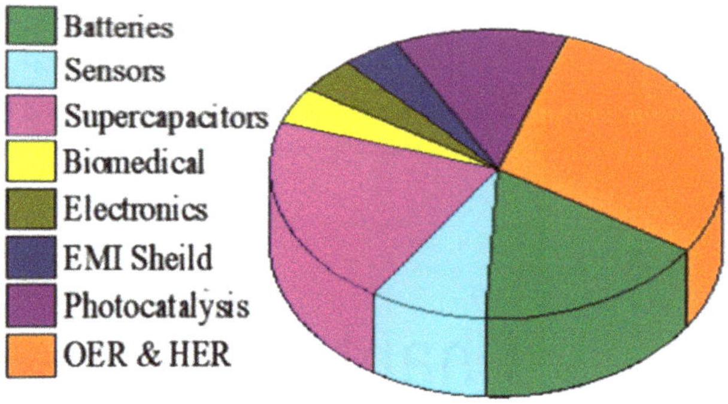

Figure 8.1. A pie chart showing the possible applications of TMNs in interdisciplinary fields with the number of publications.

attempts have been made in the investigation of new electrode materials for LIBs, SIBs, and SCs with attractive features. In the past decades, metal oxides, metal sulphides, metal phosphates, and their composites, either with carbon related materials or polymers, have been used as electrodes for energy storage devices. Recently, the transition metal nitrides (TMNs) have also become of great interest owing to their good mechanical and physicochemical properties such as high electrical conductivity, high melting point, hardness, good corrosion resistance, high specific capacity, and high thermal and chemical stability. These properties are favourable for applications in different fields such as sensors, supercapacitors, batteries, and photocatalysis (figure 8.1). Predominantly, the TMNs are suitable for LIBs, due to their capability for lithium storage through the conversion mechanism. In energy storage field, the most reported metal nitrides are VN, TiN, Mo_2N, NiN, FeN, etc [1–3].

The metal nitrides are a type of interstitial compound comprising parent metals and nitrogen atoms, where the nitrogen atoms are combined into the interstitial position. The structure of the parent metal could be altered by the intercalation of the nitrogen atoms. First, the metal nitrides offer higher power density due to their superior electrical conductivity compared to the majority of metal oxides. Second, the high energy density of the metal nitrides is due to the fact that their specific capacities are far higher than for carbon-based materials and also comparable to most of the metal oxides. Furthermore, in comparison to metal oxides, they have a higher mechanical stability. They are incredibly promising as high-performance electrodes for energy storage devices due to these characteristics [4, 5].

8.2 TMNs for supercapacitors

The cumulative effects of environmental pollution and the scarcity of fossil fuels have significantly encouraged the exploration of renewable energy sources (wind, solar, and geothermal energy). These sources cannot be used directly in electrical power networks since they require energy input from high power transistors. In this regard, supercapacitors could fulfil the role of providing transient energy due to their suitable features. Based on the storage mechanism, supercapacitors are characterized into two types namely, pseudocapacitors (PCs) and electrical double layer capacitors (EDLCs). Usually, EDLCs provide high power energy density through non-faradaic processes and pseudocapacitors offer high energy density through faradaic processes [6, 7]. Recently, great efforts have been devoted to enhancing the electrochemical behaviour of electrode materials. In general, transition metal oxides, carbon materials, and conducting polymers have been used extensively as the active electrodes of supercapacitors. The carbon related materials (graphene oxide, reduced graphene oxide, and carbon nanotubes) have high electrical conductivity and electrochemical stability, but they suffer from a low specific capacitance which makes them interesting for use in high energy density supercapacitors. Conversely, the conducting polymers and metal oxides can deliver significantly higher specific capacitances through the faradaic mechanism. However, their kinetic irreversibility and poor conductivity prevent their real-life application. With regard to their use as pseudocapacitive electrodes for SCs, TMNs have drawn a lot of attention owing to their outstanding electrical conductivity, good electrochemical activity, high capacitance, and minimal environmental impact. However, the application of metal nitrides still requires some work owing to their instability in aqueous electrolytes during cycling which leads to fast capacitance fading [8, 9].

Kumar *et al* reported nickel molybdenum nitride (Ni_2Mo_3N) by a hydrothermal method and subsequent NH_3 treatment. As a supercapacitor electrode, Ni_2Mo_3N delivered a specific capacity of 264 C g^{-1} in a half cell configuration and a specific capacity of 157 C g^{-1} in a full cell configuration with activated carbon (AC) as the anode in 6M KOH. The asymmetric cell exhibited 96% of cycling stability at a high current density of 5 A g^{-1} after 3000 cycles. The device showed a maximum energy density of 34.9 Wh kg^{-1} at a power density of 800 W kg^{-1} [10]. Gao *et al* prepared Co_3N by nitridation. It demonstrated notable electrochemical behaviour with a capacitance of 112 F g^{-1} at 0.5 A g^{-1} of current density, a good rate capability of 72%, and good cycling stability of 109% retention after 10000 cycles at 2 A g^{-1} current density. As an asymmetric supercapacitor, the assembled device (Co_3N//AC) achieved a better energy density of 12 Wh kg^{-1} at a power density of 205 W kg^{-1}, which showed that the Co_3N could be a promising electrode material for the supercapacitors [11].

Xu *et al* synthesized iron nitride and mesoporous carbon composites (Fe_2N@OMC) by a facile nano-casting method followed by ammonia calcination. The structural characterization results specified that tiny non-aggregated iron nitride nanoparticles were uniformly distributed on the exterior surface and in the pores of the carbon with a lined array of mesopores in a systematic arrangement. An electrochemical characterization was carried out in 6 M KOH electrolytes using Fe_2N@OMC as a negative electrode. The results indicated that the obtained Fe_2N@OMC-2 showed a specific capacitance of value of 520 F g^{-1} at a current density of 0.5 A g^{-1} and outstanding rate capability of 398 F g^{-1} at 20 A g^{-1} with 76% of the retention capacitance. After 1000 cycles, Fe_2N@OMC-2 retained 85% of the specific capacitance [12].

Zhu *et al* described the synthesis of Fe_2N on a graphene matrix for solid-state asymmetric supercapacitors sing an atomic layer deposition method through which the maximum usage is ensured with the coverage of GNS matrix and its connection with metal nitride particles. The asymmetric supercapacitors fabricated using a TiN/GNS cathode, Fe_2N/GNS anode, and polyvinyl alcohol–lithium chloride solid electrolyte achieved 58 F g^{-1} of irreversible capacitance with good cycling stability. After 20000 cycles, 98% of capacity retention was observed [13]. Sliwak *et al* reported an iron nitride-activated carbon composite by a hydrothermal route followed by ammonia treatment. The combination of mesoporous carbon with an iron nitride resulted in an outstanding specific capacitance of 507 F g^{-1} at a current density of 0.5 A g^{-1} and 72% of rate capability as a supercapacitor electrode [14]. Sammed *et al* synthesized nickel–cobalt nitride nanoparticles embedded on a carbon nano coils/Ni foam (CNCs/NF) matrix by the solvothermal reaction and consequent annealing process in ammonia atmosphere. The CNC/NF framework offered a favourable surface area for the growth of nickel–cobalt nitride nanoparticles and more electroactive spots for redox reactions and the required electronic conductivity. This hybrid composite electrode prepared at 600 °C resulted in a specific capacitance of 5235 F g^{-1} at a current density of 1 A g^{-1}, and 86% rate capability and 96% retention capacitance over 3000 cycles. The extraordinary electrochemical performance was mainly due to the synergistic effect of the NiCoN and CNC/NF matrix [15].

The metal embedded metal nitrides can further boost the electrical conductivity, stability, and morphology in appropriately controlled experimental conditions, where extra and complex experiment procedures would raise the total cost. In this regard, pure NH_3 gas has the ability to reduce the metal ions present in the metal oxides into metal at high temperatures. The metal nitrides and heterostructures of the metal–metal nitrides can be obtained by the nitridation process with controlled atmosphere and temperature. Liu *et al* demonstrated a nickel doped cobalt–cobalt nitride heterostructure by a method of simple room temperature synthesis and annealing in ammonia atmosphere at different temperatures. The sample prepared at 350 °C (Ni/Co-N-350) showed a better capacitance of 362 C g^{-1} at 2 mA cm^{-2}. Additionally, a full cell configuration with porous carbon showed a better energy density of 20 Wh kg^{-1} at a power density of 10 Kw kg^{-1}. This 2D nanoflake texture exhibited outstanding behaviour in an alkaline electrolyte, moreover, the good electrical conductivity was accomplished by heterostructure of metal–metal nitride. These findings offered key evidence on controlling the structure of metal nitrides and supported the metal doped metal nitride heterostructures as an innovative electrode material for supercapacitor applications [16].

Even though most of the TMNs have attracted copious consideration as electrodes for supercapacitors, their poor cycling stability limits their practical application. Lu *et al* used TiN to clarify the mechanism that triggering the loss of capacitance in the charge–discharge process. Notably, the authors proved that TiN can be steadied without losing its electrochemical activity with gel of PVA/KOH as the electrolyte. The electrolyte solution overcame the oxidation on the surface of the electrode. The electrochemical analysis reports showed that the TiN solid-state supercapacitors revealed outstanding stability up to 15 000 cycles and attained good energy density

(0.05 mWh cm^{-3}). The ability of effectually stabilizing metal nitride could show the innovative prospects in emerging high-performance supercapacitors [17]. Zhou *et al* prepared titanium nitride–vanadium nitride with core–shell assemblies by electrospinning method and subsequent annealing process. The hybrid of nitrides showed a capacitance of 248 F g^{-1} (2 mV s^{-1}) and rate capability of 161 F g^{-1} (50 mV s^{-1}). The good electrochemical results are mainly due to the combined effect of TiN (better rate capability) as well as VN (high specific capacitance) [18].

Hyoung *et al* proposed a structure comprised of vanadium nitride (VN) embedded carbon fibre (CF) matrix by electrospinning and carbonization treatments. In addition, excellent cycling stability (93% at 80 A g^{-1} after 10 000 cycles) and electrode flexibility have been observed. These synthesis tactics provided useful insight to increase the stability of metal nitrides through encapsulation of carbon, which further led to a smaller dissemination pathway due to the crumpled porous surfaces. Further, the authors claimed that the notable electrochemical properties were attributable to the high energy density of the VN phase and a high power density was obtained owing to the crumpled porous surfaces [19]. Robert *et al* provided a new viewpoint on the storage behaviour of VN films in an aqueous KOH electrolyte. The authors succeeded in tuning the VN properties for use as both a current collector as well as electrode material (bifunctional films) for on-chip micro-supercapacitors. These VN films (16 μm thickness) displayed both a high specific area and areal capacitance of 1.2 F cm^{-2} as well as 4700 F cm^{-3} of volumetric capacitance. It is notable that to verify the cycling stability, these micro-supercapacitors were made using a compatible microfabrication technique (CMOS) and characterized using *in situ* AFM methods. This study evidently proved the method for the production of miniaturized devices on a silicon wafer, and therefore for the effectual driving of next-generation micro-devices [20].

Glushenkov *et al* synthesized VN from vanadium pentoxide (V_2O_5) through ammonia reduction in a temperature-controlled experiment. The electrochemical test was performed in three different aqueous electrolytes (NaCl, H_2SO_4, KOH) with different concentrations. In all the electrolytes, the material showed a satisfactory rate capability but in KOH the high capacitance of 186 F g^{-1} was attained at 1 A g^{-1} of current density [21]. Guan *et al* developed a low temperature solution technique for the preparation of the transition metal nitrides, such as WN, δ-MoN, TN as well as VN nanosheets. When compared to the common high temperature synthesis process, the reaction temperature of this method was controlled up to 270 °C, which significantly reduces the energy consumption. The resultant δ-MoN nanosheets showed high crystallinity, thermal stability, good conductivity, and large surface area, with photothermal conversion efficacy and a strong LSPR effect. It is assumed that this method can be prolonged further to prepare other 2D metal nitrides through suitable modification [22].

As electrode materials for supercapacitors, enhancing the capacitance and cycling stability of the metal nitrides still requires much consideration. The metal nitride composites were considered by many investigators as an approach to instantaneously improve the capacitance and conductivity. They are combined with metal oxides, conducting polymers, and carbon materials in order to improve the cycling stability and electron transport. This is to reduce the capacity fading and cycling degradation

Table 8.1. The electrochemical performance of metal nitrides and their composites.

Materials	Electrolyte	Capacitance	Cycling stability (%)	Number of cycles	Reference
VN-NGr	6 M KOH	255 F g^{-1} at 10 mV s^{-1}	94.0	2000	[23]
N-PCNPs	2 M KOH	255 F g^{-1} at 1 A g^{-1}	66.0	1000	[24]
VNQD-CNF	6 M KOH	407 F g^{-1} at 0.5 A g^{-1}	75.3	1000	[25]
VN-GO	2 M KOH	110 F g^{-1} at 1 A g^{-1}	93.0	5000	[26]
TiN-VN	1 M KOH	170 F g^{-1} at 2 mV s^{-1}	89.0	500	[27]
TiN-VN	1 M KOH	248 F g^{-1} at 2 mV s^{-1}	88.0	500	[28]
Mo_3N_2	PVA/H_2SO_4	220 F g^{-1} at 0.4 A g^{-1}	98.3	1500	[29]
CC-CN-MoN	1 M LiOH	468 mF cm^{-2} at 5 mA cm^{-2}	90.0	10 000	[30]
Mo_2N-rGO	1 M H_2SO_4	142 mF cm^{-2} at 1 mA cm^{-2}	93.0	10 000	[31]
MoN_x-TiN	1 M LiOH	122 mF cm^{-2} at 0.3 mA cm^{-2}	93.8	1000	[32]
C-MoN_x	1 M H_2SO_4	251 F g^{-1}	85.7	10 000	[33]
Ni_3N	3 M KOH	320 mF cm^{-2} at 10 mV s^{-1}	93.7	2000	[34]
Fe_3N	6 M KOH	556 F g^{-1} at 1 A g^{-1}	73.6	5000	[35]
Nb_4N_5	1 M H_2SO_4	226 mF cm^{-2} at 0.2 mA cm^{-2}	70.9	2000	[36]
Nb_4N_5	1 M H_2SO_4	124 F g^{-1} at 0.2 mA cm^{-2}	80.0	1000	[37]
TiNbN	0.5 M H_2SO_4	59 mF cm^{-2} at 1 mA cm^{-2}	94.2	20 000	[38]

triggered by volume changes and can further increase the electrochemical performance [3]. The metal nitrides used for supercapacitor applications are listed in table 8.1.

8.3 Photocatalysis

Photocatalysis is a greener technology which facilitates the energy to activate a large number of chemical reactions which are very difficult to achieve under normal conditions, such as water splitting and CO_2 reduction. Photocatalysis plays a major role in energy generation, such as H_2 production from the splitting of water and CO_2 reduction to fuels [39–41]. It is important to emphasize that photocatalysis are also involved in environmental remediation, including dye degradation, and inhibition of water and air pollutants [42, 43]. The first report published on the photocatalysis process was by Fujishima and Honda investigating TiO_2 for the production of H_2 from water under solar light [44]. This greener technology has now emerged as a fascinating approach in the field of energy and environmental remediation. There is an escalating number of catalysts for photocatalytic processes, including metal organic frameworks [45], conjugated polymers [46], composite metal oxides [47], carbon nitrides [48–53], and metal nitrides [54].

The transition metal nitrides (TMNs) are emerging as a promising photocatalytic material due to their distinct physical and electronic properties and noble-metal-like properties [55]. The intercalation of nitrogen atoms enables the lattice expansion modification in the parent metal structure and the contraction of metal d-bands increasing the density of state (DOS) near the Fermi level compared to that of the parent metal [56]. They possess high electrical conductivity, high structural stability, and tuneable band gap properties, which are essential to enable light harvesting and energetic photoelectron carrying properties. TMNs act as effective semiconductors in photocatalytic reactions. They have the ability to reduce the overpotential and activation energies of the photocatalytic reactions by affording additional active sites and promoting the electron–hole separation at the interface between the co-catalyst and semiconductor [57]. The low cost of TMNs, and their thermal stability and environmentally friendly nature ensure the selection of TMNs as effective photocatalytic materials. The TMNs are reported as either photocatalysts or co-catalysts in various transformations [58, 59]. There are plenty of reports on TMNs (Ta_3N_5 [60], InN [61], and InGaN [62]) as semiconducting photocatalysts due to their simple chemical composition, stability, and tunable band gap. TiN [63], FexN [64], CoxN [65], Ni_3N [66], MoxN [67], and VN [68] are used effectively as co-catalysts due to the availability of active sites and the facilitation of the transportation of photoexcited electrons.

The following section provides insights into the transition metal nitride based photocatalysts, with a classification based on the semiconducting photocatalysts and co-catalysts. More specifically, the use of TMN based photocatalysts for emerging applications such as water splitting and CO_2 reduction is described in this chapter.

8.4 Transition metal nitrides in water splitting reactions

Water splitting reactions have emerged as one of the most prominent renewable methods for the production of H_2 powered by the photovoltaic process. H_2 production by the water splitting reactions generates an environmentally friendly and cost-effective method which uses abundant solar energy/light energy, which in turn is used to supply clean fuel for electric vehicles and automobiles [69–71].

The mechanism of the water splitting reactions involves the following steps. The first step is the absorption of photons to form electron–hole pairs. Figure 8.2 presents a schematic representation of the photocatalytic water splitting reactions by semiconductor materials.

Semiconductors are materials in which the valence band and conduction band are separated by a band gap of an appropriate width. When the energy of the incident light exceeds the band gap energy, the electrons and holes are generated in the conduction and valence bands, respectively. The redox reactions are carried out by photo-generated electrons and holes. In the water splitting reactions, the water molecules are reduced by the electrons to form H_2 and are oxidized by the holes to form O_2. The important aspect of these reactions is the suitable band gap width between the conduction and valence bands. The second step is the charge separation and migration of the photoelectron carriers and the third step involves the surface reactions in which the active sites and surface area are important factors for effective catalytic reactions. The co-catalysts are loaded to improve the active sites for the photocatalyst.

Nanoparticles are potential catalysts due to the large surface area and increased surface reaction sites of the nanoscale architectures. However, the band gap of the TMNs can be tuned which enables them to act as an efficient photocatalysts.

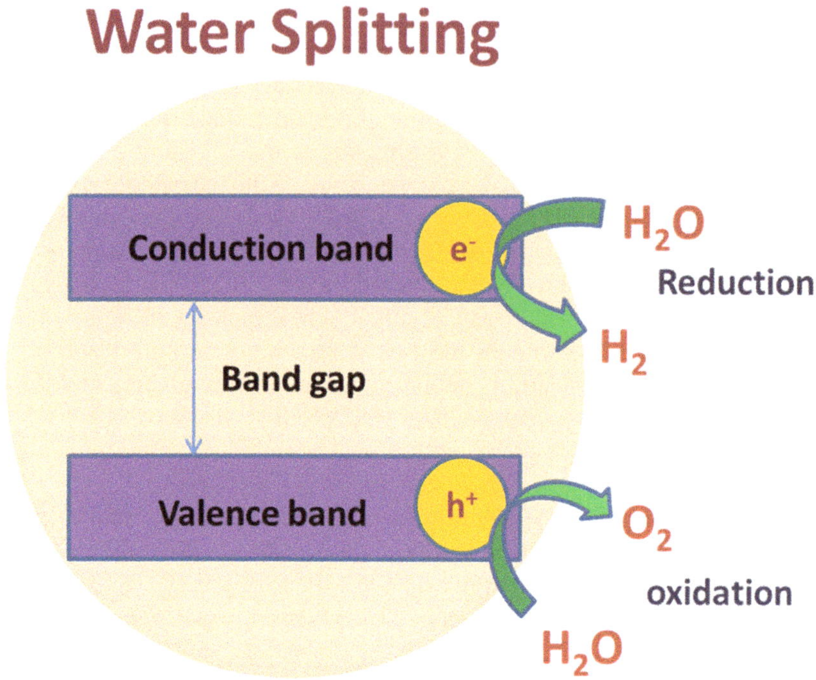

Figure 8.2. Schematic representation of the photocatalytic processes in overall water splitting.

β-Ge_3N_4 with dispersed RuO was the first successfully investigated non-oxide photocatalyst for the overall water splitting reactions and the amount of H_2 evolved was found to be 11.2 mmol over 24 h [72]. Tantalum nitride (Ta_3N_5) is a semiconductor used in photocatalysis due to its ease of synthesis and simple chemical formula with a narrow band gap of $\approx$ 2.1 eV [73], which makes it easier for photons of visible light to be absorbed and generate photoexcited electrons. Modifications in the crystal structure of the Ta-based nitrides ensures enhanced photocatalytic activity. The incorporation of a co-catalyst is one of the efficient ways to reduce the back-oxidation reaction and develop hydrogen production with better efficacy. Wang *et al* reported Ta_3N_5 supported on cubic $KTaO_3$ particles for an enhanced rate of water splitting reactions. The efficiency of the evolution of H_2 and O_2 are still better in the above-mentioned nitrides in the presence of the co-catalyst Rh/Cr_2O_3 [74]. The polymer co-catalysts PANI and Ppy for Ta_3N_5 improved the hydrogen production rates (30.2 and 28.9 mmol g^{-1} h^{-1}), and prevented the recombination of the generated electrons and holes of the Ta_3N_5 photocatalyst [75]. Wang *et al* fabricated an InGaN nanowire array coated with a Rh/CrO_x shell to improve the overall water splitting efficacy and to reduce the back reactions [76]. Chen *et al* constructed $CoPi/Co(OH)_2$-Ta_3N_5 supported on Ta foil as a conductive substrate for photoelectrochemical water splitting reactions. The $CoPi/Co(OH)_2$ shell on Ta_3N_5 facilitates the photoelectrochemical water splitting reactions faster with a reduction in self-oxidation reactions. The crystallinity of the nanomaterials also influences the catalytic performance by affecting the photocarrier generation, recombination, and migration. The mesoporous highly ordered Ta_3N_5 with a crystalline thin-wall structure showed three-fold higher activity (3.4 μmol h^{-1}) than the bulk Ta_3N_5. The small particle size, high surface area, and low density of defect sites would facilitate the increased activity [77].

Doping of metal nitrides was also done to improve the efficiency of the photocatalytic activity. Ta_3N_5 is the most explored metal nitride and its photocatalytic efficiency has been improved by the dopants Zr [78], W [79], Mg [80], Na, and K [81]. There are also quite large numbers of reports on GaN for photocatalytic applications. GaN is a bulk nitride material that support efficient water splitting reactions with a band gap of 3.4 eV [82]. The photocatalytic activity of GaN mainly depends on the crystallinity of the material and the co-catalyst $Rh_{2-y}Cr_yO_3$ employed on well-crystallized GaN resulted in the stable stoichiometric evolution of H_2 and O_2 under ultraviolet irradiation.

TMNs have also been found to act as effective co-catalysts for water splitting reactions. Chen *et al* reported that Co_3N can be used as an efficient co-catalyst on CdS nanorods for photocatalytic H_2 production under visible light irradiation. The loading of Co_3N onto CdS nanorods helps in the separation and transfer of photogenerated charge carriers, resulting in enhanced photocatalytic H_2 production. The H_2 evolution rate was found to be ~137.33 μmol mg^{-1} h^{-1} [83]. The chromium–titanium nitride ($Cr_{0.5}Ti_{0.5}N$) co-catalyst incorporated on CdS nanoparticles boosted the photo-induced charge carrier separation and migration, improving the reaction kinetics compared to TiN–CdS and CrN–CdS with a hydrogen evolution rate of 2.44 mmol g^{-1} h^{-1} [84]. Sun *et al* utilized a facile *in situ* method to grow Ni_3N on a CdS semiconductor which achieved a H_2 production rate of 888 mmol g^{-1} h^{-1}. It was observed that the efficiency was improved ten-fold compared to that of CdS nanorods [85]. Chen *et al* reported a H_2 evolution rate of 169 μmol g^{-1} h^{-1} for Ni_3N/g-C_3N_4 catalyst which was found to be higher than that of Pt modified g-C_3N_4. Ni_3N anchored on the g-C_3N_4 surface forms heterostructures which help in the separation of photogenerated electron–hole pairs. This consequently increases the active sites and photocatalytic hydrogen evolution [66]. Meng *et al* revealed that Ni_3N–Au–TiO_2 exhibited promising photocatalytic activity with a H_2 evolution rate of 87.7 μmol g^{-1} h^{-1}, which was comparatively higher than that of Ni_3N–TiO_2 [86]. There are many reports on Mo_2N as a co-catalyst for the photocatalytic H_2 evolution [87–89]. Gong *et al* reported Mo_2N anchored on ultrathin g-C_3N_4 nanosheets with a catalytic rate 148 times higher than that of the pure precursor [88]. The 2D ultrathin layer of g-C_3N_4 reduces the electron moving path, improves the charge transfer rate, and creates more active sites. The improved catalytic activity was due to the interaction between Mo_2N(111) planes and g-C_3N_4 (002) planes at the interface.

There are increasing reports on the metal nitrides as photocatalysts for effective water splitting reactions [90, 91] and they are listed in table 8.2.

8.5 Transition metal nitrides in CO_2 reduction reactions

The burning of fossil fuels leads to an increase in CO_2 emissions in the atmosphere and is the foremost contributor to greenhouse gases. The photocatalytic conversion of CO_2

Table 8.2. Photocatalytic water splitting reactions by metal nitrides.

Catalyst	Surface area ($m^2\ g^{-1}$)	Band gap (eV)	H_2 evolution rate ($mmol\ g^{-1}\ h^{-1}$)	References
TaON	58.2	2.07	381.6	[92]
Ta_3N_5	32.26	2.1	98.67	[93]
$Ta_3N_5/CdIn_2S_4$	11.86	2.06/2.08	122.6	[94]
Ta_3N_5/TiO_2	43.523	—	466.54	[95]
$TaON/Ta_2O_5/Rh$	17.4	2.04/3.86	39.41	[96]
$Ta_3N_5/Ta_2N/CoN$	21.0	2.05	75.69	[97]
Ta_3N_5/WO_3	—	2.03/2.7	2.94	[98]
Ni_3N/CdS	—	2.4	88	[99]
$Ni_3N/g\text{-}C_3N_4/Pt$	—	—	169	[100]
$Ni_3N\text{–}NiO/g\text{-}C_3N_4$	38.62	2.7	151.5×10^3	[101]
VN/CdS	13.6	—	6.24×10^3	[68]
InN/GaZnON	4.6	2.5	0.74	[61]
Nb_4N_5/TiO_2	23.4	3.2	108.13	[102]
$Fe_2N/g\text{-}C_3N_4$	81.11	2.73	88.7	[103]
GaInZnON/GaInON	102.7/4.5	2.74/2.51	603	[104]
ZnGaON	9.1	2.57	82	[105]
$Mo_2N/NC/CdS$	22.2	2.4	7.29×10^3	[87]
$Mo_2N/g\text{-}C_3N_4$	—	2.66	0.89	[88]
$Co_3N/Cd_{0.5}Zn_{0.5}S$	49.25	2.48	218.8×10^3	[106]

is a greener approach to convert CO_2 into carbon-containing fuels which helps to meet energy demands and overcome environmental hazards. The kinetics and thermodynamic factors involved in the formation of different products makes this reaction more challenging compared to water splitting reactions. The photocatalytic CO_2 reduction reactions are more important than the water splitting reactions as they can produce valuable fuels such as formic acid and alcohols. The CO_2 reduction yields products based on the number of electrons involved in the reactions. Methanol is one of the products of CO_2 hydrogenation and involves six electrons with an activated energy of −0.38 eV. Highly selective catalysts are required to avoid the formation of by-products in the CO_2 reduction reaction. The structural design, band gap, and composition tuning ability of nanocatalysts allow them to serve as promising photocatalysts for the CO_2 reduction reaction. Figure 8.3 presents a schematic representation of the photocatalytic CO_2 reduction reactions.

The metal oxides have been recognized as effective catalysts for the past few decades [107, 108]. However, the band gap of the metal oxides (>3 eV) limits light harvesting in the visible region. Instead, the metal nitrides with their narrow band gap serve as a better photocatalysts than the metal oxides. Han *et al* designed spherical TaON photocatalysts for the CH_3CHO and C_2H_5OH evolution with rates of 0.52 and 2.03 mmol g^{-1} h^{-1}, respectively, which are superior to the other commercial catalysts [109]. GaN nanowires were used as effective CO_2 reduction

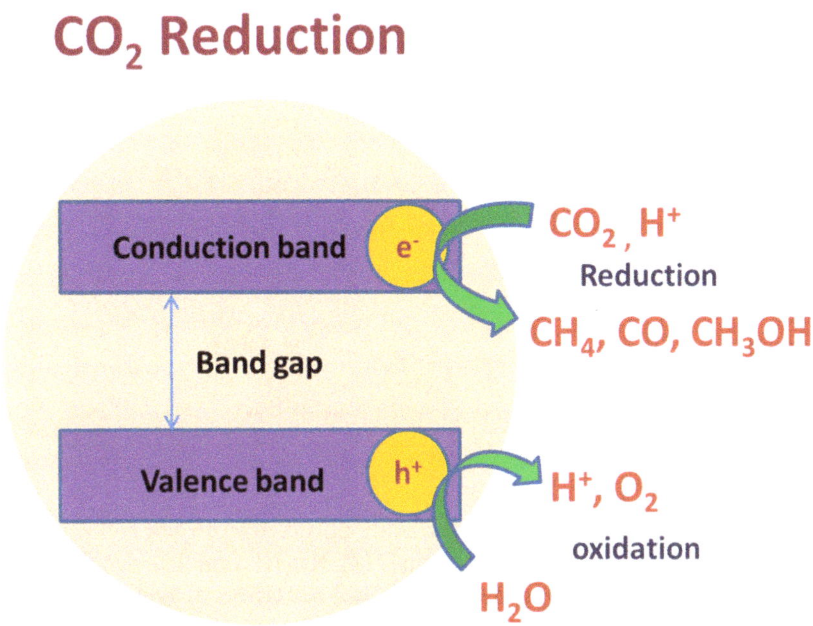

Figure 8.3. Schematic representation of the photocatalytic processes in CO_2 reduction.

photocatalysts with CH_4 and CO being the predominant products. Increased formation of CO was observed when GaN nanowires were under UV irradiation. Selective CH_4 formation was achieved by using a Rh/Cr_2O_3 core/shell co-catalyst, with an average rate of ~3.5 μmol g^{-1} h^{-1} in 24 h [110]. The same research group synthesized GaInN nanowires and reported the reduction of CO_2 to CO, CH_4, and CH_3OH with production rates of 5, 0.86, and 0.5 mmol g^{-1} h^{-1}, respectively [111]. Di *et al* demonstrated that the Co_2N co-catalyst effectively boosted the photocatalytic activity of BiOBr for the CO_2 reduction. The unique surface configuration of Co_2N generates a strong electronic interaction with BiOBr, which drives the electron transfer from the semiconducting photocatalyst and thus electrons can cross the interface and occupy the surface of the co-catalyst. The CO_2 reduction occurs on the surface of the co-catalyst (Co_2N). The effective charge separation, the surface morphology of Co_2N, and optimized Co_2N/BiOBr resulted in pronounced activity with a CO generation rate of 67.8 μmol g^{-1} h^{-1} [112].

Doping will have an influence in CO_2 reduction reactions similar to that of water splitting reactions. The synthesis of Ta_3N_5 and V-Ta_3N_5 for CO_2 conversion to generate valuable fuels, including CH_4, CO, H_2, and O_2, under visible light was reported by Dieu *et al* [113]. Vanadium (V) dopant on the Ta_3N_5 lattice (V-Ta_3N_5) reduces the band gap and increases the photocarrier separation which in turn enhances the photocatalytic efficiency of V-Ta_3N_5. However, a substantial increase in V dopant decreased the photocatalytic efficiency, thus optimizing the dopant level to be 2 wt%. The production rates of CH_4, CO, O_2, and H_2 by 2 wt% V-Ta_3N_5 photocatalytic CO_2 reduction under visible light are obtained to be 425, 236, 1003, and 56 μmol g^{-1} h^{-1}, respectively.

Table 8.3. The photocatalytic CO_2 reduction by metal nitrides and their composites.

Catalyst	Surface area ($m^2\ g^{-1}$)	Selective product	References
TiN	—	HCOOH	[117]
TaON	11.12	CH_3CHO	[109]
WN/WO_3	—	CH_4, CO	[118]
GaN	—	CH_4, CO	[110]
Ag/TaON/Ru	—	HCOOH	[115]
$Li_2LaTa_2O_6N$/Ag/Ru	2.4	HCOOH	[119]
Mg–Ga(In)N/Pt	—	CO, CH_4, CH_3OH	[111]
$CaTaO_2N$/Ag/Ru	—	HCOOH	[120]
ZnGaON	12.1	CH_4	[121]
ZnGeON	40.47	CH_4	[122]

The design of heterojunction and Z-scheme photocatalysts are reported to have a greater influence on the band gap tailoring [114]. Ru(II) dinuclear complex adsorbed on Ag-loaded TaON was designed for methanol oxidation by Sekizawa *et al.* It was the first visible light-driven Z-scheme for the photocatalytic reduction of CO_2. HCOOH and HCHO were formed as reduced and oxidized products of methanol, respectively [115]. Muraoka *et al* developed a semiconductor YTaON for a Z-scheme photocatalytic CO_2 reduction in combination with a binuclear Ru(II) complex. This hybrid photocatalyst selectively reduce CO_2 to HCOOH [116]. The few selective metal nitride catalysts for photoreduction of CO_2 are listed in table 8.3.

8.6 Outlook and conclusions

The use of transition metal nitrides as electrode materials for supercapacitors is discussed briefly and the advancements in the progress of these materials are summarized in this chapter. Recently, the metal nitrides have been adopted as high-performance materials for other applications as well, owing to their good electronic structure and physicochemical characteristics, and have been used widely in energy storage and conversion devices. As the electrode materials for supercapacitors, TMNs are a promising choice due to their good conductivity and electrochemical behaviour, but they still face major challenges in some aspects such as cost-effective synthesis methods with reproducibility, dissolution in aqueous electrolytes, and poor cycling stability, which need to be focused upon more.

Furthermore, this chapter summarizes recent reports on highly effective and selective metal nitride photocatalysts for the evolution of H_2 and the reduction of CO_2. Promising achievements have been reported using metal nitride photocatalysts for the water splitting and CO_2 reduction reactions with valuable products formed, which are very important for sustainable development. There are challenges yet to be explored in the synthesis of TMNs with controllable morphology using environmentally friendly approaches. The appropriate mechanisms of the photocatalytic reactions are not clear. Theoretical investigations and *in situ* surface chemical studies should be performed to establish improved TMNs photocatalytic systems. The morphology modification, optimization of size, different combinations of the metal nitrides, and incorporation of dopants will open up the possibility to improve photocatalytic reactions.

Acknowledgments

The authors KR, UR, PS, and PA acknowledge Fondecyt postdoctoral projects (Nos.: 3200232, 3220357and 3220176) ANID, Government of Chile, Santiago, for financial assistance.

References

[1] Liu Y, Wu Q, Liu L, Manasa P, Kang L and Ran F 2020 Vanadium nitride for aqueous supercapacitors: a topic review *J. Mater. Chem.* A **8** 8218–33

[2] Zhou Y, Guo W and Li T 2019 A review on transition metal nitrides as electrode materials for supercapacitors *Ceram. Int.* **45** 21062–76

[3] Balogun M S, Qiu W, Wang W, Fang P, Lu X and Tong Y 2015 Recent advances in metal nitrides as high-performance electrode materials for energy storage devices *J. Mater. Chem.* A **3** 1364–87

[4] Arulraj A and Nishanthi S T 2020 *2D Materials-based Flexible Supercapacitors for High Energy Storage Devices* (Amsterdam: Elsevier) pp 417–36

[5] Kumar K S, Choudhary N, Jung Y and Thomas J 2018 Recent advances in two-dimensional nanomaterials for supercapacitor electrode applications *ACS Energy Lett.* **3** 482–95

[6] Krishnan S G, Arulraj A, Khalid M, Reddy M V and Jose R 2021 Energy storage in metal cobaltite electrodes: opportunities and challenges in magnesium cobalt oxide *Ren. Sustain. Energies Rev.* **141** 110798

[7] Hong N V M, Huang H, Zhou K, Lee P S, Que W, Xu J Z and Kong L B 2017 Recent progress in layered transition metal carbides and/or nitrides (MXenes) and their composites: synthesis and applications *J. Mater. Chem.* A **5** 3039–68

[8] Syam G K, Arulraj A, Jagadish P, Khalid M, Nasrollahzadeh M, Fen R, Chun-Chen Y and Hegde G 2023 Pore size matters! a critical review on the supercapacitive charge storage enhancement of biocarbonaceous materials *Crit. Rev. Solid State Mater. Sci.* **48** 1–56

[9] Wang H, Li J, Li K, Lin Y, Chen J, Gao L and Lee J M 2021 Transition metal nitrides for electrochemical energy applications *Chem. Soc. Rev.* **50** 1354–90

[10] Kumar R, Bhuvana T and Sharma A 2020 Ammonolysis synthesis of nickel molybdenum nitride nanostructures for high-performance asymmetric supercapacitors *New J. Chem.* **44** 14067–74

[11] Gao J F, Zhang W B, Zhao Z Y and Kong L B 2018 Solid-phase synthesis and electro chemical pseudocapacitance of nitrogen-atom interstitial compound Co_3N sustain *Energy Fuels* **2** 1178–88

[12] Xu L, Sun L, Feng J, Qi L, Muhammad I, Maher J and Song W 2017 Nanocasting synthesis of an iron nitride-ordered mesopore carbon composite as a novel electrode material for supercapacitors *RSC Adv.* **7** 44619–25

[13] Zhu C, Yang P, Chao D, Wang X, Zhang X, Chen S and Fan H J 2015 All metal nitrides solid-state asymmetric supercapacitors *Adv. Mater.* **27** 4566–71

[14] Śliwak A, Moyseowicz A and Gryglewicz G 2017 Hydrothermal assisted synthesis of an iron nitride-carbon composite as a novel electrode material for supercapacitors *J. Mater. Chem.*A **5** 5680–4

[15] Abdul Sammed K, Pan L, Farid A, Javid M, Yang S and Asif M 2021 Carbon nanocoil-supported three-dimensional structure of nickel–cobalt nitrides as the electrode material for supercapacitors *ACS Appl. Energy Mater.* **4** 6678–87

[16] Liu X, Zang W, Guan C, Zhang L, Qian Y, Elshahawy A M and Wang J 2018 Ni-doped cobalt–cobalt nitride heterostructure arrays for high-power supercapacitors *ACS Energy Lett.* **3** 2462–9

[17] Lu X, Wang G, Zhai T, Yu M, Xie S, Ling Y, Liang C, Tong Y and Li Y 2012 Stabilized TiN nanowire arrays for high-performance and flexible supercapacitors *Nano Lett.* **12** 5376–81

[18] Zhou X, Shang C, Gu L, Dong S, Chen X, Han P and Cui G 2011 Mesoporous coaxial titanium nitride–vanadium nitride fibers of core–shell structures for high-performance supercapacitors *ACS Appl. Mater. Interfaces* **3** 3058–63

[19] An G H, Lee D Y and Ahn H J 2017 Vanadium nitride encapsulated carbon fibre networks with furrowed porous surfaces for ultrafast asymmetric supercapacitors with robust cycle life *J. Mater. Chem.*A **5** 19714–20

[20] Robert K, Stiévenard D, Deresmes D, Douard C, Iadecola A, Troadec D and Lethien C 2020 Novel insights into the charge storage mechanism in pseudocapacitive vanadium nitride thick films for high-performance on-chip micro-supercapacitors *Energy Environ. Sci.* **13** 949–57

[21] Glushenkov A M, Hulicova-Jurcakova D, Llewellyn D, Lu G Q and Chen Y 2009 Structure and capacitive properties of porous nanocrystalline VN prepared by temperature-programmed ammonia reduction of V_2O_5 *Chem. Mater.* **22** 914–21

[22] Guan H, Yi W, Li T, Li Y, Li J, Bai H and Xi G 2020 Low temperature synthesis of plasmonic molybdenum nitride nanosheets for surface enhanced Raman scattering *Nat. Commun.* **11** 3889

[23] Liu H H, Zhang H L, Xu H B, Lou T P, Sui Z T and Zhang Y 2018 *In situ* self-sacrificed template synthesis of vanadium nitride/nitrogen-doped graphene nanocomposites for electrochemical capacitors *Nanoscale* **10** 5246–53

[24] Yang Y, Shen K, Liu Y, Tan Y, Zhao X, Wu J and Ran F 2017 Novel hybrid nanoparticles of vanadium nitride/porous carbon as an anode material for symmetrical supercapacitor *Nanomicro Lett.* **9** 6

[25] Wu Y and Ran F 2017 Vanadium nitride quantum dot/nitrigen-doped microporous carbon nanofibers electrode for high-performance supercapacitors *J. Power Sources* **344** 1–10

[26] He T, Wang Z, Li X, Tan Y, Liu Y, Kong L, Kang L, Chen C and Ran F 2019 Intercalation struture of vanadium nitride nanoparticles growing on graphene surface toward high negative active material for supercapacitor utilizaton *J. Alloys Compd.* **781** 1054–8

[27] Dong S *et al* 2011 TiN/VN composites with core/shell structure for supercapacitors *Mater. Res. Bull.* **46** 835–9

[28] Zhou X *et al* 2011 Mesoporous coaxial titanium nitride-vanadium nitride fibers of core–shell structures for high-performance supercapacitors *ACS Appl. Mater. Interfaces* **3** 3058–63

[29] Xiong Z, Yang J, Gao Z, Yang Q and Shi D 2020 Orthorhombic Mo_3N_2 nanobelts with improved electrochemical properties as electrode material for supercapacitors *Results Phys.* **16** 102941

[30] Lv F, Ma H, Shen L, Jiang Y, Sun T, Ma J, Geng X, Kiran A and Zhu N 2021 Wearable helical molybdenum nitride supercapacitors for self-powered healthcare smartsensors *ACS Appl. Mater. Interfaces* **13** 29780–7

[31] Ma G *et al* 2015 Multilayered paper-like electrodes composed of alternating stacked mesoporous Mo_2N nanobelts and reduced graphene oxide for flexible all-solidstate supercapacitor *J. Mater. Chem.* A **3** 14617–24

[32] Xie Y and Tian F 2017 Capacitive performance of molybdenum nitride/titanium nitride nanotube array for supercapacitor *Mater. Sci. Eng.* **215** 64–70

[33] Tan Y, Meng L, Wang Y, Dong W, Kong L, Kang L and Ran F 2018 Negative electrode materials of molybdenum nitride/N-doped carbon nano-fiber via electrospinning method for high-performance supercapacitors *Electrochim. Acta* **277** 41–9

[34] Prasad S, Durai G, Devaraj D, AlSalhi M S, Theerthagiri J, Arunachalam P, Gurulakshmi M, Raghavender M and Kuppusami P 2018 3D nanorhombus nickel nitride as stable and cost-effective counter electrodes for dye-sensitized solar cells and supercapacitor applications *RSC Adv.* **8** 8828–35

[35] Kumaresan L, Shanmugavelayutham G and Saravanan P 2022 Single-phase ferromagnetic iron nitride (ε-Fe_3N) nanoparticles synthesized by thermal plasma method for oxygen evolution and supercapacitor applications *Appl. Phys.* A **128** 1073

[36] Cui H, Zhu G, Liu X, Liu F, Xie Y, Yang C, Lin T, Gu H and Huang F 2015 Niobium nitride Nb_4N_5 as a new high-performance electrode material for supercapacitors *Adv. Sci.* **2** 1500126

[37] Gao B, Xiao X, Su J, Zhang X, Peng X, Fu J and Chu P K 2016 Synthesis of mesoporous niobium nitride nanobelt arrays and their capacitive properties *Appl. Surf. Sci.* **383** 57–63

[38] Wei B, Ming F, Liang H, Qi Z, Hu W and Wang Z 2021 All nitride asymmetric super capacitors of niobium titanium nitride–vanadium nitride *J. Power Sources* **481** 228842

[39] Zou Z, Ye J, Sayama K and Arakawa H 2001 Direct splitting of water under visible light irradiation with an oxide semiconductor photocatalyst *Nature* **414** 625–7

[40] Tu W, Zhou Y and Zou Z 2014 Photocatalytic conversion of CO_2 into renewable hydrocarbon fuels: state-of-the-art accomplishment, challenges, and prospects *Adv. Mater.* **26** 4607–26

[41] Li P, Zhou Y, Zhao Z, Xu Q, Wang X, Xiao M and Zou Z 2015 Hexahedron prism-anchored octahedronal CeO_2: crystal facet-based homojunction promoting efficient solar fuel synthesis *J. Am. Chem. Soc.* **137** 9547–50

[42] Shen L, Huang L, Liang S, Liang R, Qin N and Wu L 2014 Electrostatically derived self-assembly of NH_2-mediated zirconium MOFs with graphene for photocatalytic reduction of Cr(VI) *RSC Adv.* **4** 2546–9

[43] Chen H, Nanayakkara C E and Grassian V H 2012 Titanium dioxide photocatalysis in atmospheric chemistry *Chem. Rev.* **112** 5919–48

[44] Fujishima A and Honda K 1972 Electrochemical photolysis of water at a semiconductor electrode *Nature* **238** 37–8

[45] Liu S, Zhang C, Sun Y, Chen Q, He L, Zhang K, Zhang J, Liu B and Chen L F 2020 Design of metal–organic framework-based photocatalysts for hydrogen generation *Coord. Chem. Rev.* **413** 213266–83

[46] Woods D J *et al* 2020 Side-chain tuning in conjugated polymer photocatalysts for improved hydrogen production from water *Energy Environ. Sci.* **13** 1843–55

[47] El-Maghrabi H H, Nada A A, Bekheet M F, Roualdes S, Riedel W, Iatsunskyi I, Coy E, Gurlo A and Bechelany M 2021 Coaxial nanofibers of nickel/gadolinium oxide/nickel oxide as highly effective electrocatalysts for hydrogen evolution reaction *J. Colloid Interface Sci.* **587** 457–66

[48] Hao Q, Jia G H, Wei W, Vinu A, Wang Y, Arandiyan H and Ni B J 2020 Graphitic carbon nitride with different dimensionalities for energy and environmental applications *Nano Res.* **13** 18–37

[49] Talapaneni S N, Singh G, Kim I Y, AlBahily K, Al-Muhtase A H, Karakoti A S, Tavakkoli E and Vinu A 2019 Nanostructured carbon nitrides for CO_2 capture and conversion *Adv. Mater.* **32** 1904635–156

[50] Cha W, Kim I Y, Lee J M, Kim S, Ramadass K, Gopalakrishnan K, Premkumar S, Umapathy S and Vinu A 2019 Sulfur-doped mesoporous carbon nitride with an ordered porous structure for sodium-ion batteries *ACS Appl. Mater. Interfaces* **11** 27192–9

[51] Kim I Y, Kim S, Jin X Y, Premkumar S, Chandra G, Lee N S, Mane G P, Hwang S J, Umapathy S and Vinu A 2018 Ordered mesoporous C_3N_5 with a combined triazole and triazine framework and its graphene hybrids for the oxygen reduction reaction (ORR) *Angew. Chem. Int. Ed.* **57** 17135–40

[52] Benzigar M R, Talapaneni S N, Joseph S, Ramadass K, Singh G, Scaranto J, Ravon U, Al-Bahily K and Vinu A 2018 Recent advances in functionalized micro and mesoporous carbon materials: synthesis and applications *Chem. Soc. Rev.* **47** 2680–721

[53] Park D H, Lakhi K S, Ramadass K, Kim M K, Talapaneni S N, Joseph S, Ravon U, Al-Bahily K and Vinu A 2017 Energy efficient synthesis of ordered mesoporous carbon nitrides with a high nitrogen content and enhanced CO_2 capture capacity *Chem. Eur. J.* **23** 10753–7

[54] Han N, Liu P, Jiang J, Ai L, Shao Z and Liu S 2018 Recent advances in nanostructured metal nitrides for water splitting *J. Mater. Chem.* A **6** 19912–33

[55] Urbankowski P, Anasori B, Hantanasirisakul K, Yang L, Zhang L, Haines B, May S J, Billinge S J L and Gogotsi Y 2017 2D molybdenum and vanadium nitrides synthesized by ammoniation of 2D transition metal carbides (MXenes) *Nanoscale* **9** 17722–30

[56] Sullivan M M, Chen C J and Bhan A 2016 Catalytic deoxygenation on transition metal carbide catalysts *Catal. Sci. Technol.* **6** 602–16

[57] Ran J, Zhang J, Yu J, Jaroniec M and Qiao S Z 2014 Earth-abundant cocatalysts for semiconductor-based photocatalytic water splitting *Chem. Soc. Rev.* **43** 7787–812

[58] Xiao Y, Hwang J Y and Sun Y K 2016 Transition metal carbide-based materials: synthesis and applications in electrochemical energy storage *J. Mater. Chem.*A **4** 10379–93

[59] Qi W, Zhou Y, Liu S, Liu H, Hui L S, Turak A, Wang J and Yang M 2020 Oxidized impurity in transition metal nitride for improving the hydrogen evolution efficiency of transition metal nitride-based catalyst *Appl. Mater. Today* **18** 100476

[60] He Y, Ma P, Zhu S, Liu M, Dong Q, Espano J, Yao X and Wang D 2017 Photo-induced performance enhancement of tantalum nitride for solar water oxidation *Joule* **1** 831–42

[61] Hou X, Jiang S, Li Y Y, Xiao J and Li Y Y 2015 Ni-doped InN/GaZnON composite catalyst for overall water splitting under visible light irradiation *Int. J. Hydrogen Energy* **40** 15448–53

[62] Ebaid M, Min J W, Zhao C, Ng T K, Idriss H and Ooi B S 2018 Water splitting to hydrogen over epitaxially grown InGaN nanowires on a metallic titanium/silicon template: reduced interfacial transfer resistance and improved stability to hydrogen *J. Mater. Chem.* A **6** 6922–30

[63] Zhou X, Zolnhofer E M, Nguyen N T, Liu N, Meyer K and Schmuki P 2015 Stable co-catalyst-free photocatalytic H_2 evolution from oxidized titanium nitride nanopowders *Angew. Chem. Int. Ed.* **54** 13385–587
[64] Cheng Z, Saad A, Adimi S, Guo H, Liu S, Thomas T and Yang M 2020 Metal organic framework-derived porous Fe_2N nanocubes by rapid-nitridation for efficient photocatalytic hydrogen evolution *Mater. Adv.* **1** 1161–7
[65] Liang L *et al* 2020 Efficient infrared light induced CO_2 reduction with nearly 100% CO selectivity enabled by metallic CoN porous atomic layers *Nano Energy* **69** 104421
[66] Chen L *et al* 2018 Noble-metal-free Ni_3N/gC_3N_4 photocatalysts with enhanced hydrogen production under visible light irradiation *Dalton Trans.* **47** 12188–96
[67] Ma B, Zhang J, Lin K, Li D, Liu Y and Yang X 2019 Improved photocatalytic hydrogen evolution of CdS using Earth-abundant cocatalyst Mo_2N with rod shape and large capacitance *ACS Sustainable Chem. Eng.* **7** 13569–75
[68] Tian L, Min S, Wang F and Zhang Z 2019 Metallic vanadium nitride as a noble-metal-free cocatalyst efficiently catalyzes photocatalytic hydrogen production with CdS nanoparticles under visible light irradiation *J. Phys. Chem.* C **123** 28640–50
[69] Praveen Kumar M, Sasikumar M, Arulraj A, Rajasudha V, Murugadoss G, Rajesh Kumar M, Gouse Peera S and Mangalaraja R V 2022 NiFe layered double hydroxide electrocatalyst prepared via an electrochemical deposition method for the oxygen evolution reaction *Energies* **12** 1470
[70] Arulraj A, Thachnatharen N, Rajkumar C and Mangalaraja R V 2022 Engineering 2D chalcogenides for energy and environmental remediation *Nanostructured Materials for Sustainable Energy and Environmental Remediation* (Bristol: IOP Publishing) pp 2–28
[71] Arulraj A, Praveen Kumar M, Alejandra T Z and Mangalaraja R V 2023 Nanoarchitectonics of layered metal chalcogenides-based ternary electrocatalyst for water splitting *Energies* **16** 1669
[72] Sato J, Saito N, Yamada Y, Maeda K, Takata T, Kondo J N, Hara M, Kobayashi H, Domen K and Inoue Y 2005 RuO_2-loaded β-Ge_3N_4 as a non-oxide photocatalyst for overall water splitting *J. Am. Chem. Soc.* **127** 4150–1
[73] Hitoki G, Takata T, Kondo J N, Hara M, Kobayashi H and Domen K 2002 (Oxy) nitrides as new photocatalysts for water splitting under visible light irradiation *Electrochemistry* **70** 463–5
[74] Wang Z, Inoue Y, Hisatomi T, Ishikawa R, Wang Q, Takata T, Chen S, Shibata N, Ikuhara Y and Domen K 2018 Overall water splitting by Ta_3N_5 nanorod single crystals grown on the edges of $KTaO_3$ particles *Nat. Catal.* **1** 756–63
[75] Dao V D *et al* 2019 Superior stability and photocatalytic activity of Ta_3N_5 sensitized/protected by conducting polymers for water splitting *J. Alloys Compd.* **775** 942–9
[76] Wang Y, Wu Y, Sun K and Mi Z 2019 A quadruple-band metal-nitride nanowire artificial photosynthesis system for high efficiency photocatalytic overall solar water splitting *Mater. Horizons* **6** 1454–62
[77] Yuliati L, Yang J-H, Wang X, Maeda K, Takata T, Antoniettic M and Domen K 2010 Highly active tantalum(V) nitride nanoparticles prepared from a mesoporous carbon nitride template for photocatalytic hydrogen evolution under visible light irradiation *J. Mater. Chem.* **20** 4295–8
[78] Wang Y, Zhu D and Xu X 2016 Zr-doped mesoporous Ta_3N_5 microspheres for efficient photocatalytic water oxidation *ACS Appl. Mater. Interfaces* **8** 35407–18

[79] Grigorescu S, Barhausen B, Wang L, Mazare A, Yoo J E, Hahn R and Schmuki P 2015 Tungsten doping of Ta_3N_5-nanotubes for band gap narrowing and enhanced photo-electrochemical water splitting efficiency *Electrochem. Commun.* **51** 85–8
[80] Chen Y, Xia H, Feng X, Liu Y, Zheng W, Ma L and Li R 2021 Synergy of porous structure and cation doping in Ta_3N_5 photoanode towards improved photoelectrochemical water oxidation *J. Energy Chem.* **52** 343–50
[81] Kado Y, Lee C Y, Lee K, Muller J, Moll M, Spiecker E and Schmuki P 2012 Enhanced water splitting activity of M-doped Ta_3N_5 (M = Na, K, Rb, Cs) *Chem. Commun.* **48** 8685–7
[82] Maeda K, Teramura K, Saito N, Inoue Y and Domen K 2007 Photocatalytic overall water splitting on gallium nitride powder *Bull. Chem. Soc. Jpn.* **80** 1004–10
[83] Chen H, Jiang D, Sun Z, Irfan R M, Zhang L and Du P 2017 Cobalt nitride as an efficient cocatalyst on CdS nanorods for enhanced photocatalytic hydrogen production in water *Catal. Sci. Technol.* **7** 1515–22
[84] Meng X, Qi W, Kuang W, Adimi S, Guo H, Thomas T, Liu S, Wang Z and Yang M 2020 Chromium-titanium nitride as efficient co-catalyst for photocatalytic hydrogen production *J. Mater. Chem.*A **8** 15774–81
[85] Sun Z, Chen H, Zhang L, Lu D and Du P 2016 Enhanced photocatalytic H_2 production on cadmium sulfide photocatalysts using nickel nitride as a novel cocatalyst *J. Mater. Chem.*A **4** 13289–95
[86] Meng X, Kuang W, Qi W, Cheng Z, Thomas T, Liu S, Yang C and Yang M 2020 Ultra-low loading of Au clusters on nickel nitride efficiently boosts photocatalytic hydrogen production with titanium dioxide *Chem. Cat. Chem.* **12** 2752–9
[87] Mu R, Ao Y, Wu T, Wang C and Wang P 2020 Synergistic effect of molybdenum nitride nanoparticles and nitrogen-doped carbon on enhanced photocatalytic hydrogen evolution performance of CdS nanorods *J. Alloys Compd.* **812** 151990
[88] Gong S, Jiang Z J, Shi P, Fan J, Xu Q and Min Y 2018 Noble-metal-free heterostructure for efficient hydrogen evolution in visible region: molybdenum nitride/ultrathin graphitic carbon nitride *Appl. Catal.*B **238** 318–27
[89] Ma B, Liu Y, Li J, Lin K, Liu W and Zhan H 2016 Mo_2N: an efficient non-noble metal cocatalyst on CdS for enhanced photocatalytic H_2 evolution under visible light irradiation *Int. J. Hydrogen Energy* **41** 22009–16
[90] Rao T, Cai W, Zhang H and Liao W 2021 Nanostructured metal nitrides for photocatalysts *J. Mater. Chem.*C **9** 5323–42
[91] Cheng Z, Qi W, Pang C H, Thomas T, Wu T, Liu S and Yang M 2021 Recent advances in transition metal nitride-based materials for photocatalytic applications *Adv. Funct. Mater.* **2100553**
[92] Wang Z, Hou J, Yang C, Jiao S, Huang K and Zhu H 2013 Hierarchical metastable γ-TaON hollow structures for efficient visible-light water splitting *Energy Environ. Sci.* **6** 2134–44
[93] Guo Q, Zhao J, Yang Y, Huang J, Tang Y, Zhang X, Li Z, Yu X, Shen J and Zhao J 2020 Mesocrystalline Ta_3N_5 superstructures with long-lived charges for improved visible light photocatalytic hydrogen production *J. Colloid Interface Sci.* **560** 359–68
[94] Peng Z Y, Jiang Y, Xiao Y, Xu H, Zhang W and Ni L 2019 $CdIn_2S_4$ surface-decorated Ta_3N_5 core–shell heterostructure for improved spatial charge transfer: *in-situ* growth, synergistic effect and efficient dual-functional photocatalytic performance *Appl. Surf. Sci.* **487** 1084–95

[95] Jiang Y, Jing X, Zhu K, Peng Z, Zhang J, Liu Y, Zhang W, Ni L and Liu Z 2018 Ta_3N_5 nanoparticles/TiO_2 hollow sphere (0D/3D) heterojunction: facile synthesis and enhanced photocatalytic activities of levofloxacin degradation and H_2 evolution *Dalton Trans.* **47** 13113–25
[96] Jiang H, Zhang W, Zang S and Zhang W 2019 Rh nanospheres anchored TaON@ Ta_2O_5 nanophotocatalyst for efficient hydrogen evolution from photocatalytic water splitting under visible light irradiation *Int. J. Hydrogen Energy* **44** 24218–27
[97] Jiang H, Li X, Zang S and Zhang W 2021 Mixed cobalt-nitrides Co_xN and Ta_2N bifunction-modified Ta_3N_5 nanosheets for enhanced photocatalytic water-splitting into hydrogen *J. Alloys Compd.* **854** 155328
[98] Liu Y, Mishra M, Chen Y H and Perng T P 2018 On the processes of photo-induced hydrogen generation from methanol solution by Ta_3N_5 and WO_3 *Int. J. Hydrogen Energy* **43** 23255–61
[99] Sun Z, Chen H, Zhang L, Lu D and Du P 2016 Enhanced photocatalytic H_2 production on cadmium sulfide photocatalysts using nickel nitride as a novel cocatalys *J. Mater. Chem.* A **4** 13289–95
[100] Chen L, Huang H, Zheng Y, Sun Z, Zhao Y, Francis P S and Wang X 2018 Noble-metal-free Ni_3N/gC_3N_4 photocatalysts with enhanced hydrogen production under visible light irradiation *Dalton Trans.* **47** 12188–96
[101] Qi W, Zhou Y, Liu S, Liu H, Hui L S, Turak A, Wang J and Yang M 2020 Oxidized impurity in transition metal nitride for improving the hydrogen evolution efficiency of transition metal nitride-based catalyst *Appl. Mater. Today* **18** 100476
[102] Ji Y, Yang R, Wang L, Song G, Wang A, Lv Y, Gao M, Zhang J and Yu X 2020 Visible light active and noble metal free Nb_4N_5/TiO_2 nanobelt surface heterostructure for plasmonic enhanced solar water splitting *Chem. Eng. J.* **402** 126226
[103] Qi W, Liu S, Li F, Jiang H, Cheng Z, Zhao S and Yang M 2019 Prussian blue derived Fe_2N for efficiently improving the photocatalytic hydrogen evolution activity of g-C_3N_4 nanosheets *Catal. Sci. Technol.* **9** 2571–7
[104] Fu W, Guan X, Huang Z, Liu M and Guo L 2019 Efficient photocatalytic overall water splitting over a core–shell GaInZnON@GaInON homojunction *Appl. Catal.* B **255** 117741
[105] Yang J H, Pei Y R, Kim S J, Choi G, Vinu A and Choy J H 2018 Highly enhanced photocatalytic water-splitting activity of gallium zinc oxynitride derived from flux-assisted Zn/Ga layered double hydroxides *Ind. Eng. Chem. Res.* **57** 16264–71
[106] Jin Z, Wei T, Li L, Li F, Tao R and Xu L 2019 Loading Co_3N nanoparticles as efficient cocatalysts over $Zn_{0.5}Cd_{0.5}S$ for enhanced H_2 evolution under visible light *Dalton Trans.* **48** 2676–82
[107] Yu J, Low J, Xiao W, Zhou P and Jaroniec M 2014 Enhanced photocatalytic CO_2-reduction activity of anatase TiO_2 by coexposed {0 0 1} and {1 0 1} facets *J. Am. Chem. Soc.* **136** 8839–42
[108] Wu Y A *et al* 2019 Facet-dependent active sites of a single Cu_2O particle photocatalyst for CO_2 reduction to methanol *Nat. Energy* **4** 957–68
[109] Han Q, Zhou Y, Tang L, Li P, Tu W, Li L, Li H and Zou Z 2016 Synthesis of single-crystalline, porous TaON microspheres toward visible-light photocatalytic conversion of CO_2 into liquid hydrocarbon fuels *RSC Adv.* **6** 90792–6
[110] Alotaibi B, Fan S, Wang D, Ye J and Mi J 2015 Wafer-level artificial photosynthesis for CO_2 reduction into CH_4 and CO using GaN Nanowires *ACS Catal.* **5** 5342–8

[111] Alotaibi B *et al* 2016 Photochemical carbon dioxide reduction on Mg-doped Ga(In)N nanowire arrays under visible light irradiation *ACS Energy Lett.* **1** 246–52
[112] Di J *et al* 2021 Cobalt nitride as a novel cocatalyst to boost photocatalytic CO_2 reduction *Nano Energy* **79** 105429
[113] Nguyen T D C, Nguyen T P L C, Mai H T T, Dao V-D, Nguyen M P and Nguyen V N 2017 Novel photocatalytic conversion of CO_2 by vanadium-doped tantalum nitride for valuable solar fuel production *J. Catal.* **352** 67–74
[114] Low J, Yu J, Jaroniec M, Wageh S and Al-Ghamdi A A 2017 *Heterojunction Photocatalysts* (Weinheim: Wiley) vol 29
[115] Sekizawa K, Maeda K, Domen K, Koike K and Ishitani O 2013 Artificial Z-scheme constructed with a supramolecular metal complex and semiconductor for the photocatalytic reduction of CO_2 *J. Am. Chem. Soc.* **135** 4596–9
[116] Muraoka K, Kumagai H, Eguchi M, Ishitani O and Maeda K 2016 A Z-scheme photocatalyst constructed with an yttrium–tantalum oxynitride and a binuclear Ru(II) complex for visible-light CO_2 reduction *Chem. Commun.* **52** 7886–9
[117] Kuk S K *et al* 2019 Continuous 3D titanium nitride nanoshell structure for solar-driven unbiased biocatalytic CO_2 reduction *Adv. Energy Mater.* **9** 1900029
[118] Liu D, Xu Y, Sun M, Huang Y, Yu Y and Zhang B 2020 Photothermally assisted photocatalytic conversion of CO_2–H_2O into fuels over a WN–WO_3 Z-scheme heterostructure *J. Mater. Chem.* A **8** 1077–83
[119] Oshima T *et al* 2018 Undoped layered perovskite oxynitride $Li_2LaTa_2O_6N$ for photocatalytic CO_2 reduction with visible light *Angew. Chem.* **130** 8286–90
[120] Yoshitomi F, Sekizawa K, Maeda K and Ishitani O 2015 Selective formic acid production via CO_2 reduction with visible light using a hybrid of a perovskite tantalum oxynitride and a binuclear ruthenium(II) complex *ACS Appl. Mater. Interfaces* **7** 13092–7
[121] Zhou P, Gao H L, Yan S C and Zou Z G 2016 The Kirkendall effect towards oxynitride nanotubes with improved visible light driven conversion of CO_2 into CH_4 *Dalton Trans.* **45** 3480–5
[122] Liu Q, Xu M, Zhou B, Liu R, Tao F and Mao G 2017 Unique zinc germanium oxynitride hyperbranched nanostructures with enhanced visible-light photocatalytic activity for CO_2 reduction *Eur. J. Inorg. Chem.* **18** 2195–200

Chapter 9

Recent developments in metal nanoparticle-based electrochemical biosensors for medicinal applications

Victor Vinoth, Rubina Michael, Nalandhiran Pugazhenthiran, Sambandam Anandan and R V Mangalaraja

The unique physicochemical properties of nanoparticles are well-suited for designing novel and advanced sensing devices, in particular electrochemical biosensors. Recently, nanomaterial-based electrochemical biosensors have played an important role in the development of public health due to their rapid detection, excellent sensitivity, and selectivity in clinical diagnosis. Many types of nanoparticles, such as metal-based nanoparticles, are used to develop electrochemical biosensors, and these nanoparticles play a significant role as mediators in various analytical processes. The nanomaterials offer significant functions, including the catalysis of electrochemical reactions, immobilization of biomolecules, and enhanced electron transfer rates between biomolecules and electrode surfaces. In addition, nanomaterials enable the functional labeling of biomolecules with precisely designed signal tags and highly selective and improved sensitivity in the wider application of electrochemical devices. In this chapter we summarize the different strategies used to attach metal nanoparticles to the surface of electrodes and highlight the electrochemical detection of biomolecules.

9.1 Introduction

The development of nanomaterial-based analytical techniques has attracted much interest for various applications, particularly in the fundamental biological research, pharmaceutical analysis, health monitoring, clinical diagnosis, food safety, ecological monitoring, defense, security, etc. At this scale (often between 1 and 100 nm),

doi:10.1088/978-0-7503-5724-1ch9

nanoparticles attract much attention because they exhibit unique optical, magnetic, electrical, and other characteristics that differ from those of bulk materials [1]. Nanoparticles have captured attention due to their advanced facile synthesis techniques and functionalization, chemical stability, low intrinsic toxicity (biocompatibility), and adjustable optical, electronic (absorption, fluorescence, and conductivity), and catalytic properties [2, 3]. The unique characteristics of nanoparticles have found use in the fabrication of new sensing technology, particularly in sensors. Considering their sensitivity, adaptability, and miniaturization, these sensors might provide a novel pattern for the analytical devices used in clinical analytical tools. Numerous nanoparticles, including metal, metal oxide, semiconductor, and composite nanoparticles, have been employed extensively in biosensors in recent years. Moreover, different kinds of nanoparticles and occasionally the same kind can play a variety of roles in various biosensor systems. In particular, noble metal nanoparticles, such as gold (Au), silver (Ag), and platinum (Pt), are of tremendous interest because of their distinctive size, biocompatibility, low cytotoxicity, stability, conductivity, and catalytic and thermal properties [4–6]. The strong correlations that exist between the optoelectronic properties of the noble metal nanoparticles and their distinctive size, shape, surface contact, aggregation behavior, and biocompatibility, are of major significance [7]. The noble metal nanoparticles are now being investigated for a wide range of the technological applications in addition to their general scientific significance. The many technological applications of metal nanoparticles that have been researched so far in the development of new fields of study include drug delivery, catalysis, solar cells, chemical sensing, nano-electronics, medical diagnostic tools, and complex nano-assemblies as components [8–13]. The development of biosensors is depending increasingly on nanotechnology. The biosensor's response times, sensitivity, selectivity, compactness, and other properties can be enhanced by incorporating recent advances in nanotechnology into their design. There are several potential applications for metal nanoparticles in biosensors. For example, metal nanoparticles with better optical, magnetic, and electrical properties are now being used in biosensors to detect and amplify various signals [14].

Biosensors based on optical, piezoelectric, electrochemical, and acoustic wave metal nanoparticles are a few other examples [15–18]. The development of electrochemical sensors based on nanoparticles for biomedical applications has been investigated using several electrochemical approaches. They may be used to develop structures for transducers. Some nano- and micro-fabrication techniques are well-established and employed [19]. Two main effects, (i) the Volta effect (voltammetry) and (ii) galvanic effect (amperometry), are used in electrochemical sensors. The voltammetric techniques have proven to be quite effective in measuring blood levels, drug excretion in urine after modest dosages, and metabolites. It may be applied as a quick way to assess a molecule's potential after an electron transfer. Because they may be used to elucidate redox mechanisms and electrode processes, cyclic voltammetry (CV), and linear sweep voltammetry (LSV) are receiving a lot of interest [20]. In comparison to the time-consuming chromatographic methods, differential pulse voltammetry (DPV), stripping analysis, and square-wave

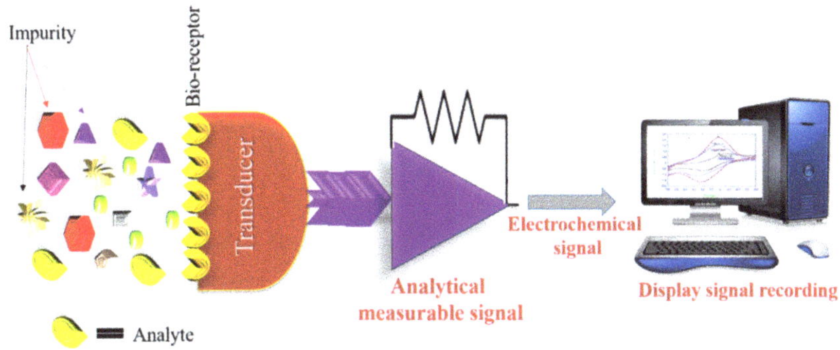

Figure 9.1. Schematic representations illustrating the basic elements of an electrochemical biosensor.

voltammetry (SWV) are more sensitive, making it possible to detect the trace amounts of analytes electroactive compounds in clinical and biological fluids with shorter analytical times [21]. Stripping voltammetry is extensively applied for the quantitation of ultra-trace analytes in complex clinical, industrial, and environmental samples [22]. The amperometry (i-t) sensors offer the ability to distinguish selectively between many electroactive species in solutions by the judicious selection of the applied potential and/or choice of the electrode materials. Nanomaterials can be used to enhance the performance of both types of sensors [23, 24]. The development of high-throughput analytical instruments for the ultrasensitive detection of trace amounts of various analytes in the environmental and clinical applications has recently received much attention. These devices use functionalized nanoparticles in combination with a variety of electrochemical methods to generate several signal amplification techniques [25] (figure 9.1).

9.2 Synthesis approaches for metal nanoparticles

Figure 9.2 illustrates the various techniques used to produce metal nanoparticles, which may be categorized into two groups: 'top-down' and 'bottom-up' techniques [26–30]. The key distinction between the two methods is the raw material used to produce the nanoparticles: in the top-down methods, the starting material is the bulk substance, which is reduced to nanoparticle size by a variety of physical, chemical, and mechanical processes; in bottom-up methods, atoms or molecules are the initial substance [26, 27]. The bottom-up method is particularly intriguing and practical for sensor applications because it allows for the control of the size, shape, stoichiometry, and surface area [31].

9.2.1 General approaches for the production of metal nanoparticles

This section discusses the physical and chemical processes used to produce the metal nanoparticles. Typically, the required metal salts are reduced or precipitated to produce ultra-fine metal particles with a limited size distribution [32]. They may be used under moderate conditions and are quite effortless to use. The sensors may be arranged in ordered arrays on the substrate/transducer surface for applications in a

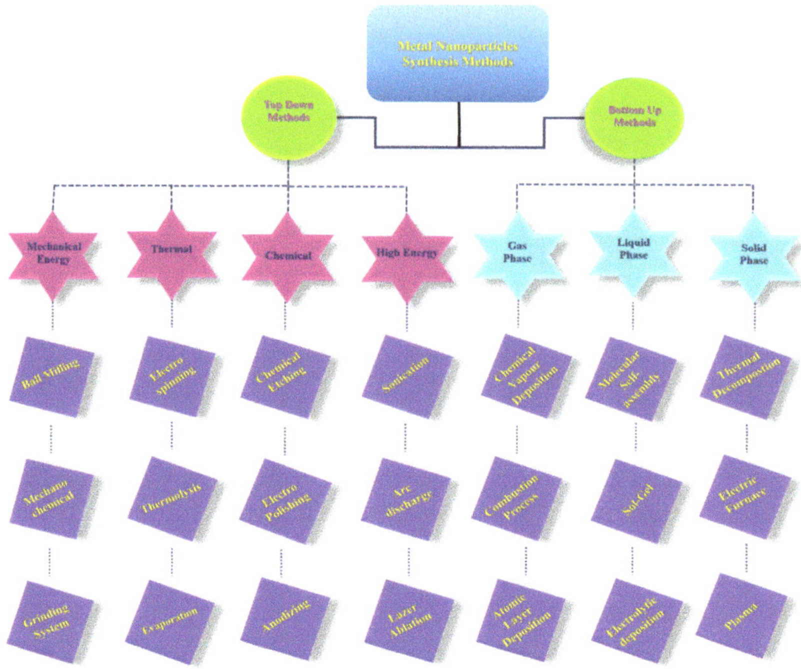

Figure 9.2. Fundamental synthesis approaches for the formation of metal nanoparticles.

sensing system, or they may be suspended in a liquid or gas phase (such as colloidal suspensions). The arrangement of such nanoparticles on transducers and other substrates plays a crucial role in the development of nanotechnology-based sensors and nano/micro devices [33, 34]. The solution-phase bottom-up process has been used to produce a variety of metal nanoparticles with various structures, including nanowires [35], nanospheres [36], nanodisks [37], nanorods [38], nanocubes [39], nanorings [40], and branching nanocrystals [41]. The recommended preparation technique must be determined by the kind of nanoparticle, its surface, and the functional medium to which it is linked.

The conventional physical methods (grinding, laser ablation, and physical vapor deposition) and the chemical methods (photochemical reduction, chemical reduction, thermal decomposition, hydrolysis, hydrothermal, sol–gel, combustion, precipitation, and co-precipitation) can be used to produce a variety of metal nanoparticles. Noble metal nanoparticles (gold, silver, platinum, and palladium) have promising uses in electronics, sensing, optics, medicine, and other disciplines [42–45]. The metallic nanoparticles can be used as catalysts, optical resonators, biomaterial labels, and a variety of other functional devices depending on the sensor type and intended application [46]. The noble metals are typically not harmful to biomolecules because of their inert nature and, as a result, they are gaining popularity in biosensing and biotechnology. Recent developments in the colloidal chemistry-based synthesis of nanoparticles have reached a peak of innovative accomplishment. The size of the nanoparticles may now be well-ordered at the atomic level, retaining more homogeneity, with a noticeably narrower size distribution. The particle sizes may now be down to cluster limits of less than one nanometer [47–49]. It has been possible to make complicated shapes, such as icosahedron, decahedron, octahedron, tetrahedron, cube, and sphere, as well as one- or two-dimensional shapes, using flexible, colloidal synthesis techniques (wire, rod, disc, plate, triangular- and hexagonal-plate, and more complex bipyramid, tetrapods, and hyper-branched structures) (table 9.1) [50–53]. Since the features of nanoparticles depend heavily on their shape and size, a synthesis procedure that regulates the growth and morphology of the nanoparticles is essential and desired [54].

Table 9.1. The construction of different shaped metal nanocrystals.

Structures	Shapes	Schematic drawings	Metals
Single crystal	Cube	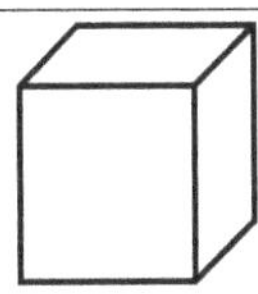	Au, Ag, Pt, Pd, Cu, and Rh
	Octahedron	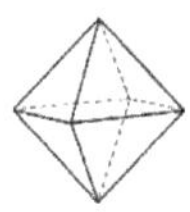	Au, Ag, Pt, and Pd
	Tetrahedron	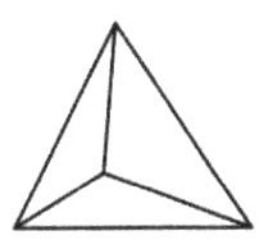	Au, Ag, Pt, and Rh
	Rectangular bar	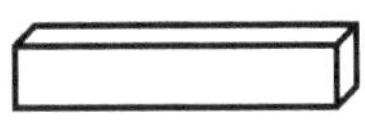	Ag, Pt, and Pd
	Octagonal rod	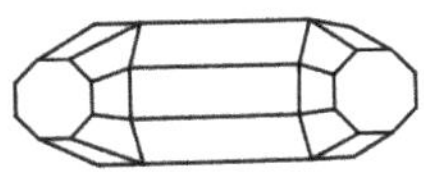	Au, Pd, Fe, Co, and Ni
Singly twinned	Bipyramid		Ag and Pd
	Beam	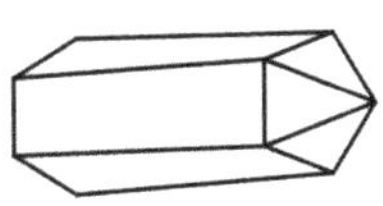	Ag
Multiply twinned	Decahedron	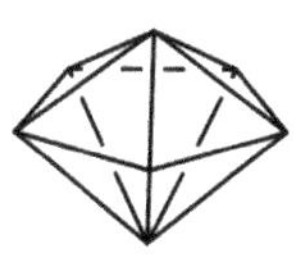	Au, Ag, and Pd

(*Continued*)

Table 9.1. (*Continued*)

Structures	Shapes	Schematic drawings	Metals
	Icosahedron	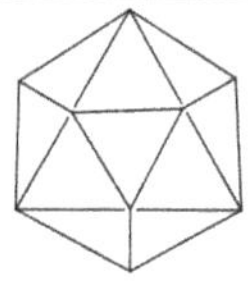	Au and Pd
	Five-fold twinned pentagonal rod	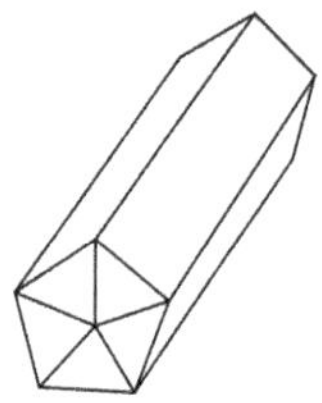	Au, Ag, Pd, and Cu
	Triangular/hexagonal plates	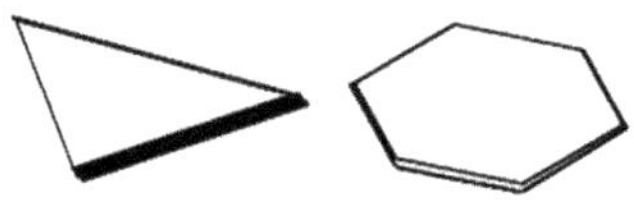	Au, Ag, Pd, Pb, Bi, Cu, Co, and Ni
	Disc		Sn and Co

The evaluation of modified electrodes is essential for the development of next-generation electroanalysis systems with higher sensitivity and selectivity since they provide fascinating properties the help in the specific recognition and/or pre-concentration of analytes. The primary goal of these modified electrode investigations has been to better understand the nature of the charge transfer and charge transport mechanisms within the thin films [55]. The homogeneous formation of a thin coating on the electrode surface and by-product desorption can achieve nanoparticle adsorption attachment. A variety of techniques, including electroplating, electropolymerization, physical absorption, chemical covalent bonding, redox polymers, and others, can keep metal nanoparticles as sensing components fixed on the working electrode surface. The local microenvironment of metal nanoparticles can be controlled by cross-linking elements in a permeable film produced by electrodes modified with multilayers of conductive nanoparticles [56].

The electroless metallic coating is the process of substrate material deposition by a chemical process or aqueous metal ions deposited by autocatalytic reduction on a substrate material without any external power [57]. Electrochemical deposition does not require an expensive experimental set-up; in recent years this technique has been used increasingly to generate the electrodes in micro-analysis devices [58, 59].

The mechanical and electrical properties of the electrodes generated by electrodeposition are appropriate for this application. However, several nanomaterials, such as gold, silver, platinum, etc [60, 61], are used to make electrodes for microanalytical devices. The drop-casting 'manual deposition' method is used widely to modify the surfaces of the electrodes using nanomaterials for sensor construction due to its simplicity and ease-of-use. Also, different methods of deposition of electrodes are reported, such as dip coating, spray coating, and spin coating [62]. However, this process uses binders with the nanoparticles (i.e. chitosan, nafion, dimethylformamide, butyl carbitol acetate, polyvinyl alcohol, glutaraldehyde, and cellulose) [63]. To do so, nanomaterials and solvents or appropriate binders in certain ratios are suspended and compose a uniform suspension or slurry using either physical mixing or sonication-based methods. The nanomaterials formed on the electrodes have been employed effectively for several sensor applications [64]. Metal nanoparticles have been utilized effectively to modify the surfaces of electrodes using the self-assembled monolayer formation of silane-active molecules and alkanethiol molecules using sulfur–metal (–S–M) and amine–metal (–NH_2–M) binding. The nanoparticle-modified electrode materials greatly increase the electrochemical response [65].

9.2.1.1 Gold nanoparticles

Over the decades, there has been significant progress in the field of biology and nanomedicine with the use of gold nanoparticles. These capabilities are based on their characteristic properties such as adjustable optical properties, large surface area, and efficiency for modification of the surface. Due to their high stability and absolute recovery in biochemical redox processes, gold nanoparticles are used as catalysts in diverse biomedical applications. In addition, gold nanoparticles are a redox-active material, which opens up the possibility of miniaturization of nanoscale sensitivity devices, with excellent opportunities for chemical and biological sensing. These noble metal nanoparticles are extensively recognized as the ultimate contributor to the development of electrochemical sensors. The use of gold nanoparticles in both macro- and nanoscale medicine is a fast-growing field with massive potential to achieve positive effects on human health. Fang *et al* reported an ultrasensitive and incubation-free electrochemical biosensor using a gold-nanocatalyst label mediating outer-sphere-response-philic and inner-sphere-response-philic species [66], with the sensitive measurement of creatine kinase-muscle brain (CK-MB). Sun *et al* reported a gold nanoparticle-based electrochemical method for the detection of protein kinase with a peptide-like inhibitor as the bioreceptor [67]. Chen *et al* described an electrochemical sandwich apta-sensor based on a signal amplification system involving two thrombin (TB) aptamers (TBA1 and TBA2), gold nanoparticles (AuNPs) as aptamer carriers, and $[Ru(NH_3)_6]^{3+}$ for signal conversion. In the presence of the target thrombin, TBA1 and TBA2 specifically bind to TB, and the TBA1–TB–TBA2 complexes cause the formation of a sandwich structure, meaning more $[Ru(NH_3)_6]^{3+}$ can be adsorbed on the negatively charged phosphate backbone of the aptamers, increasing DPV current and a limit of detection of 0.1429 fM ($S/N = 3$) for TB [68].

9.2.1.2 Silver nanoparticles

The development of the silver nanoparticle (AgNP)-based sensor platform has had a significant influence on the field of nanoscience, due to its excellent biocompatibility, high conductivity, and enhanced electrochemical signal. Over the last few decades, significant efforts have been undertaken to develop innovative analytical techniques for environmental monitoring and food safety based on AgNPs and their nanocomposites. They are also considered as one of the best classes of nanomaterials for biosensing and electrochemical sensing due to their excellent sensitivity and specificity. AgNP-based electrodes have made it possible to develop novel analytical platforms for growing environmental organic, inorganic, and biological pollutants [69]. Nantaphol *et al* modified the surfaces of glassy carbon electrodes made using electrochemical deposition, which served in the determination of cholesterol based on the coupling of the enzymatic assay, up to 3.9 to 773.4 mg dl^{-1} with a detection limit of 0.99 mg dl^{-1} [70]. Yazdanparast *et al* used a dual-aptamer to attach aptamer-labeled AgNPs to the electrode surface to establish a sensitive and selective approach for detecting human breast cancer cells and MUC1 biomarkers [71]. Similarly, AgNPs modified with Cys-Arg-Gly-Asp-Ser (CRGDS) were demonstrated by Meng *et al* to identify cancer cells on a gold electrode with a DNA tetrahedron nanostructure formed [72]. Metal–S interaction, hydrogen bonding, and certain chemical reactions are used as bridges to immobilize AgNPs on the surface of electrodes. The melamine-functionalized AgNPs (M-AgNPs) were introduced by Miao *et al* as electroactive probes for the electrochemical method of clenbuterol sensing (figure 9.3) [73]. Through the hydrogen bonding interactions, clenbuterol and melamine may interact with the melamine-modified gold electrodes and further trap M-AgNPs on the electrode hydrogen, forming a sandwich structure. This method was able to obtain a broad detection range of 10 pM to 100 nM with a low detection limit of 10 pM.

9.2.1.3 Platinum nanoparticles

Platinum (Pt) nanoparticles have drawn a lot of attention in the discipline of electrochemical sensors and biosensors for diverse biomedical and environmental

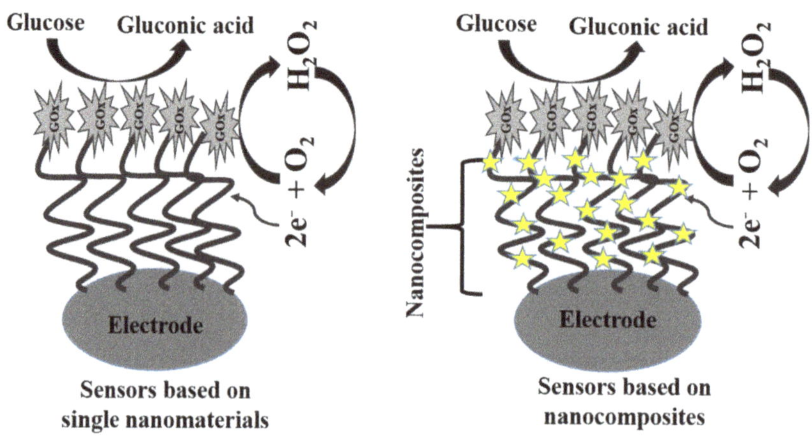

Figure 9.3. Mechanics of single nanomaterial and nanocomposite-based glucose biosensors for glucose detection.

applications because of their distinctive size- and shape-dependent electrocatalytic and electronic properties. The Pt nanoparticles can alter the biomolecules to improve their immobilization on the electrode surface as well as to target them more precisely. The performance of the sensors detecting the elements is affected significantly in terms of their detection range, sensitivity, stability, response time, linear range, selectivity, and repeatability. Due to their high specific surface area, outstanding chemical stability, superior electrical and thermal conductivity, and acceptable biocompatibility, noble metal nanoparticles have been employed as sensing components in electrochemical biosensors. The platinum nanoparticles may be produced utilizing a variety of methods with a wide range of physical and chemical properties. They exhibit volume, surface, and quantum size effects, as well as macroscopic quantum tunneling effects [74–76].

9.3 Metal NP-based biosensor applications in the medical field

The metal nanoparticle (MNP)-based biosensors are vital for the medical field. The requirements of clinical care cannot be addressed using traditional testing methods, which involve challenging procedures and long cycles. A rapid test may be created thanks to the biosensors. MNP-based biosensors can detect lactic acid, uric acid, and glutamic acid, as well several carcinogens, in bodily fluids.

9.3.1 Detection of glucose

Blood glucose levels significantly influence the diagnosis and prognosis of diabetes [77]. The food, beverage, and fermentation sectors also need to know the glucose concentration [78]. Glucose oxidase (GO_x) was crucial in the creation of glucose biosensors. Several clinical and experimental tests for glucose sensing exist, and the production of rapid, accurate, and affordable biosensors is a promising research area in numerous labs all over the world. Glucose biosensors can detect glucose either directly or by encouraging its conversion to other measurable electroactive molecules. Until now, for the determination of glucose both enzyme-based and enzyme-independent techniques have been used. MNPs have a wide range of uses in the biomedical sciences. By increasing the surface area and aiding in the electron transfer from an enzyme to the electrode, MNPs have recently found extensive use in glucose biosensors, and have improved the detection signals. Li *et al* investigated the peroxidase- and catalase-like properties of gold, palladium, platinum, and silver [79]:

$$H_2O_{2(ads)} \leftrightarrow H_{(ads)} + HO_{2(ads)},$$

where the species adsorbed on the metal is indicated by the subscript (ads).

On the other hand, a metal displays peroxidase-like activity at low pH and with H groups pre-adsorbed, decomposing H_2O_2 in a base-like based manner:

$$H_2O_{2(ads)} \leftrightarrow H_2O_{(ads)} + O_{(ads)}.$$

Additionally, it was demonstrated that the latter mechanism has a lower energy barrier at neutral pH, making it more advantageous. Shen *et al* [80] investigated the nature of the oxidase-like activity of gold, silver, platinum, and palladium.

Biosensors have typically become miniaturized as a result of the incorporation of nanomaterials. Several MNPs, including Au, Ag, Pt, and Pd, have been used in glucose biosensors as moderators, labeling agents, or immobilizers. MNPs are often used as electrode modifiers and immobilizers in glucose biosensors because of their electrocatalytic activity [81, 82]. Furthermore, the ability of MNPs to bind to a wide range of biological molecules, such as superantigens, antibodies, and nucleic acids [83–85], is advantageous for developing biosensors. In MNP-based glucose biosensors, the incorporation of composites or films and polymers (such as chitosan and mesoporous silica) have been utilized to improve the surface stability, prevent enzyme leakage, and protect against considerably extreme conditions. German *et al* recently tested and compared various glucose biosensors based on GO_x immobilized on bare carbon electrodes modified with AuNPs to biosensors based on GO_x immobilized on bare carbon electrodes [86]. PtNPs are used in a wide range of applications in nanotechnology, catalysis, and medicine [87] because of their antioxidant qualities. PtNPs can significantly increase the catalytic activity of glucose biosensors that employ hydrogen peroxide to detect the glucose [88]. Li *et al* [89] developed a biosensor based on PtNPs supported on CNTs under optimal conditions that showed a linear response in the range of 1.2×10^{-6}–2.0×10^{-3} M with a limit of detection of 0.4 μM. Many different kinds of biosensors, including glucose sensors, have had their sensitivity increased using AgNPs [90]. In glucose biosensors AgNPs can also serve as redox markers. The activity of free Ag^+ released from AgNPs is being monitored to detect glucose using AgNPs [91]. Hydrogen peroxide is produced during the enzyme–substrate interaction between β-D-glucose and GO_x, and oxidizes the AgNPs into free Ag^+ ions. The concentration of glucose is correlated with the number of free Ag^+ ions. Activated carbon (PtNi/AC) loaded with PtNi nanocomposites was developed by Savk *et al* in 2020 [92]. To develop an enzyme-free electrochemical glucose biosensor, the biosensors were modified on the glassy carbon (GC) electrode's surface using PtNi/AC. In the absence of an enzyme, the oxidation of glucose may be efficiently catalyzed by PtNi/AC. The current intensity was significantly changed with the addition of glucose. In 2 s the current density stabilized and increased steadily. The linear range was 0.025–12 mM and the detection limit was 0.052 μM with a sensitivity of 40.9 mA mM^{-1} cm^{-2}.

9.3.2 Detection of glutamate

One of the primary components of mammalian neurotransmitters is glutamate acid, and irregularities in its concentration can lead to chronic conditions including Parkinson's and Alzheimer's disease [93]. Therefore, measuring glutamate levels in the extracellular space of brain tissues is a crucial use for assessing a variety of neurological conditions, including epilepsy, stroke, traumatic brain injury, Alzheimer's disease, and Parkinson's disease. These techniques need specialized expertise in addition to being time and labor-intensive [94]. The electrochemical approach for glutamate detection is simple, economical, and provides real-time results promptly [95]. Graphene oxide (Crbxl-RGO) and Pt nanoparticles were used to develop an electrochemical L-glutamic acid biosensor. The synthesized

Crbxl-RGO was then immobilized with the composite material in the sensing area. Subsequently, treatment with 1-ethyl-3-(3-dimethylaminopropyl)-carbodiimide (EDC) and activation of the –COOH group were used to increase the stability of the immobilized glutamate oxidase. The sensor's linear range was 0.004–0.9 mM and its sensitivity was 973 ± 4 μA mM^{-1} cm^{-2}, with a detection limit of 0.1 μm, demonstrating high specificity and selectivity.

In 2020 Nguyen *et al* [96] produced current on the electrode surface by integrating Pt particles, commercially activated carbon, and glutamate oxidase to convert glutamic acid into H_2O_2. The glutamate as a consequence has a quick response time, great sensitivity, and low detection limit. It has astrocytes, which are extremely glutamate-sensitive. In particular, it has a linear range of 1–925 mm, a detection limit of 0.03 mm, and a response time of <1 s, with a sensitivity of 5.73 ± 0.078 nA μM^{-1} mm^{-2}.

9.3.3 Detection of H_2O_2

The reactive oxygen species (ROS) are crucial intracellular signaling molecules that play a role in many important cellular processes, protein synthesis, and cell death. However, they also call for particular skills [97]. Nevertheless, an excessive concentration of ROS can result in oxidative stress, which in turn fuels the development of aging, heart diseases, diabetes, Alzheimer's disease, neurological disorders, and cancer [98]. In the cellular environment, hydrogen peroxide (H_2O_2) is a typical ROS. We may comprehend H_2O_2's crucial role in cell physiology in detail through a focused quantitative investigation of the substance and an assessment of its dynamic release mechanism [97].

Due to the high oxidability of H_2O_2, the precise and real-time detection of H_2O_2 is significant in the medicinal, industrial, and military applications. For the detection of H_2O_2, an electrode surface is formed with an electrolytically active film containing Au nanoparticles (AuNPs) embedded in a layer of a metal–organic framework (UiO-66). As a result, the enhanced AuNPs/UiO-66 film model demonstrates a substantial electrochemical response for the detection of H_2O_2, with a linear range (0.2–23 mM), high sensitivity (329 μA mM^{-1} cm^{-2}), and low detection limit (0.045 μM, $S/N = 3$) [98]. Loza *et al* [99] observed that the addition of strong oxidizing species such as H_2O_2 can enhance the dissolution of AgNPs. By using the modified AgNP electrode as a sensing interface, H_2O_2 will decompose into hydroxyl radicals and superoxide anions, which will change into O_2 and H_2O. The mechanism is as follows:

$$Ag_n(\text{solid}) + 3H_2O_2 + 2H^+ \rightarrow Ag_{n-2}(\text{solid}) + 3H_2O + 2Ag^+ + O_2.$$

An electrochemical biosensor modified with AgNPs was investigated for monitoring H_2O_2 secretion in living cells by taking advantage of this feature.

9.3.4 Detection of NADH

Nicotinamide adenine dinucleotide (NAD+) serves as a cofactor in several enzymatic reactions and is a crucial substrate in a variety of biological, chemical, and medicinal procedures. Systems for dehydrogenase amperometric sensitization were developed as a result of effective bio-electrocatalysis of NADH electrochemical oxidation [100]. Similarly, by utilizing poly(L-cysteine), electrochemically reduced graphene oxide (ERGO), and glassy carbon electrodes modified with gold–silver bimetallic nanoparticles (Au–AgNPs), a new NADH sensor were developed. Due to the synergistic effects of the Au–AgNPs, polymer, and ERGO, the composite electrode demonstrated strong electrocatalytic reactivity towards NADH at a low oxidation potential (+ 0.35 V) and reduction of surface contamination. The improved sensors enabled the detection of NADH under optimal conditions with a wide linear range of 0.083 M to 1.05 mM and a low detection limit of 9.0 nM ($S/N = 3$) [101].

Jimenez *et al* [102] developed new amperometric biosensors using covalent cross-linking, and alcohol dehydrogenase (ADH) was immobilized on the electrodes after polymer membranes were electrodeposited on Pt or PtNPs/Pt electrodes. A linear concentration range of 25 mM (R = 0.9979) and 21 mM (R = 0.9984) may be tolerated by the polymer/PtNPs/Pt electrode. The electrode has a sensitivity of 68.24 and 40.21 $\mu A\ mM^{-1}\ cm^{-2}$ and a detection limit of 4.78 and 6.18 μM for PDAMS/PtNPs/Pt and PMDUS/PtNPs/Pt, respectively. Its linear range is 30 mm, and its sensitivity to ADH/PDAMS/PtNPs/Pt and ADH/PMDUS/PINPSP is 0.957 or 0.756 $\mu A\ mM^{-1}\ cm^{-2}$, respectively.

9.3.5 Detection of xanthine

A precursor to uric acid is 3,7-dihydro-purine-2,6-dione or xanthine (XA). High levels of XA in biological fluids from humans can result from hyperuricemia, xanthinuria, gout, renal failure, and other conditions [103]. Therefore, clinical diagnosis must keep track of the XA level in human bodily fluids. The oxidation of H_2O_2 or dielectric redox processes serves as the basis for the electrochemical study of xanthine, which is based on the following electrochemical reactions [104, 105]:

$$\text{xanthine} + H_2O + O_2 \xrightarrow{\text{Xanthineoxidase}} \text{uric acid} + H_2O_2$$

$$H_2O_2 \xrightarrow{\text{high potential}} O_2 + 2H^+ + 2e^-.$$

Dervicevic *et al* [105] produced a novel xanthine biosensor built on a hybrid bio-nanocomposite platform for the electrochemical detection of the adenosine-3-phosphate degradation product. To improve the electron transport and offer strong enzyme affinity, the chitosan-polypyrrole-gold nanoparticles were synthesized using an *in situ* chemical synthesis approach placed on a glassy carbon electrode surface in the design of the amperometric xanthine biosensor. The biosensor demonstrated a good linear range of 1–200 μM, a low detection limit of 0.25 μM with an average reaction time of 8 s, and other impressive characteristics.

Additionally, xanthine, biomolecules with a high and close oxidation over-potential on traditional glassy carbon electrodes, such as hypoxanthine, dopamine, uric acid, and ascorbic acid, are easily absorbed along with their oxidation products on the electrode surface, resulting in the pollution that negatively affects the analytical effect of the detection.

9.3.6 Detection of ascorbic acid

Ascorbic acid (AA) is a strong antioxidant due to its ability to donate a hydrogen atom and form the relatively stable ascorbyl free radical. Superoxide radical ions, hydrogen peroxide, the hydroxyl radical, singlet oxygen, and nitric oxide are only a few examples of the reactive oxygen and nitrogen oxide species. Because these substances have an unpaired electron, which makes them highly reactive, they harm people at the molecular level. This is because of how they interact with lipids, proteins, and nucleic acids. Ascorbate is initially converted to monodehydroascorbate and subsequently to dehydroascorbate by reactive oxygen species. The oxidized forms of ascorbate are comparatively stable and unreactive and do not harm cells, whereas the reactive oxygen species are reduced to water.

Many studies have been published in the literature in the previous few decades on the electrochemical sensing of AA. Despite the enormous volume of articles written on the subject, it is receiving continued attention due to its significance. Recently, electrodes modified with MNPs have been used for the reliable and accurate detection of AA.

Yang *et al* [106] developed Pt nanoparticle-modified nanoporous AuSn(Pt@NP-AuSn) alloy on Ni buffered flexible carbon fiber paper (CFP) for the detection of AA. As synthesized, the Pt@NP-AuSn/Ni/CFP had a hierarchical pore structure, high specific surface area, and outstanding catalytic activity. The Pt@NP-AuSn/Ni/CFP exhibited high sensitivity in detecting ascorbic acid, with a sensitivity of 0.14 $\mu A\ \mu M^{-1}\ cm^{-2}$, under a concentration ranging from 200 to 2000 μM for AA. This is due to the bi-functions of both the large surface area of nanoporous metal and macroporous carbon fiber paper facilitating the mass transfer. Zou *et al* [107] produced PdAu/graphene (RGO) nanocomposites which were made using an easy, environmentally friendly process, and PdAu/RGO/GC was utilized to detect AA electrochemically. Its good linearships were observed for the AA molecule in the range of 12.5–700 μM, and the detection limit was 12.5 μM. The modified PdAu/RGO/GC electrode works quite well when real sample analysis is performed to determine the ascorbic acid level.

9.3.7 Detection of urea

Blood urea is a low-molecular-weight waste product of tissue protein turnover and dietary protein catabolism. It has an inverse relationship with the decline in the glomerular filtration rate (GFR). Freely filtered urea is recycled between the tubule and interstitium in the renal medulla, where a variable quantity (between 30% and 70%) is reabsorbed mostly in the proximal tubule. The typical range for serum or blood urea nitrogen is 1.8–7.2 mmol l^{-1} (or 5–20 mg dl^{-1}). The blood urea concentration will also rise as a result of any cause that enhances the tubular reabsorption of urea, such as decreased effective vascular volume (i.e. impaired renal perfusion) and/or blockage of urine drainage. A significant accumulation of urea in

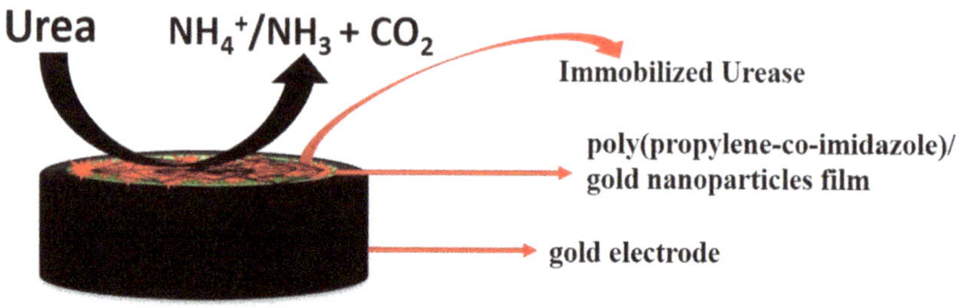

Figure 9.4. Schematic representation of the detection of urea on the working electrode surface.

the serum may lead to shock, dehydration, and renal problems. As a result, precise urea detection is essential for the treatment of many problems. Given their excellent sensitivity, specificity, and long-term performance, biosensors provide a quick option for urea detection.

Korkut *et al* [108] produced a novel amperometric urea biosensor based on poly (propylene-co-imidazole) coated with gold nanoparticles to measure urea by utilizing differential pulse voltammetry, the ammonium was electro-oxidized on the gold electrode with the help of gold nanoparticles at +0.2 V versus Ag/AgCl after being hydrolyzed by the urease adsorbed on the polymeric film to ammonium and bicarbonate ions. The biosensor offered a linear current response to urea concentration between 0.1 and 30 mM, with a relative standard deviation of 2.43% ($n = 18$), and a detection limit of 36 μM. A schematic representation of the biosensor is shown in figure 9.4.

The nanosized tin oxide was synthesized hydrothermally at 150 °C for 6 h. Then, the thin layers were made using the electrophoretic method on an electropolished aluminum substrate. Tetragonal rutile particles with a diameter of 30–50 nm are seen with partial aggregation when observed under an electron microscope. The films that were deposited have been employed for enzyme-free urea sensing at concentrations of 1–20 mM in buffer solution. To achieve a cyclic voltammogram (CV) that was dependent on both urea concentration and scan rate, the sensors performed electrochemical characterization. The observed limit of detection was 0.6 mM, and the sensitivity was reported as 18.9 μA mM^{-1} for below 5 mM and 2.31 μA mM^{-1} above 5 mM [109].

9.3.8 Detection of DNA

The most significant molecule in life, deoxyribonucleic acid (DNA), is the largest and most distinct naturally occurring molecule [110]. It encodes the genetic code and regulates the biological synthesis of enzymes and proteins through the replication and transcription of genetic code in living species and many viruses. Guanine (G), adenine (A), thymine (T), and cytosine (C) are the nucleotides that make up the sequence that encodes the genetic information. To develop new and more effective DNA-targeted medications, it is critical to comprehend the structural characteristics of DNA, the causes of various diseases, the mutation of genes, and the mechanisms of action of some anticancer and antiviral treatments. Traditional DNA detection techniques need a lot of time and work. However, the development of rapid, inexpensive, and sensitive miniaturized devices is necessary for widespread genetic testing. As a result, biosensors present a possible replacement for nucleic acid detection techniques that are faster, less expensive, and simpler. These biosensors rely mainly on immobilizing a probe made of single-stranded DNA, or double-stranded DNA oligonucleotides onto a transducer surface to hybridize with its corresponding target sequence. Although DNA is a very simple molecule, it can be exceedingly difficult to identify the complementary sequence that carries the required information and to differentiate between perfect and imperfect matches, which are very difficult challenges. A full-match DNA hybridization must have high specificity and sensitivity to be successful.

In DNA detection, electrochemical transducers have drawn a lot of attention due to their small size, real-time analysis, practically reagent-less operation, simple pretreatment techniques, low cost of manufacturing, and ease-of-use. The electrochemical DNA biosensors have been developed widely for chemical, biochemical, medicinal, agricultural, and environmental monitoring (figure 9.5) [111].

In 2021 Wang *et al* [112] developed a novel label-free electrochemical DNA biosensor that relies on electron transport mediated by gold nanoparticles (AuNPs) via a DNA monolayer-modified gold electrode surface. They found that varying sizes of AuNPs and the various forms of DNA monolayers (single-strand or triple-helix DNA) have a substantial impact on the electron transfer via the DNA-modified Au electrodes. This approach to DNA biosensing minimizes the stages involved in electrode pretreatment and avoids the difficult manufacturing procedure for DNA capture probes. Further, this approach has a detection limit of 1.47 pM and can detect the target DNA with high sensitivity and accuracy in the range of 10 pM to 100 nM.

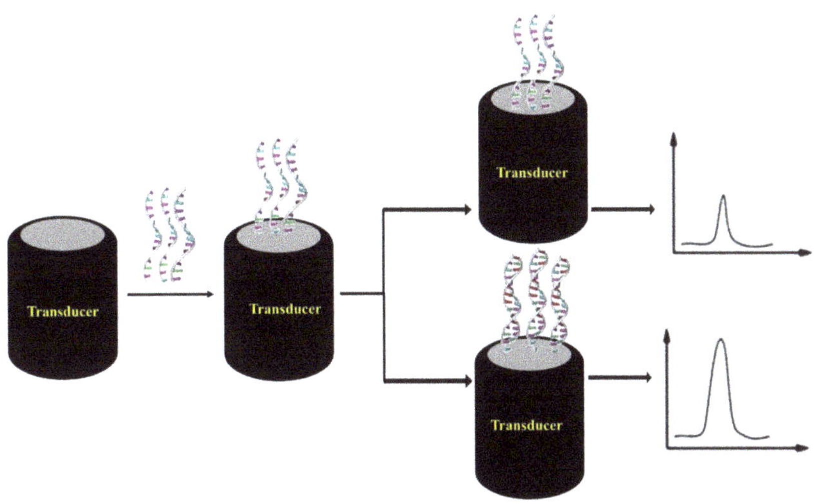

Figure 9.5. A general configuration of an electrochemical DNA biosensor.

In 2021 Chen *et al* [113] developed an electrochemical biosensor that is highly effective for the rapid determination of ctDNA in whole blood. The detection of ctDNA is based on hybridization in a network of magnetic nanoparticles with gold coatings that have been modified with DNA (DNA-Au@MNPs). Short- and long-strand DNA targets may be selectively found by this DNA-Au@MNP biosensor. For 22 nucleotide DNA targets, it has a wide dynamic range (2 aM to 20 nM) and an extremely low detection limit of 3.3 aM. A dynamic range of 200 aM to 20 nM for 101 nucleotides ctDNA target was attained with a detection limit of 5 fM. A promising way to obtain a 20 min reaction time and minimally invasive early cancer detection is through the use of this DNA-Au@MNP-based sensor.

Electrochemical biosensor-based biosensing technologies offer several benefits for environmental analysis, which include inexpensive, rapid, simple, powerful, small, sensitive, selective, and reliable instruments for the identification of hazardous pollutants. These kinds of technology may effectively monitor contaminants, improving environmental and public health protection.

9.4 Conclusion and future perspectives

In this chapter we summarized the recent developments in metal nanoparticle-based electrochemical biosensors for medical applications. Notably, metal nanoparticles have been developed as attractive candidates for the assembly of biosensors due to their superior electrical conductivity, stable structures, large specific surface area, and good biocompatibility. The benefits of metal NPs allowed for the orchestration of an extensive variety of sensors; however, new techniques should be investigated to improve the sensitivity, selectivity, and adaptability to meet the needs of practical applications.

The electrochemical methods are now emerging as a strong option in terms of sensitivity; however, sensitivity would be affected greatly by the detection conditions. By chemically modifying metal nanoparticles with physical characteristics that relate directly to their size, shape, and composition, as well as functionalizing them with molecular, polymer, biomaterial, or other nanomaterial coatings, it is possible to make various electrochemical sensors that are specifically adapted to the various electrode surfaces mentioned above. We might functionalize various nanoparticles with various modifiers based on the characteristics of particular analytes and can develop the relevant electrochemical sensors. The entire structural stability of the metal particles, and their conductive aspect, enable the development of electroanalytical and bio-electroanalytical devices with predesigned and controlled electrochemical functions. Nanoparticles can provide ideal labels for the amplified electrochemical analysis of analytes such as DNA, RNA, and other analytes because of their special target binding properties.

The covalent bonding between the target and metal NPs is now the most often employed target principle in terms of selectivity. The lack of specificity in the bonding might cause false positives or misdetections. This should be improved by using more specific bonding molecules, such as antigen and cyclodextrin, antibody, avidin or adamantane and biotin. The metal-based nanoparticle biosensors are simple to use and should be regulated in the future to increase the utilization rates in terms of portability and comfort. Accessibility may be increased by merging electrochemical sensors with regularly used electronic devices such as laptop computers, mobile phones, and tablets.

Acknowledgments

The authors V Vinoth and N Pugazhenthiran gratefully acknowledge Universidad Técnica Federico Santa Maria, Valparaíso, Chile, for financial support.

References

[1] Zhang Z Y and Lagally M G 1997 Atomic processes in the early stages of thin-film growth *Science* **276** 377–83

[2] Senthilkumar N, Arulraj A, Nandhakumar E, Ganapathy M, Vimalan M and Vetha Potheher I 2018 Green mediated synthesis of plasmonic nanoparticle (Ag) for antireflection coating in bare mono silicon solar cell *J Mater. Sci. Mater. Electron.* **29** 12744–53

[3] Shipway A N, Katz E and Willner I 2000 Nanoparticle arrays on surfaces for electronic, optical, and sensor applications *Chem. Phys. Chem.* **1** 18–52

[4] El-Sayed M A 2001 Some interesting properties of metals confined in time and nanometer space of different shapes *Accounts Chem. Res.* **34** 257–64

[5] Wang J, Boriskina S V, Wang H Y and Reinhard B M 2011 Illuminating epidermal growth factor receptor densities on filopodia through plasmon coupling *ACS Nano* **5** 6619–28

[6] Eustis S and El-Sayed M A 2006 Why gold nanoparticles are more precious than pretty gold: noble metal surface plasmon resonance and its enhancement of the radiative and nonradiative properties of nanocrystals of different shapes *Chem. Soc. Rev.* **35** 209–17

[7] Zhang Z and Lin P C 2018 Emerging applications of nanoparticles and architecture nanostructures/noble metal nanoparticles: synthesis, and biomedical implementations *Emerging Applications of Nanoparticles and Architecture Nanostructures* (Amsterdam: Elsevier) pp 177–233

[8] Dreaden E C, Alkilany A M, Huang X, Murphy C J and El-Sayed M A 2012 The golden age: gold nanoparticles for biomedicine *Chem. Soc. Rev.* **41** 2740–79

[9] Senthilkumar N, Ganapathy M, Arulraj A, Meena M, Vimalan M and Vetha Potheher I 2018 Two step synthesis of ZnO/Ag and ZnO/Au core/shell nanocomposites: structural, optical and electrical property analysis *J. Alloys Comp.* **750** 171–81

[10] Preethi S, Arulraj A, Mangalaraja R V, Ravichandran V and Subramanian N 2022 Biomass-derived carbon-based materials for microbicidal applications *Biomass-Derived Carbon Materials* ed A Pandikumar, P Rameshkumar and P Veerakumar (New York: Wiley) pp 63–92

[11] Sephra P J, Baraneedharan P and Arulraj A 2021 Nanoelectronics devices (field-effect transistors, electrochromic devices, light-emitting diodes, dielectrics, neurotransmitters) *Advances in Hybrid Conducting Polymer Technology* ed S Shahabuddin *et al* (Berlin: Springer) pp 77–100

[12] Hicks J F, Zamborini F P, Osisek A J and Murray R W 2001 The dynamics of electron self-exchange between nanoparticles *J. Am. Chem. Soc.* **123** 7048–53

[13] Elghanian R, Storhoff J J, Mucic R C, Letsinger R L and Mirkin C A 1997 Selective colorimetric detection of polynucleotides based on the distance-dependent optical properties of gold nanoparticles *Science* **277** 1078–81

[14] Farka Z, Juřík T, Kovář D, Trnková L and Skládal P 2017 Nanoparticle-based immunochemical biosensors and assays: recent advances and challenges *Chem. Rev.* **117** 9973–10042

[15] Malekzad H, Zangabad P S, Mirshekari H and Karimi M 2017 Noble metal nanoparticles in biosensors: recent studies and applications *Nanotechnol. Rev.* **6** 301–29

[16] Pramanik S, Murphy B P and Osman N A 2013 Developments of immobilized surface modified piezoelectric crystal biosensors for advanced applications *Int. J. Electrochem. Sci.* **8** 8863–92

[17] Hasanzadeh M, Shadjou N, Eskandani M, Soleymani J, Jafari F and Guardia M L 2014 Dendrimer-encapsulated and cored metal nanoparticles for electrochemical nanobiosensing *Trends Anal. Chem.* **53** 137–49

[18] Zhang Y, Yang F, Sun Z, Li Y T and Zhang G J 2017 A surface acoustic wave biosensor synergizing DNA-mediated *in situ* silver nanoparticle growth for a highly specific and signal-amplified nucleic acid assay *Analyst* **142** 3468–76

[19] Zadeh K K and Fry B 2008 Nano fabrication and patterning techniques *Nanotechnology-Enabled Sensors* (New York: Springer)vol 1 p 135

[20] Meng Y, Aldous L, Belding S R and Compton R G 2012 The formal potentials and electrode kinetics of the proton/hydrogen couple in various room temperature ionic liquids *Chem. Commun.* **48** 5572–4

[21] Erk N 2004 Voltammetric behaviour and determination of moxifloxacin in pharmaceutical products and human plasma *Anal. Bioanal. Chem.* **378** 1351–6

[22] Achterberg E P, Barriada J L and Braungardt C B 2005 Voltammetry: Cathodic stripping *Encyclopedia of Analytical Science* (Amsterdam: Elsevier) 2nd edn pp 203–11

[23] Zadeh K K and Fry B 2008 Transduction platforms *Nanotechnology-Enabled Sensors* (New York: Springer)vol 1 p 79

[24] Holade Y, Tingry S, Servat K, Napporn T W, Cornu D and Kokoh K B 2017 Nanostructured inorganic materials at work in electrochemical sensing and biofuel cells *Catalysts* **7** 31

[25] Lei J and Ju H 2012 Signal amplification using functional nanomaterials for biosensing *Chem. Soc. Rev.* **41** 2122–34

[26] Arulraj A, Govindan S, Vadivelu S and Subramanian B 2017 Photovoltaic performance of TiO_2 using natural sensitizer extracted from *Phyllanthus reticulatus J Mater. Sci. Mater. Electron.* **28** 18455–62

[27] Rajput N 2015 Methods of preparation of nanoparticles: a review *Int. J. Adv. Eng. Technol.* **7** 1806

[28] Swathy B A 2014 Review on metallic silver nanoparticles *IOSR J. Pharm.* **4** 38–44

[29] Horikoshi S A and Serpone N I 2013 Introduction to nanoparticles *Microwaves in Nanoparticle Synthesis: Fundamentals and Applications* (New York: Wiley) pp 1–24

[30] Ahmed S, Ahmad M, Swami B L and Ikram S 2016 A review on plants extract mediated synthesis of silver nanoparticles for antimicrobial applications: a green expertise *J. Adv. Res.* **7** 17–28

[31] Egerton T A and Tooley I R 2004 Femtosecond microscopic solvation dynamics of aqueous solutions *J. Phys. Chem.* B **108** 5066–72

[32] Minsker L K, Semagina N and Renken A 2008 Size-controlled Pd nanoparticles anchored to carbon fiber fabrics: novel structured catalyst effective for selective hydrogenation *Metal Nanoclusters in Catalysis and Materials Science* (Amsterdam: Elsevier) pp 293–9

[33] Arduini F, Cinti S, Scognamiglio V and Moscone D 2020 Nanomaterial-based sensors *Handbook of Nanomaterials in Analytical Chemistry* (Amsterdam: Elsevier) pp 329–59
[34] Zhang Y C, Chu W D, Foroushani A D, Wang H B, Li D, Liu J Q, Barrow C J, Wang X and Yang W R 2014 New gold nanostructures for sensor applications: a review *Materials* **7** 5169–201
[35] Takai A, Doi Y, Yamauchi Y and Kuroda K 2010 Soft-chemical approach of noble metal nanowires templated from mesoporous silica (SBA-15) through vapor infiltration of a reducing agent *J. Phys. Chem.* C **114** 7586–93
[36] Ding S, Chen J S, Qi G, Duan X, Wang Z, Giannelis E P, Archer L A and Lou X W 2011 Formation of SnO_2 hollow nanospheres inside mesoporous silica nanoreactors *J. Am. Chem. Soc.* **113** 21–3
[37] Kim B H and Lee J S 2015 One-pot photochemical synthesis of silver nanodisks using a conventional metal-halide lamp *Mater. Chem. Phys.* **149–150** 678–85
[38] Gao C, Zhang Q, Lu Z and Yin Y 2011 Templated synthesis of metal nanorods in silica nanotubes *J. Am. Chem. Soc.* **13** 19706–9
[39] Xiong Y, Chen J, Wiley B, Xia Y, Yin Y and Li Z Y 2005 Size-dependence of surface plasmon resonance and oxidation for Pd nanocubes synthesized via a seed etching process *Nano Lett.* **5** 1237–42
[40] Ghosh T, Satpati B and Senapati D 2014 Characterization of bimetallic core–shell nanorings synthesized via ascorbic acid-controlled galvanic displacement followed by epitaxial growth *J. Mater. Chem.* C **2** 2439–47
[41] Imran Hussain S, Karthick S, Arulraj A and Mangalaraja R V 2022 *Nanocrystals for Electrochemical Energy Storage Devices* (Amsterdam: Elsevier) ch 22 pp 409–26
[42] Sau T K and Rogach A L 2010 Nonspherical noble metal nanoparticles: colloid-chemical synthesis and morphology control *Adv. Mater.* **22** 1781–804
[43] Doria G, Conde J, Viegas B, Giestas L, Almeida C, Assunção M, Rosa J and Baptista P V 2012 Noble metal nanoparticles for bio-sensing applications *Sensors* **12** 1657–87
[44] Arulraj A, Ramesh M, Subramanian B and Senguttuvan G 2018 *In-situ* temperature and thickness control grown 2D-MoS_2 via pulsed laser ablation for photovoltaic devices *Sol. Energy* **174** 286–95
[45] Arulraj A, Subramanian B, Ramesh M and Senguttuvan G 2019 Effect of active sites in pulsed laser deposited bimetallic $NiMoS_2$ thin films for solar energy conversion *Mater. Lett.* **241** 132–5
[46] Bellan L M, Wu D and Langer R S 2011 Current trends in nanobiosensor technology *Wiley Interdiscip. Rev. Nanomed. Nanobiotechnol.* **3** 229–46
[47] Gross E, Krier J, Heinke L and Somorjai G A 2012 Polymer-encapsulated metallic nanoparticles as a bridge between homogeneous and heterogeneous *Catalysis, Catal. Lett.* **55** 13–23
[48] Jungemann A H *et al* 2019 The role of ligands in the chemical synthesis and applications of inorganic nanoparticles *Chem. Rev.* **119** 4819–80
[49] An K and Somorjai G A 2012 Size and shape control of metal nanoparticles for reaction selectivity in catalysis *Chem. Cat. Chem.* **4** 1512–24
[50] Jungck J R, Brittain S, Plante D and Flynn J 2023 Self-assembly, self-folding, and origami: comparative design principles *Biomimetics* **8** 12
[51] Lim B and Xia Y 2011 Metal nanocrystals with highly branched morphologies *Angew. Chem. Int. Ed.* **50** 76–85

[52] Xia Y, Xiong Y, Lim B and Skrabalak S E 2009 Shape-controlled synthesis of metal nanocrystals: simple chemistry meets complex physics? *Angew. Chem.* **121** 62–108

[53] Boles M A, Engel M and Talapin D V 2016 Self-assembly of colloidal nanocrystals: from intricate structures to functional materials *Chem. Rev.* **116** 11220–89

[54] Caschera D, Federici F, Zane D, Focante F, Curulli A and Padeletti G J 2009 Gold nanoparticles modified GC electrodes: electrochemical behaviour dependence of different neurotransmitters and molecules of biological interest on the particles size and shape *Nanopart. Res.* **11** 1925–36

[55] Zen J M, Kumar A S and Tsai D M 2003 Recent updates of chemically modified electrodes in analytical chemistry *Electroanalysis* **15** 1073–87

[56] Wang F and Hu S 2009 Electrochemical sensors based on metal and semiconductor nanoparticles *Microchim. Acta* **165** 1–22

[57] Sudagar J, Tamilarasan R, Sanjith U, Rajendran R and Kumar R 2017 Electroless deposition of nanolayered metallic coatings *Nanoscaled Films and Layers* (London: IntechOpen) pp 27–50

[58] Losey M W, Kelly J J, Badgayan N D, Sahu S K and Sreekanth P S R 2017 Electrodeposition ed S Hashmi *Reference Module in Materials Science and Materials Engineering* vol 13 (Oxford: Elsevier) pp 1–20

[59] Pan Z, Sun F, Zhu X, Chen Z, Lin X, Zheng Y, Zhong W, Zhuang Z and Gu F 2019 Electrodeposition-based *in situ* construction of a ZnO-ordered macroporous film gas sensor with enhanced sensitivity *J. Mater. Chem.* A **7** 1287–99

[60] Wang H, Tian T, Zhang Y, Pan Z, Wang Y and Xiao Z 2008 Sequential electrochemical oxidation and site-selective growth of nanoparticles onto AFM probes *Langmuir* **24** 8918–22

[61] Chen G, Xia D, Nie Z, Wang Z, Wang L, Zhang L and Zhang J 2007 Facile Synthesis of Co–Pt Hollow Sphere *Chem. Mater.* **19** 1840–4

[62] Zadeh K K and Fry B 2008 *Nanotechnology-Enabled Sensors* (New York: Springer) vol 32 pp 182–4

[63] Ibrahim A A, Ahmad R, Umar A, Al-Assiri M S, Al-Salami A E, Kumar R, Ansari S G and Baskoutas S 2017 Two-dimensional ytterbium oxide nanodisks based biosensor for selective detection of urea *Biosens. Bioelectron.* **98** 254–60

[64] Maduraiveeran G and Jin W 2017 Nanomaterials based electrochemical sensor and biosensor platforms for environmental applications *Trends Environ. Anal. Chem.* **13** 10–23

[65] Saei A A, Ezzati J, Dolatabadi N, Marandi P N, Abhari A and Guardia M L 2013 Electrochemical biosensors for glucose based on metal nanoparticles *Trends Anal. Chem.* **42** 216–27

[66] Fang C S, Oh K H, Oh A, Lee K, Park S, Kim S, Park J K and Yang H 2016 An ultrasensitive and incubation-free electrochemical immunosensor using a gold-nanocatalyst label mediating outer-sphere-reaction-philic and inner-sphere-reaction-philic species *Chem. Commun.* **52** 5884–7

[67] Sun K, Chang Y, Zhou B, Wang X and Liu L 2017 Gold nanoparticles-based electrochemical method for the detection of protein kinase with a peptide-like inhibitor as the bioreceptor *Int. J. Nanomed.* **12** 1905–15

[68] Chen Y, Xiang J, Liu B, Chen Z and Zuo X 2020 Gold nanoparticle-engineered electrochemical aptamer biosensor for ultrasensitive detection of thrombin *Anal. Methods* **12** 3729–33

[69] Kumar J V, Karthik R, Chen S M, Muthuraj V and Karuppiah C 2016 Fabrication of potato-like silver molybdate microstructures for photocatalytic degradation of chronic toxicity ciprofloxacin and highly selective electrochemical detection of H_2O_2 *Sci. Rep.* **6** 34149

[70] Nantaphol S, Chailapakul O and Siangproh W 2015 Sensitive and selective electrochemical sensor using silver nanoparticles modified glassy carbon electrode for determination of cholesterol in bovine serum *Sensors Actuators* B **207** 193–8

[71] Yazdanparast S, Benvidi A, Banaei M, Nikukar H, Tezerjani M D and Azimzadeh M 2018 An optoelectronic tongue based on an array of gold and silver nanoparticles for analysis of natural, synthetic and biological antioxidants *Microchim. Acta* **185** 405

[72] Meng F, Han K, Wang B, Liu T, Liu G, Li Y and Miao P 2016 Nanoarchitectured electrochemical cytosensor for selective detection of cancer cells *ChemistrySelect* **1** 1515

[73] Kokkinos C 2019 Electrochemical DNA biosensors based on labeling with nanoparticles *Nanomaterials (Basel)* **9** 1361

[74] Yu H, Yu J, Li L, Zhang Y, Xin S, Ni X, Sun Y and Song K 2021 Recent progress of the practical applications of the platinum nanoparticle-based electrochemistry biosensors *Front. Chem.* **9** 677876

[75] Choi G, Kim E, Park E and Lee J H 2017 A cost-effective chemiluminescent biosensor capable of early diagnosing cancer using a combination of magnetic beads and platinum nanoparticles *Talanta* **162** 38–45

[76] Wang L, Liu D Q, Sun Y L, Su J J, Jin B W, Geng L, Song Y Y, Huang X and Yang M 2018 Signal-on electrochemiluminescence of self-ordered molybdenum oxynitride nanotube arrays for label-free cytosensing *Anal. Chem.* **90** 10858–64

[77] Kavakiotis I, Tsave O, Salifoglouc A, Maglaveras N, Vlahavas I and Chouvarda I 2017 Machine learning and data mining methods in diabetes research *Comput. Struct. Biothechnol. J.* **15** 104–16

[78] Sharma R, Garg P, Kumar P, Bhatia S K and Kulshrestha S 2020 Microbial fermentation and its role in quality improvement of fermented foods *Fermentation* **6** 106

[79] Li J, Liu W, Wu X and Gao X 2015 Mechanism of pH-switchable peroxidase and catalase-like activities of gold, silver, platinum and palladium *Biomaterials* **48** 37–44

[80] Shen X, Liu W, Gao X, Lu Z, Wu X and Gao X 2015 Mechanisms of oxidase and superoxide dismutation-like activities of gold, silver, platinum, and palladium, and their alloys: a general way to the activation of molecular oxygen *J. Am. Chem. Soc.* **137** 15882–91

[81] Haruta M 1997 Size- and support-dependency in the catalysis of gold *Catal. Today* **36** 153

[82] Malik P, Gupta R, Malik V and Ameta R K 2021 Emerging nanomaterials for improved biosensing *Meas.: Sens.* **16** 100050

[83] El-Sayed I H, Huang X and El-Sayed M A 2005 Surface plasmon resonance scattering and absorption of anti-EGFR antibody conjugated gold nanoparticles in cancer diagnostics: applications in oral cancer *Nano Lett.* **5** 829

[84] Copland J A, Eghtedari M, Popov V L, Kotov N, Mamedova N, Motamedi M and Oraevsky A A 2004 Bioconjugated gold nanoparticles as a molecular based contrast agent: implications for imaging of deep tumors using optoacoustic tomography *Mol. Imag. Biol.* **6** 341

[85] Bai Y, Yang H, Yang W, Li Y and Sun C 2007 Gold nanoparticles-mesoporous silica composite used as an enzyme immobilization matrix for amperometric glucose biosensor construction *Sensors Actuators* B **124** 179

[86] German N, Ramanavicius A, Voronovic J and Ramanaviciene A 2012 Glucose biosensor based on glucose oxidase and gold nanoparticles of different sizes covered by polypyrrole layer *Colloids Surf.* A **413** 224
[87] Bloch K, Pardesi K, Satriano C and Ghosh S 2021 Bacteriogenic platinum nanoparticles for application in nanomedicine *Front. Chem.* **9** 624344
[88] Fiérrez F J, Sánchez M I G, Pérez R J, Iniesta J and Valero E 2020 Glucose biosensor based on disposable activated carbon electrodes modified with platinum nanoparticles electrodeposited on poly(Azure A) *Sensors* **20** 4489
[89] Li W J, Yuan R, Chai Y Q, Zhong H A and Wang Y 2011 Study of the biosensor based on platinum nanoparticles supported on carbon nanotubes and sugar–lectin biospecific interactions for the determination of glucose *Electrochim. Acta* **56** 4203
[90] Naresh V and Lee N A 2021 Review on biosensors and recent development of nanostructured materials-enabled biosensors *Sensors* **21** 1109
[91] Ma J L, Yin B C, Wu X and Ye B C 2017 Simple and cost-effective glucose detection based on carbon nanodots supported on silver nanoparticles *Anal. Chem.* **89** 1323–8
[92] Savk A, Aydin H, Cellat K and Sen F 2020 A novel high performance non-enzymatic electrochemical glucose biosensor based on activated carbon-supported Pt–Ni nanocomposite *J. Mol. Liq.* **300** 112355
[93] Barman S C, Hossain M F, Yoon H and Park J Y 2018 Carboxyl terminated reduced graphene oxide (Crbxl-RGO) and Pt nanoparticles based ultra-sensitive and selective electrochemical biosensor for glutamate detection *J. Electrochem. Soc.* **165** B296–301
[94] Knittl E T, Rusakov D A, Korotkova E I, Dorozhko E V, Voronova O A, Plotnikov E V, Sozuer I T and Linert W 2015 Characterization of a novel dioxomolybdenum complex by cyclic voltammetry *Anal. Lett.* **48** 2369–79
[95] Barman S C, Hossain M F and Park J Y 2017 Gold nanoparticles assembled chemically functionalized reduced graphene oxide supported electrochemical immunosensor for ultrasensitive prostate cancer detection *J. Electrochem. Soc.* **164** B234–9
[96] Nguyen T N H *et al* 2020 Fabrication and *ex vivo* evaluation of activated carbon–Pt microparticle based glutamate biosensor *J. Electroanal. Chem.* **866** 114136
[97] Zhang Y, Wu C Y, Zhou X J, Wu X C, Yang Y Q, Wu H X, Guo S and Zhang J 2013 Graphene quantum dots/gold electrode and its application in living cell H_2O_2 detection *Nanoscale* **5** 1816–9
[98] Wang Q, Zhang X, Chai X, Wang T, Cao T, Li Y, Zhang L, Fan F, Fu Y and Qi W 2021 An electrochemical sensor for H_2O_2 based on Au nanoparticles embedded in UiO-66 metal–organic framework films *ACS Appl. Nano Mater.* **4** 6103–10
[99] Loza K *et al* 2014 The dissolution and biological effects of silver nanoparticles in biological media *J. Mater. Chem.* B **2** 1634–43
[100] Govindhan M, Amiri M and Chen A 2015 Au nanoparticle/graphene nanocomposite as a platform for the sensitive detection of NADH in human urine *Biosens. Bioelectron.* **66** 474–80
[101] Tığ G A 2017 Highly sensitive amperometric biosensor for determination of NADH and ethanol based on Au–Ag nanoparticles/poly(L-cysteine)/reduced graphene oxide nanocomposite *Talanta* **175** 382–9
[102] Jimenez A, Armada M P G, Losada J, Villena C, Alonso B and Casado C M 2014 Amperometric biosensors for NADH based on hyperbranched dendritic ferrocene polymers and Pt nanoparticles *Sensors Actuators* B **190** 111–9

[103] Mu S L and Shi Q F 2013 Xanthine biosensor based on the direct oxidation of xanthine at an electrogenerated oligomer film *Biosens. Bioelectron.* **47** 429–35
[104] Joon A, Ahlawat J, Aggarwal V, Jaiwal R and Pundir C S 2021 An improved amperometric determination of xanthine with xanthine oxidase nanoparticles for testing of fish meat freshness *Sens. Bio-Sens. Res.* **33** 100437
[105] Dervisevic M, Dervisevic E, Cevik E and Senel M 2017 Novel electrochemical xanthine biosensor based on chitosan-polypyrrole-gold nanoparticles hybrid bio-nanocomposite platform *J. Food Drug Anal.* **25** 510–9
[106] Yang H, Zhao J, Qiu M, Sun P, Han D, Niu L and Cui G 2019 Hierarchical bi-continuous Pt decorated nanoporous Au–Sn alloy on carbon fiber paper for ascorbic acid, dopamine and uric acid simultaneous sensing *Biosens Bioelectron.* **124–125** 191–8
[107] Manu P, Al-Dhaher Z, Khan S, Kane J M and Correll C U 2014 Elevated blood urea nitrogen and medical outcome of psychiatric inpatients *Psychiatr.* Q. **85** 111–20
[108] Korkut S, Uzuncar S, Kilic M S and Hazer B 2019 Electrochemical determination of urea using a gold nanoparticle-copolymer coated-enzyme modified gold electrode *Instrum. Sci. Technol.* **47** 1–18
[109] Ansari S G, Fouad H, Shin H S and Ansari Z A 2015 Electrochemical enzyme-less urea sensor based on nano-tin oxide synthesized by hydrothermal technique *Chem. Biol. Interact.* **242** 45–9
[110] Movilli J, Kolkman R W, Rozzi A, Corradini R, Segerink L I and Huskens J 2020 Increasing the sensitivity of electrochemical DNA detection by a micropillar-structured biosensing surface *Langmuir* **36** 4272–9
[111] Campuzano S, Pedrero M, Gamella M, Serafín V, Sedeño P Y and Pingarrón J M 2020 Beyond sensitive and selective electrochemical biosensors: towards continuous, real-time, antibiofouling and calibration-free devices *Sensors* **20** 3376
[112] Wang J, Zhou H, Liu J, He J, Liu J and Yang W 2021 Electrochemical detection of DNA by formation of efficient electron transfer pathways through adsorbing gold nanoparticles to DNA modified electrodes *Microchem. J.* **169** 106581
[113] Chen D, Wu Y, Hoque S, Tilley R D and Gooding J J 2021 Rapid and ultrasensitive electrochemical detection of circulating tumor DNA by hybridization on the network of gold-coated magnetic nanoparticles *Chem. Sci.* **12** 5196–201

Chapter 10

Two-dimensional (2D)-nanolayer based functional materials in the advanced oxidation process

Mehdi Abedi, Zahra Rahiminezhad, Marzieh Mehrzadeh, Vahid Alimardani, Younes Ghasemi, Divya Chauhan, Mohammad Ashfaq, R V Mangalaraja and Neetu Talreja

The advent of two-dimensional (2D)-nanolayers (2D-NLs) has caused a revolution in the field of materials science and nanomaterials to develop the emerging functional materials. The 2D-NLs are unremittingly gaining the interest of researchers globally due to their exceptional characteristics such as high reactivity, large surface area, and mechanical, optical, and electrical properties. These characteristics provide new insight toward resolving numerous associated issues that directly relate to the society, mainly energy production and water treatment technology. Numerous water treatment technologies have been used to treat environmental pollution, including advanced oxidation processes (AOPs) such as photocatalysis. AOPs/photocatalysis is an emerging process to treat/degrade environmental pollution using photocatalyst materials. 2D-NL-based functional material, such as WS_2, TiO_2, MXene, MoS_2, etc, based photocatalyst materials have been used successfully to treat innumerable environmental pollutants. In this chapter, we focus on synthesizing 2D-NLs and their functional materials, and how tuning the surface texture, surface functional group, and band gap might improve the photocatalytic performance. Finally, we discuss the applications of 2D-NLs and their potential suitability for next-generation technology.

doi:10.1088/978-0-7503-5724-1ch10

10.1 Introduction

Water pollutants contribute to a hazardous environment that harms human health. The quality of water and its availability are deteriorating due to changes in the hydrological cycle, and indirect/direct contamination (Naresh *et al* 2010). Even though the world is swimming with water, much of it is either too salty for human consumption, as are 97.2% of the world's oceans and seas, or too frozen (2.15%) to be used in their current condition. Major portions of the world's freshwater supplies are contaminated due to variety of natural and manmade causes, including the exploitation of water resources for residential, industrial, and agricultural purposes, leaving just a small percentage of the world's water as usable for human consumption (0.65%) (Kurian 2021). Industrial processes, particularly those in the chemical, pharmaceutical, and cosmetic industries, are connected to generating a wide variety of pollutants that threaten aquatic ecosystems and freshwater supplies for human life (Musolff *et al* 2010). The wide-ranging issues of water pollution, diversity, and non-biodegradable waste have evolved into a challenge that the natural cleansing cycle cannot address and resolve (Bayati *et al* 2010, Tajeddine *et al* 2010). Therefore, there is an immediate requirement for the widespread availability of effective water purification and environmental remediation technologies. Physical, chemical, and biological treatments are the most frequent and widely utilized techniques for treating wastewater, as shown in figure 10.1. The larger grains and granules can be removed using physical methods such as filtration, sedimentation, drainage, adsorption, and membrane separation (Lin *et al* 2017, Mustafa and Asmatulu 2020, Yang *et al* 2018). However, the conventional physical treatment is inadequate for eradicating organics from the water (Bartolomeu *et al* 2018). Such issues can be resolved by biological therapies such as biofilm and activated sludge (Sgroi *et al* 2021, Zhang *et al* 2020b). While the biological treatment procedure has many advantages, it also has several drawbacks, including high initial and ongoing costs, a propensity for sludge swelling, and lengthy pre-preparation cycles (Zhang *et al* 2019). The organic contaminants in water may be oxidized and degraded quickly using chemical processes, which are effective water treatment approaches (George *et al* 2015). There are several chemical treatments employed for the organic wastewater, but advanced oxidation processes (AOPs) are often the most effective (Chávez *et al* 2019).

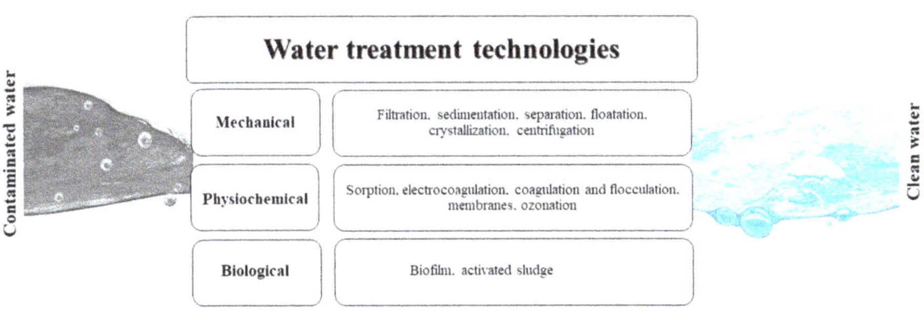

Figure 10.1. The various existing technologies for wastewater treatment.

AOPs, which entail the *in situ* production of severely oxidizing HO•, were first used for drinking water treatment in the 1980s (Glaze 1987, Glaze *et al* 1987). With its oxidation potential of 2.80 eV, the produced HO• can oxidize non-biodegradable and toxic organic pollutants into relatively harmless compounds such as H_2O, CO_2, or HCl under ambient conditions, which is more effective than other oxidants used in the oxidation process in water (Devi and Kavitha 2013). Several novel heterogeneous and homogeneous oxidation techniques have been explored for application in wastewater treatment during the last decades. Ultraviolet (UV) or visible light, different oxidizing agents (H_2O_2, O_2, or O_3), or catalytic materials are the various viable options for activating AOPs (Ameta and Ameta 2018). While different AOP approaches are now in use, they can be categorized generally as either photochemical or non-photochemical processes (Kanakaraju *et al* 2018). The conventional water treatment methods affect the environment negatively and are ineffective in removing new pollutants and endocrine disruptors. The traditional water treatment methods have environmental drawbacks and may not be effective enough to rid water of new pollutants such as endocrine chemicals. Even though some of the available water treatment technologies are efficient, they are not always the best options for removing certain contaminants, in particular those present in low concentrations and those that need to use of a lot of energy and result in a lot of hazardous waste (Adeleye *et al* 2016).

Nanomaterials, including those with dimensions from zero-dimensional (0D) to three-dimensional (3D), are useful for the detection and elimination of a wide range of chemical and biological contaminants (Kouhbanani *et al* 2019, Shaker Ardakani *et al* 2021, Jani *et al* 2020). The purification rates may be increased by using nanomaterials in AOP design by the strategic manipulation of nanoscale components. Nanomaterials possess a higher surface area to volume ratio thereby greatly improving the adsorption characteristics of the targeted materials, suggesting their usage in the AOP might increase its purification rates (Bethi *et al* 2016). Many studies have been conducted to improve water treatment using nanotechnology and AOPs (Kurian 2021). The low production of environmentally harmful by-products is a major benefit of adopting nanotechnology in wastewater treatment. Nanotechnology is sometimes viewed as costly. However, it may offer more cost-effective solutions than the conventional approaches under some conditions. The cost of the products based on nano-enabled technology may be lowered through industrial production and by creating innovative methods that use less expensive raw materials and less energy (Adeleye *et al* 2016). Many nanomaterials may be reused in nanotechnology-based systems, reducing energy consumption, improving treatment efficiency, and reducing waste (Cardoso *et al* 2021).

Among the different families of nanomaterials under research and development, two-dimensional (2D)-nanolayers (2D-NLs) have garnered the most interest in water treatment due to the unique qualities that their structure affords. When contrasted with 0D and 1D nanomaterials, the evident advantages of 2D-NLs can be seen in architectural modification, light harvesting, charge separation, and tunability (Zhang 2015). In more specific terms, the large lateral dimension of 2D-NLs makes it possible to include many active sites capable of adsorbing pollutants and undergoing surface reactions (Sun *et al* 2015). Hence, 2D-NLs have recently received much attention in AOPs for treating wastewater. This chapter explains how these 2D functional materials are synthesized, modified, and work, and how they can be used in AOPs for cleaning up the environment. Next-generation technologies are also detailed.

10.2 Nanomaterial-based AOPs

Nanomaterials should have at least one distinctive dimension between 1 and 100 nm (Alimardani *et al* 2021). All the zero-dimensional (0D), one-dimensional (1D), two-dimensional (2D), and three-dimensional (3D) nanostructures exhibit distinctive structural, physicochemical, optical, and electrochemical properties that make them promising candidates for the removal or degradation of wastewater pollutants, as shown in figure 10.2 (Monga *et al* 2021). A nanoparticle's surface and interface properties tend to predominate over its bulk properties. The high surface energy and large available surface area of nanoparticles are responsible for their remarkable catalytic activity. More reactions can occur because more of the reactant's surface area is in contact with the particles. Increasing selectivity is achieved by using nano-materials (nano-catalysts) that allow reactions to occur at lower temperatures, decreased side reactions, and increased recycling rates. The active surface sites interact with different oxidants (H_2O_2, O_3) to produce hydroxyl radicals, which are essential for effective degradation of pollutants in AOPs. Super-paramagnetism is another property that may be used to separate the nanomaterials from bulk materials (Abedi *et al* 2019). These will help with the fundamental and practical problems that have impeded the widespread application of AOPs in water treatment.

Polymers, metal/metal oxides, ceramics, and carbon-based materials are some of the most common types of nanomaterials employed today for degrading/removing pollutants in the wastewater (Adeleye *et al* 2016, Sasidharan *et al* 2021). The polymeric nanoparticles can remove organic/inorganic pollutants. The metallic nanoparticles such as iron (Fe), silver (Ag), gold (Au), zirconium (Zr), and

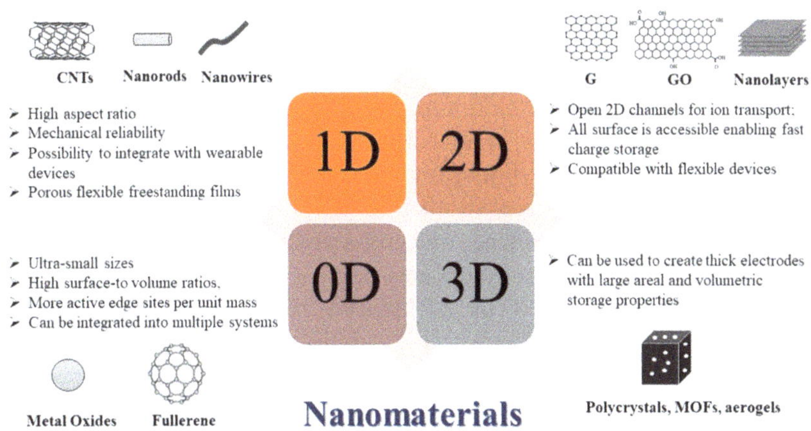

Figure 10.2. Schematic overview of 0D, 1D, 2D, and 3D nanomaterials.

palladium (Pd) have also been used to clean the wastewater (Shaker Ardakani *et al* 2021, Kouhbanani *et al* 2019, Naseem and Durrani 2021). The decomposition of trichloroethane in the groundwater was achieved using gold nanoparticles infused with palladium. Compared to pure palladium, gold nanoparticles impregnated in palladium exhibit 2 200 times higher catalytic activity (Tobiszewski and Namieśnik 2012). When different materials, such as metal oxides, are combined into single nanostructures, they work better together and have more catalytic activity than when used alone (Kulesza *et al* 2013). Metal oxide nanoparticles, such as titanium dioxide (TiO_2), zinc oxide (ZnO), and cerium oxide (CeO_2), were shown to be particularly effective at degrading organic pollutants in an aqueous system (Naseem and Durrani 2021). TiO_2, ZnO, and copper oxide (Cu_2O) nanoparticles have all been used in electro- and photocatalytic-assisted oxidation processes. Under UV light, these nanoparticles were reported to be effective at removing chemical oxygen demand and oxidizing water. Inorganic, nonmetallic ceramic nanoparticles primarily consist of carbides, oxides, phosphates, and carbonates of metals and metalloids, including titanium, calcium, and silicon (Thomas *et al* 2015). They have found widespread use in fields such as catalysis and photocatalysis due to their easy production with the desired structure, size, and porosity, as well as their extreme stability (at a given pH and temperature). These nanomaterials can be combined with various materials for efficient photocatalysis and water purification (Horovitz *et al* 2020). The ceramic-based nanomaterials, which permit and improve catalytic redox processes, have garnered much attention in recent years for wastewater treatment (Singh *et al* 2020). Carbon-based nanomaterials, made up almost entirely of carbon atoms, are extremely helpful in treating emerging contaminants and in energy conversion and storage due to their exceptional surface area and high thermal and electrical conductivity (Dai *et al* 2012). Notable nonmetal catalysts that have seen extensive usage for a variety of AOPs include fullerenes, graphene, carbon nanotubes (CNTs), and multiwalled CNTs (MWCNTs) (González-González *et al* 2022). Allotropic fullerenes are nanomaterials with a hollow spherical cage structure made from carbon and the benefits of strong thermal strength, electrical conductivity, and flexibility make these materials (fullerenes) more interesting. CNTs can be tubular or rod-shaped, and thanks to the surface modification, they are now colloidally stable, enabling efficient degradation of hexavalent chromium (Gu *et al* 2013). The surface adsorption capacity and the removal rate of MWCNTs can be improved due to their unique structures and semiconductor properties. Graphene is a monolayer of carbon atoms bonded by sp^2 bonds, which provides it with exceptional mechanical strength, thermal conductivity, and electronic abilities (Borandeh *et al* 2021). Graphene and its derivatives can inactivate bacteria independent of light due to wrapping, insertion of sharp edges, stress extraction, and oxidative stress generated by G-induced reactive oxygen species (ROS) (Alimardani *et al* 2019, Zou *et al* 2016). Additionally, the effective synthesis of graphene has prompted the development of 2D-NLs with the thinnest structures and highest surface-to-volume ratios, which promise quicker processing times and greater treatment efficiency than the chemical and biological remediation techniques (see next section) (Zhang 2015). Overall, the nanomaterial's combination with AOPs might lead to a revolution in the water treatment technology. This combination is one of the most advanced contaminant removal techniques, which can the replace conventional environmental treatment methods.

10.3 2D-NL in AOPs

The rapid synthesis of 2D-NLs has been made possible by efficiently separating graphene monolayers from graphite (Zhang 2015). Because of the unique qualities linked to structural features, functional 2D-NLs for photocatalysis have attracted much interest. 2D-NLs outperform 0D and 1D nanomaterials in manipulating architecture, charge separation, tenability, and light harvesting. More specifically, the enormous lateral sizes of 2D-NLs provide many active sites for pollutant adsorption and surface reactions (Sun *et al* 2015). The ultra-thin thickness of 2D-NLs also promotes charge diffusion to the material surface due to their small transit distance (Deng *et al* 2016b). A significant proportion of low-coordination 2D layers allows more exposed surfaces for light harvesting (Di *et al* 2018). Furthermore, the individual active facets might be completely revealed or tuned to a precise ratio by adjusting the manufacturing conditions rationally. Lastly, 2D-NL band gaps and light absorption may be adjusted by changing the number and thickness of the nanolayers, which may differ from 0D and 1D structures (Yu *et al* 2014b, Liu *et al* 2014). To date, numerous attempts have been made to create extremely effective photocatalysts using 2D-NLs as well as to investigate their potential for water remediation, all of which have been motivated by the special optical, electrical, and physicochemical features of 2D-NLs. In order to study this subject, a wide variety of 2D-NLs have been used, such as MXenes, graphene and its derivatives, metal oxides (such as TiO_2), and transition metal dichalcogenides (TMDs).

10.3.1 TiO_2

2D TiO_2-NLs, due to their unique properties, are considered as suitable materials for developing different photocatalytic and microelectronic devices and are of great interest to researchers to investigate their catalytic, optical, and electronic structure and properties. The majority of these uses depend on TiO_2's exceptional electrical characteristics. Controlling the properties is crucial for appropriate application, which depend on film thickness, preparation methods, impurities, preparation temperature, substrate, and preparation atmosphere (Yildiz *et al* 2010). By considering the electronic structure, TiO_2 film exhibited a high energy band gap of about 3.2 eV that can serve as a semiconductor, and electron transition occurs in the visible region, but by using some impurity band gap change significantly. The present impurity role of an interstice band gap in the electrical conductivity is neglectable; even at high temperatures, impurity plays an essential role in jumping electrons between different band gaps (Yildiz *et al* 2008). In addition, the electrical conductivity of this material depends strongly on temperature. To add corrosion-resistant properties to metal, TiO_2 is frequently used as a thin film (Rajib *et al* 2021). The ideal film thickness is typically required for the best performance when producing high-quality optoelectronic devices based on TiO_2. Since this factor is the most crucial determinant of the film's physical properties, there is also a lot of research studying the effect of the film thickness on the property of the prepared film, such as optical and electrical performance. Since it is a key factor influencing the films' electrical and optical properties, the film thickness is significant. Higher transparency and resistivity are the characteristics of thinner films. It will be especially difficult to transmit light in the visible range as the film thickness increases (Zeribi *et al* 2022). The different techniques, including CVD, ultrasonic spraying, precipitation method, sol–gel, spray pyrolysis, solvothermal method, radio frequency magnetron sputtering, etc, can be used to create thin film TiO_2 (Zhang and Lu 2018). The primary challenge to TiO_2's use in practical applications is the quick recombination of electron–hole pairs produced by photons which causes a reduction in photocatalytic activity and reduced sensitivity to TiO_2 to visible light (Li *et al* 2020, Cai *et al* 2017). The main drawback of pure TiO_2 is its interstice band gap, which limits its wavelength adsorption under 400 nm, which makes up only about 5% of sunlight (Tian *et al* 2014).

One of the most extensively researched topics in photocatalysis is modifying TiO_2 to address the issues with using pure TiO_2. Through various mechanisms, the modified TiO_2 can enhance the photocatalytic activity of TiO_2-based materials such as reducing electron–hole interactions and band gap reduction. Various strategies have been used recently to increase TiO_2's photocatalytic activity. Ion doping TiO_2 or coupling it to other semiconductor materials for creating a composite semiconductor material are the two most popular techniques for reducing TiO_2's band gap (Nzaba *et al* 2019, Hlekelele *et al* 2019). Another method for improving the photocatalytic activity is preventing electron–hole recombination by adding some electron acceptor agents, including Ag, Au, etc, on the surface of TiO_2 film (Yoshida *et al* 2020, Deng *et al* 2019). Recently, numerous studies on TiO_2 property enhancement and modification with different metallic and nonmetallic materials have been done. Some researchers immobilize TiO_2 on a nonmetallic substrate such as polymer, glass, or carbon. Suarez *et al* immobilized TiO_2 in the pores of porous zeolite and prepared a TiO_2/zeolite hybrid material for pollutant degradation. Due to the composite catalyst's high surface area, pollutants are attracted to it more readily, which increases the efficiency and speed of catalytic activity while lowering the probability of aggregation (Suárez *et al* 2019). Ling *et al*, by combining montmorillonite with TiO_2, created a composite catalyst that increased absorption capacity in a range of UV–vis light from 70% to 87% (Liang *et al* 2017). The findings Ni *et al* indicated that the degradation rate of methylene blue was 97% after 150 min for immobilized TiO_2 on polyvinyl alcohol–ethylene–ethylene nanofiber (Ni *et al* 2019). TiO_2's photoresponse range can be expanded by doping nonmetal atoms. The valence band of TiO_2 is restructured by doping nonmetallic ions, increasing and shortening the gap's width. Recently, it has been discovered that doping TiO_2 with nonmetallic elements such as carbon, sulfur, iodine, brome, and fluoride can successfully enlarge its optical response range to the visible light region. Doping with nonmetal elements involved substituting non-metals for oxygen or partial oxygen in the TiO_2 (Fang *et al* 2017, Kovalevskiy *et al* 2020). Table 10.1 shows TiO_2-based material with photocatalytic degradation activity.

Table 10.1. TiO_2-based material with photocatalytic degradation activity.

Method	Substrate	Oxidized layer	Application	Results	References
Sol–gel	Glass substrates	TiO_2	Photocatalyst	Thin film with visible light transparency and photocatalytic activity comparable to Degussa P-25, a common commercial TiO_2 catalyst.	Nobuaki *et al* (1995)
Sol–gel	Glass substrates	TiO_2	Photocatalyst	Compared to Degussa P-25, semitransparent TiO_2 demonstrated higher photocatalytic activity for acetaldehyde gas with quantum efficiencies greater than 100%.	Sopyan *et al* (1996)
Dip-coating process	Silica coated glass plates	TiO_2	Photocatalyst	Degussa P-25-like photocatalytic activity for NO_x removal in the visible range, NO_x emissions strongly depend on film thickness, and efficiency decreases as film thickness decreases.	Negishi *et al* (1998)
Template method	Ce^{3+}/TiO_2	TiO_2	Photocatalyst	UV-A radiation in thin films of cerium-doped TiO_2 effectively breaks down amoxicillin and tetracycline due to inhibition from electron–hole recombination, and cerium presence increases the photocatalytic activity.	Lalliansanga *et al* (2022)
Anodic oxidation	Indium tin oxide glass	$BiFeO_3/TiO_2$	Photocatalyst	The prepared composite sample exhibited excellent photocatalytic activity and removed 97.6% of the chromium (VI) when illuminated with visible light.	Huo *et al* (2022)
Sol–gel	Glass	Fe^{3+}–TiO_2	Photocatalyst	Effective optical band gap was decreased by Fe doping, and degradation of sulfamethoxazole increased with film thickness.	Abedi *et al* (2019)
Sol–gel	Glass	TiO_2	Photocatalyst	Diclofenac, propranolol, albuterol, ofloxacin, ciprofloxacin, azithromycin, erythromycin, hydrochlorothiazide, and furosemide are just a few of the pharmaceutically active substances that are detected in wastewater at levels higher than 50%. *Escherichia coli* in wastewater was effectively inactivated.	Rueda-Márquez *et al* (2020)
Magnetron sputtering	TiO_2	TiO_2	Photocatalyst	Fe-doped TiO_2 films showed 100% antimicrobial ability against *E. coli.*	Lou *et al* (2022)
Sol–gel	Glass	TiO_2	Photocatalyst	Thin films of TiO_2 that had been Sm_3, Eu_3, and Dy_3 doped demonstrated greater photocatalytic activity than TiO_2 that had not been doped for methyl orange degradation under UV–vis light.	Radha *et al* (2022)
Magnetron sputtering	WO_3/TiO_2	TiO_2	Photocatalyst	Degrades 96.49% RhB in 2 h under UV–vis light. Performance of WO_3/TiO_2 as a catalyst was strengthened with O_2 flux.	Zhang *et al* (2022a)

10.3.2 MXenes

MXenes are a new class of nanomaterial with a 2D structure discovered in 2011 (Naguib *et al* 2011, Arulraj *et al* 2022). These nanomaterials are composed of inorganic compounds such as nitrides, carbides, or carbonitrides of transition metals. MXenes are 2D-NLs with several atom thicknesses and a planar size in the order of several micrometers (Mashtalir *et al* 2013, Ling *et al* 2014, Kim *et al* 2021, Naguib *et al* 2014). They have several distinctive properties that make them appropriate and attractive for various applications (Lukatskaya *et al* 2017, Shahzad *et al* 2016, Naguib *et al* 2012, Talreja *et al* 2023). MXenes possess a large specific surface area, fine mechanical, electronic, and physicochemical properties, and good biocompatibility (Huang *et al* 2020, Wu *et al* 2021). The MXenes general formula is $M_{n+1}AX_n$, in which M stands for an early transition metal, A stands for any element from IIIA or IVA groups, X stands for carbon, nitrogen, or carbonitride, and n is 1–4 (Jimmy and Kandasubramanian 2020). MAX phases are applied as the main precursor material for the synthesis of MXenes. The formula $M_{n+1}AX_nT_x$ is also sometimes used for MXenes in which T is added to represent the surface functional groups, for instance, F, O, OH, H_2O, and Cl (Zhang *et al* 2021). Their precedent $M_{n+1}AX_n$ phases possess a hexagonal structure. Commonly, two basic structural components in their simplest units can be distinguished: octahedral units (M_3X units), which have common edges and are identical to the NaCl cubic face-centered lattice of the type of binary nitrides or carbides, as well as element layers placed between them. In this case, the X atoms are located in the space of octahedral interstices. It should be noted that M–X chemical bonds have a combination of metallic, ionic, and covalent properties, whereas M–A chemical bonds only have metallic properties. As the covalent bond is stronger than the metal one, a suitable etching reagent can break the M–A bonds and remove the elements of the A layer (Xiao *et al* 2020). Consequently, selective etching of elements A (such as In, S, Al, Ga, and Si) results in the layered M formation with an 'accordion-like' structure.

The properties of the MXenes may differ depending on the synthesis strategy, the starting materials used, and surface modification technologies (Pogorielov *et al* 2021). The first MXene was obtained from MAX phases via selective etching of element A (Naguib *et al* 2011). These materials are the main component for synthesizing 2D carbides/sodium/carbonitrides of transition metals. It was found that new elements adding to the structure of the MAX phases at the M or A positions can lead to combining their advantages in the systems, such as high-temperature resistance to corrosion and oxidation (Zhou *et al* 2008).

10.3.2.1 Strategies of MXene synthesis

Two main strategies have been applied to synthesize MXenes: the 'top-down' and the 'bottom-up' methods. The top-down approach includes separating the bulk materials (MAX phases) into several layers. The bottom-up approach applies deposition technology, and the material forms from the bottom to the top. One popular method for synthesizing MXenes is to etch a specific layer from a bulk primary substance, and MXenes are synthesized from top to bottom due to the different chemical bond strengths in MAX phases. It is possible to synthesize 2D MXenes from layered precursors by selectively etching certain elements using other etching reagents (Pogorielov *et al* 2021). As mentioned before, the M–X bonds are stronger than the M–A bonds, so the elements of the A layer can be selectively removed without destroying the M–X bonds. Top-down methods for MXene synthesis typically involve two steps. The first step is to selectively remove the elements from layer A by placing the precursor material in an etching reagent for a certain time. Then, sonication of the treated materials is used to dissociate them into separate nanolayers to obtain ultra-thin 2D MXenes. The ultrasonic devices used for this purpose are bath and probe/tip sonication systems. MXenes with smaller flake sizes can be produced with probe sonicators, while bath sonication is more suitable for producing MXenes with larger flakes. Power, amplitude, and frequency are critical parameters in the sonication process (Malaki *et al* 2019).

In addition to the top-down approaches, bottom-up methods have also been developed for 2D carbide fabrication (Soundiraraju and George 2017). The bottom-up approach is based on the MXenes' formation through epitaxially controlled deposition of multilayer films. The top-down systems require ultrasonic treatment to decrease the MXene size, which can lead to structural defect formation. In contrast, the top-down approach leads to large transverse-dimension MXenes with significantly fewer defects in their structures (Xu *et al* 2015).

10.3.2.2 MXene properties and application

Due to the unique properties of MXenes, they are suitable for use in areas such as wireless communication, electromagnetic shielding interference, energy-storage systems, chemical sensors, catalysis, optoelectronics, and flexible electronics. These properties include chemical stability, high thermal and electrical conductivity, hydrophilicity, and excellent biocompatibility (Pogorielov *et al* 2021). It was found that the MXene characteristics can be controlled by adjusting their layer distances, changing their functional groups, MAX phase doping in the M, A, or X position, and changing their composition. This controllability makes MXenes more attractive than other existing 2D materials. Adjusting MXene's characteristics can be addressed by processing, functionalization, and doping, while graphene can only be functionalized (Zhu *et al* 2021).

MXene's stability in water and oxygen is limited, and its oxidation can be accelerated in light. The MXene surface is oxidized at high temperatures, and the transition metal oxidation forms gradually. Hence, MXenes oxidation can be limited by isolating them from oxygen and water and using a vacuum, low temperatures, and an inert gas atmosphere. The various structural phase transformations of MXenes are also possible. The transformation temperature and structures depend on MXene's composition and storage conditions. Thus, the MXenes suspensions should be stored in a refrigerated, oxygen-free, dark environment (Zhu *et al* 2021). New MXenes with various chemical compounds, structures, and characteristics are rapidly discovered. Due to the variety of inherent MXene properties, they have been successfully applied for different applications. For instance, their notable electrical and optical characteristics have made them suitable for sensor design (Szuplewska *et al* 2020). Recently, electrical power systems have been developed that require new dielectrics with a high dielectric constant and low dielectric loss to be effective for electrostatic film capacitors with flexibility, suitable energy density, and acceptable breakdown strength. The combination of polymers and MXene nanosheets has led to the optimal composition with a high dielectric constant and low dielectric loss simultaneously (Li *et al* 2019b).

MXenes are actively studied for their potential medical applications. They also displayed low cell toxicity and appropriate biocompatibility, which make them good candidates for use as drug delivery systems (Huang *et al* 2018). Their ultrafine planar structure provides a large surface area for therapeutic loading agents. MXene nanosheets have inherently strong near-infrared (NIR) absorption and excellent photothermal conversion properties. Therefore, they can be further used as photothermal agents for cancer treatment via hyperthermia exposure to external NIR laser irradiation. These benefits make MXenes suitable for the photothermal drug delivery systems, bioimaging, theranostic systems, antibacterial substances, and an electroconductive base in biomaterial science.

The photothermal therapy (PTT) is a non-invasive and light-triggered cancer treatment strategy (Kotagiri *et al* 2015). In this approach, the NIR light is absorbed by a photothermal agent that transforms light into thermal energy and generates heat minimally invasively. PTT is regarded as a potential substitute for the conventional clinical cancer treatments (Chen *et al* 2014). MXenes possess a wide-spectrum photothermal conversion range (750–1350 nm), allowing maximal penetration into the living tissues (Han *et al* 2018). In addition to PTT, the MXenes can be modified with specific anti-cancer antibodies to provide a targeted cancer PTT, leading to accumulation at the target sites and decreasing side effects. The active MXene surface with high capacity can also be loaded with anti-cancer and ROS agents simultaneously to combine chemotherapy and photodynamic therapy. Despite much research, some challenges about the MXenes' biological effects need to be addressed, such as their tissue distribution and metabolism, long-term biological effects, and their influence on immune cells.

Another medical application of MXenes is as an antibacterial agent. The antibacterial activity of MXene was reported for the first time in 2016 (Rasool *et al* 2016). MXenes showed more than 98% effectiveness against both strains of Gram-positive bacteria and Gram-negative bacteria at 100 $\mu g\ ml^{-1}$ concentration within 4 h. It was observed that MXenes have dose-dependent effectiveness against both strains of bacteria. However, they have a more substantial influence on gram-positive strains. This could be due to the fact that the wall of this strain has a more negatively charged surface. MXene's antibacterial effects are thought to be caused by three mechanisms: (a) sharp MXene invading bacteria walls and disintegrating their cell membrane; (b) forming a conductive bridge over the bacteria's lipid bilayer and blocking ion transports; and (c) ablating bacteria via photothermal effects caused by NIR irradiation (Rasool *et al* 2017, Mayerberger *et al* 2018, Jastrzębska *et al* 2019, Arabi Shamsabadi *et al* 2018). In addition, it is not necessary to sink photothermal materials in the water; they can be placed directly at the water–air interface, reducing heat loss to bulk water. Recent studies have revealed that

MXene, with its high efficiency in solar energy absorption, has tremendous potential for applications in solar steam generators. In 2017 the MXenes were studied as a solar photothermal material for the first time (Cai *et al* 2017). Furthermore, the MXene-based membranes with non-wetting characteristics can naturally repel water, effectively purify water from contamination, and have environmentally friendly operation in the long term (Al-Hamadani *et al* 2020).

Due to MXene's economy, simplicity, and absence of secondary pollution, it is also used for purifying water in an adsorption approach (Yao *et al* 2021). Adsorption is regarded as an effective approach for eliminating ions of heavy metals because other strategies, such as chemical reactions and biological processing, cannot metabolize or decompose them. MXenes' inherent properties, such as their large surface area, an abundance of active sites, good reducibility and dispersibility, and intercalation of many cations between their layers, make them promising candidates for water purification via heavy metal ions, organic dyes, and radionuclide adsorption (Kim *et al* 2020). One of the other interesting fields where MXenes are implemented effectively is wearable electronics with multifunctional operations. Wearable and flexible smart MXene-based textiles have been applied in healthcare electronics (Lee *et al* 2020), pressure sensors (Li *et al* 2019a), electromagnetic interference shielding (Liu *et al* 2019), antibacterial medical devices (Liu *et al* 2020b, Kyrylenko *et al* 2020), human motion monitoring (Lou *et al* 2020), and so forth. However, these textiles have some limitations, such as difficulties integrating necessary functions and maintaining their intrinsic properties (Zheng *et al* 2021, Du *et al* 2020). MXenes, as conductive materials, enable the creation of wearable electronic smart textiles to maintain warmth, protect from electromagnetic irradiation, and sense human temperature and movements during daily activities.

10.3.2.3 MXene-based photocatalysts for environmental and water treatment

Photocatalysis is a most safe and economical approach for removing toxic pollutants (Ihsanullah 2022, Velempini *et al* 2021, Siwal *et al* 2022). MXenes are used extensively to remove environmentally hazardous pollutants and as co-catalysts for enhancing photocatalytic degradation in combination with other substances (Sreedhar and Noh 2021, Assad *et al* 2022). Several studies have investigated the photocatalytic ability of MXenes to degrade heavy metals, dyes, and pharmaceutical pollutants. For example, MXene/TiO_2 nanocomposites were constructed for the photocatalytic degradation of organic pollutants (Biswal *et al* 2022). For instance, MXene (Ti_3C_2) nanosheets modified with CuS particles were used to enhance the catalytic persulfate activation under light (Wang *et al* 2022). CuS@MXene composites showed notable reactivity for the rapid elimination of orange II and acceptable stability, even for reusability in four cycles.

It was found that MXene efficiency in photocatalytic degradation can be improved by hybridization techniques (Iravani and Varma 2022). Several studies have hybridized MXenes with other substances, such as graphite-based materials, metal organic frameworks (MOFs), polymers, carbon-based materials, etc (Zhang *et al* 2020c, Yang *et al* 2019, Mostafavi and Iravani 2022). MXenes' hybridization with graphite-based materials has been investigated widely due to their interesting properties. For instance, graphitic layers were attached to the graphitic carbon nitride and TiO_2 and were then used for photocatalyst fabrication by applying Ti_3C_2 MXenes via *in situ* calcination reaction (Han and Wu 2019). These hybridized photocatalysts exhibited outstanding degradation efficiency for organic compounds (rhodamine B and bisphenol A) and pharmaceutical pollutants (tetracycline and ciprofloxacin) under visible light.

Additionally, this hybridization yielded a reusable and stable photocatalyst that enhanced light adsorption and carrier separation. In this system, the photocatalytic degradation rate was remarkably improved thanks to the synergistic effects of graphene, g-C_3N_4, and TiO_2 in the separation and accumulation of electrons and oxidation ability (Wu *et al* 2020b). It was reported that incorporating carbon nanotubes with MXene composites led to the higher photocatalyst degradation efficiency in successive cycles than pure MXenes (Thirumal *et al* 2022). Ti_3C_2 MXene/O-doped g-C_3N_4 Schottky-junction (Thirumal *et al* 2022) and $Ti_3C_2T_x$/Ti_3AlC_2@Ag (Lv *et al* 2022) composite are other examples of hybridized MXenes to obtain more efficient photocatalytic activity using synergy between their compounds.

In addition, a composition of multi-layered MXene nanosheets and transition metal oxide was fabricated for the photocatalytic elimination of pollutants with high adsorption capacity. Ti_3C_2 MXene nanosheets composited with bismuth molybdate, zinc oxide, and tin dioxide were synthesized for the catalytic removal of high-concentration methylene blue (150 mg l^{-1}) and 4-chlorophenol solution (20 mg l^{-1}) under visible light irradiation (Zhang *et al* 2022b). According to this study, hydroxyl and superoxide free radicals were the major reactive species produced in this degradation process. This study showed that coupling MXene and transition metal oxide could prevent the recombination of photo-generated electron–hole pairs. Due to the synergistic effect of eliminating pollutants and strong adsorption, these MXene-based nanocomposites suggest promising photocatalysts to control environmental pollution.

Due to the triggered interfacial electron transfer, the heterostructure design of MXene-based materials can improve their photocatalytic performance (Assad *et al* 2022). The hierarchical ZnO/MXene heterostructures composed of the chimeric ZnO nanosheets on the accordion-shaped surfaces of Nb_2C and V_2C MXenes were fabricated as photocatalysts for degrading methylene blue (Zhou *et al* 2022). These structures exhibited good stability and photocatalytic activity, with a degradation rate of 62.62% and 99.53% for ZnO/Nb_2C and ZnO/V_2C, respectively, after UV irradiation for 2 h. After four cycles, the degradation rate was only 22.0% for pure ZnO, reaching 58.6% and 97.3% for ZnO/Nb_2C and ZnO/V_2C, respectively. As a result, the ZnO/MXene hierarchical structures may optimize the carrier transfer pathway and facilitate photo-generated electron transfer from the ZnO conduction band to the MXenes conduction band, increasing the photocatalytic degradation activity. This strategy enhances the catalytic efficiency of 2D photocatalytic materials, paving the way for generating high-yielding photocatalysts.

Recently, hybridizations of MXenes and MOFs have been developed for the photocatalytic degradation of pollutants. As MXenes typically cannot be used directly as photocatalysts, and this issue may limit their photocatalytic applications (especially in water treatment), their hybridization with MOFs can yield improvements in their characteristics and performance. Ti_3C_2 MXenes/NH_2-MIL-88B(Fe) composites were designed by Long *et al* (2021). This system showed 90%, 83.41%, and 44.26% photocatalytic removal activity for methylene blue (50 ppm, 30 min), rhodamine B (25 ppm, 120 min), and MO (20 ppm, 120 min), respectively. The MXenes could help to inhibit the MOFs' aggregation, increasing their effective area and acting as a trap for photo-generated electrons in the photocatalytic process, resulting in an upgrade to the MOF's charge separation efficiency (Long *et al* 2021). Table 10.2 lists more examples of MXene-based materials that have been investigated in the photocatalytic degradation process.

Despite the excessive development of MXene applications, some challenges still need to be addressed before MXene-based products can enter the market. Their first challenge is using toxic fluorine-containing acidic reagents for MXene synthesis. Although the bottom-up method could partially solve this problem, a green synthesis technology with up-scaleability is still needed because MXenes' primary materials are not easily available for mass production, and their price is still high. Mxene oxidation can occur during storage, leading to changes in the characteristics. New technologies and solutions have significantly improved MXene synthesis and properties and opened wide perspectives for their applications.

10.3.3 Graphene-based 2D-NLs

Graphene is a 2D sheet composed of carbon atoms with sp^2 hybridization in hexagonal conformation arranged in honeycomb-like lattices (Priyadarsini *et al* 2018). Due to its specific structure, it possesses many desirable properties such as high crystallinity, great mechanical strength, extraordinary transparency, electrical conductivity, and molecular barrier abilities (Lee *et al* 2008, Kuilla *et al* 2010, Cui *et al* 2016, Ghulam *et al* 2022). However, using pristine graphene has some limitations, such as having no band gap, difficult synthesis via bottom-up approaches, high hydrophobicity, poor water-solubility, and aggregation in solution by van der Waals interactions (Kuilla *et al* 2010, Zhu *et al* 2010, Niyogi *et al* 2006). An approach to solve these challenges and develop graphene applications is to prepare its derivatives. Therefore, its derivative compounds have been fabricated in similar structures that can be synthesized via top-down approaches to possess the advantages of pristine graphene and improve their water dispersibility. For example, graphene oxide (GO) was produced by treating graphite or other carbon sources with strong oxidants to functionalize graphite layers with hydroxyl, carbonyl, epoxy, and carboxyl groups. This surface modification with a polar oxygen-containing group provides some advantages for GO, including simple synthesis, high hydrophilicity, and excellent dispersibility in water and other solvents. In addition, these oxygen-containing functionalities can act as reactive sites for GO to undergo more chemical modification to develop GO-based composites (Razaq *et al* 2022).

Although oxygen-containing functionalization can solve some limitations of graphene, it also causes some disadvantages for GO. Despite a unique set of

Table 10.2. Some examples of MXene-based composites with photocatalytic degradation activity.

MXene-based material	Pollutants	Advantages	References
α-$Fe_2O_3/ZnFe_2O_4/Ti_3C_2$	Rhodamine B and Cr (VI)	Improved photocatalytic activity.	Zhang *et al* (2020a)
Ti_3C_2/MoS_2 nanocomposites	Methyl orange	Excellent removal of pollutants.	Yao *et al* (2020)
Ti_3C_2-bridged Ag/Ag_3PO_4 composites	MO and Cr (VI)	Improved photocatalytic efficiency associated with the synergetic effects of Ag, Ag_3PO_4, and MXenes, promoting visible light absorption.	Sun *et al* (2021)
(Bi/BiOCl) microspheres/ Ti_3C_2	Ciprofloxacin	High photocatalytic efficiency; excellent reusability (five cycles).	Wu *et al* (2020a)
$CuFe_2O_4/Ti_3C_2$	Sulfamethazine	The synergistic degrading activity under visible light; detecting of S–N bonds breaking.	Wu *et al* (2020a)
Ti_3C_2/g-C_3N	Phenol	Excellent removal efficiency; using the MXenes leads to 32.1% phenol degradation under dark conditions.	Liu *et al* (2020a)
Magnetic-$Ti_3C_2T_X$	Diclofenac	Enhanced degradation efficiency; excellent stability and reusability (up to seven cycles).	Jang *et al* (2020)
TiO_2/Ti_3C_2 modified with Ag particles	Nitrate	The improved photoelectron density and carrier separation by applying Ag particles; high elimination rate (96.1%) with good selectivity to nitrogen (92.6%); remarkable photocatalytic efficiency (89%) up to five cycles.	Chen *et al* (2022)
Ti_3C_2/Ag_2ZnGeO_4 Schottky heterojunctions	Acid red 1	Enhanced photocatalytic activity under visible light with high charge separation.	Zhong *et al* (2021)
La/Mn-co-doped $BiFeO_3$/ Ti_3C_2 composites	Congo red	Fast and high-yield degradation (92%–93%).	Iqbal *et al* (2019)
$TiO_2/Ti_3C_2T_x$	Carbamazepine	High photocatalytic degradation (98.67%) under UV light.	Shahzad *et al* (2018)
Rod-like Nb_2O_5/Nb_2CT_x	Rhodamine B and tetracycline	High photocatalytic degradation under visible light irradiation; high cycle stability.	Cui *et al* (2021)
Carbon/Ti-based oxide nanocomposites/$Ti_3C_2T_x$	Methylene blue	High removal efficiency (90%) under visible light within 60 min; high charge separation with enhanced absorption ability.	Liu *et al* (2022)

mechanical, colloidal stability, and optical characteristics, oxygen-containing functionalization causes some disadvantages for GO. In the adopted methods for GO synthesis, strong oxidants, and acids are used, resulting in several defects in its network (Tarcan *et al* 2020). The reduced graphene oxide (rGO) can be produced by reducing GO through different methods such as thermal, chemical, and photo-irradiation. The major advantages of rGO are its facile synthesis and processing, which lead to many applications (Tarcan *et al* 2020). The production of rGO by chemical GO reduction could restore its conductivity to some extent. rGO still carries some oxygen-containing functional groups, which results in its good dispersibility in many solvents. Furthermore, the electrical properties and dispersibility of rGO can be adjusted by controlling the remaining number of oxygen-containing functional groups. The properties of rGO produced by chemical reduction are approximately similar to those of pristine graphene (Khan *et al* 2016, Eda and Chhowalla 2010).

10.3.3.1 Synthesis of graphene-based 2D-NLs

10.3.3.1.1 GO

Brodie prepared GO by adding potassium chlorate to graphite under fuming nitric acid for the first time in 1859. Staudenmaie revised this procedure to produce highly oxidized GO using concentrated sulfuric acid in 1898. Hummer's method was then developed in 1958. Hummer's method uses potassium permanganate and sodium nitrate in concentrated sulfuric acid to produce GO. This reaction occurs in a shorter time with high efficiency and provides more safety by using potassium permanganate instead of potassium chlorate and sodium nitrate instead of fuming nitric acid (Alam *et al* 2017). Nevertheless, Hummer's method has some drawbacks, such as low yield, releasing harmful gases (nitrogen dioxide and dinitrogen tetroxide), and generating ions (sodium and nitrosonium), whose elimination from wastewater is difficult (Alam *et al* 2017, Yu *et al* 2016). Currently, several strategies have been developed to improve Hummer's method and refine these drawbacks: (i) the sodium nitrate removal; (ii) adding a peroxidation step before potassium permanganate oxidation without sodium nitrate; (iii) increasing the amount of potassium permanganate instead of sodium nitrate; and (iv) removing sodium nitrate and replacing potassium permanganate with potassium ferrate (Yu *et al* 2016). Marcano *et al* reported that removing sodium nitrate leads to more potassium permanganate. While accomplishing the reaction in a mixture of sulfuric acid, phosphoric acid (ratio of 9:1) improved the oxidation efficiency and hydrophilic product yield. In addition, this modification avoided harmful gas production, and the temperature was more controllable (Marcano *et al* 2010). Various strategies have been explored for GO synthesis in the past few years, but many are costly and sophisticated.

For this reason, Hummer's method or its modified versions are the most commercially available synthesis methods for GO synthesis (Guo and Dong 2011). GO was also prepared by direct oxidation of sugarcane bagasse in a muffle furnace at 300 °C for 10 min under atmospheric conditions (Somanathan *et al* 2015). It has been reported that the sugarcane bagasse was completely oxidized into GO.

10.3.3.1.2 rGO

Exhaustive studies have been conducted to reduce the oxygen-containing functional groups of GO to produce materials with characteristics as close as possible to pristine graphene (Stankovich *et al* 2007, De Silva *et al* 2017). GO reduction can be achieved through different strategies, including thermal, chemical, and electrochemical, leading to other products in terms of electrical properties, morphology, etc. The important factors in GO reduction are the carbon/oxygen ratio of rGO, the selectivity reduction of one type of oxygen-containing group, improving the rGO surface defects, choosing green reducing reagents, and improving or maintaining the desired characteristics of GO. The thermal GO reduction is done by decomposing the oxygen-containing functional groups into carbon dioxide and carbon monoxide gases at high temperatures (Pei and Cheng 2012). It has been reported that the fast release of these gases delaminates individual GO nanosheets. A thermal reduction can be accomplished by different techniques, such as thermal annealing at high temperatures in an oxygen-free environment, microwaving GO powders, and flash reducing GO sheets in high-intensity light. The chemical reduction is done by adding reducing agents to the GO dispersion (Chua and Pumera 2014). The common reducing agents, supported well by the literature, include hydrazine, metal hydrides, or hydrohalic acids (Pei and Cheng 2012, Stankovich *et al* 2007). GO reduction can be prepared under photocatalyzed conditions. Williams *et al* reduced GO by applying TiO_2 as a catalyst in the presence of UV light (Williams *et al* 2008). An electrochemical approach is also used to minimize GO which does not require chemical agents. This method obtains rGO via electron exchange between GO and the electrodes in a typical electrochemical cell. Recently, using green reducing agents for synthesizing rGO, such as sugars, ascorbic acid, amino acids, and even microorganisms, has been increasing (De Silva *et al* 2017). These different methods provide various advantages in the end product properties, energy usage, scalability, and the amount of chemical waste produced.

10.3.3.2 Properties of GO

With the progress of science and technology, more new materials with fascinating characteristics have been obtained that can be used in many fields. Graphene-based 2D-NLs are among these materials that have received much attention due to their fascinating properties, such as stability, high mechanical strength, electrical properties, charge storage capacity, etc. It was reported that pristine monolayer graphene has a break strength of 42 N m^{-1}, a Young's modulus of 1.0 TPa, an intrinsic tensile strength of 130.5 GPa, and a fracture toughness of 4.0 ± 0.6 MPa $m^{1/2}$. To achieve these properties, GO and rGO blends are produced. The values of these properties were discovered to vary depending on the number of surface functional groups and structural defects produced during oxidation or other reduction processes. Monolayer GO was fabricated by a modified Hummer's method with an order-of-magnitude drop in Young's modulus (207.6 ± 23.4 GPa), but the Young's modulus is still notably high (Suk *et al* 2010). In the same way, the rGO monolayer was produced by Hummer's method, followed by thermal reduction with Young's modulus of 250 ± 150 TPa (Gómez-Navarro *et al* 2008). GO and its derivatives have been used widely in polymer nanocomposites to modify their mechanical properties. For example, GO has been applied as a filler in polyvinyl alcohol (PVA) films fabricated by solution casting (Cheng-an *et al* 2017). It was observed that GO (20% filler content) could increase the tensile strength of the GO-PVA nanocomposite by more than five times the strength of the pure PVA film. In another study, the incorporation of GO nanosheets (0.1 wt.%) with polyurethane (PU) in a vulcanizing press could improve the tensile strength by up to 16.4% (Jiang *et al* 2019). rGO/polylactic acid (PLA) nanocomposites were prepared by melt blending on a twin-roller mill (Bao *et al* 2012). In this study, GO was first synthesized from graphite powder under pressurized oxidation conditions and then chemically reduced to rGO. The inclusion of rGO (2 wt.%) in PLA nanocomposite resulted in an increment of the storage modulus from 2.56 to 4.04 GPa.

Graphene is electrically conductive (6500 S m^{-1}) with high electron mobility (25 m^2 V^{-1} s^{-1}) (Novoselov *et al* 2012, Park and Ruoff 2009). It has been shown that G can improve the electrical conductivity of polymers greatly at low filler contents. For example, it could increase polystyrene's conductivity to 0.1 S m^{-1} at 1 vol% (Stankovich *et al* 2006). However, during GO synthesis, the sp^2 bonding orbitals of graphene were disrupted, and various surface functional groups were added, inhibiting its electrical conductivity and making GO electrically resistive (1.64×104 Ω m) (Wang *et al* 2012b, Tang *et al* 2012). Due to the high resistivity of GO, reduction techniques have been explored to form rGO. When GO is reduced, electrical conductivity is greatly improved (Pei *et al* 2010). Nevertheless, the rGO still contains residual sp^3 bonding orbitals that disrupt the electron movement via the remaining sp^2 bonding orbitals, leading to their transportation by hopping (Bagri *et al* 2010). Regarding the improved electrical conductivity of rGO, it has been used as a conductive filler added to polymers for many materials. Kim *et al* have reported fabricating a highly flexible supercapacitor of polyaniline (PANI) with rGO with a capacitance of 431 F g^{-1} that has a higher electrical conductivity over its components (Kim *et al* 2014). More compact PANI chains achieved this improved performance via interactions with rGO. Hou *et al* developed an electrically conductive paper composed of rGO and cellulose nanofiber (CNF) by sandwiching CNF between rGO layers (Hou *et al* 2018). In addition to the example mentioned above, rGO has also been applied to improve electrical properties in combination with various other polymers such as polyethylene terephthalate (Zhang *et al* 2010), polydimethylsiloxane (Song *et al* 2016), polyvinyl alcohol (Li *et al* 2015), polyurethane (Xing *et al* 2017), etc.

Graphene is known to have one of the highest in-plane thermal conductivities with a thermal conductivity of ~3000–5000 W m^{-1} K^{-1} (Zhu *et al* 2010). On the other hand, the synthesized GO from graphite has a low thermal conductivity (0.5–1 W m^{-1} K^{-1}) that is not ideal for most thermal applications (Renteria *et al* 2015). Hence, GO reduction is necessary to improve its thermal conductivity (Kim *et al* 2018). It was shown that reducing GO films to obtain rGO via annealing at high temperatures can remarkably enhance the in-plane thermal conductivity (Renteria *et al* 2015). The rGO combination with polymers has significantly improved thermal conductivity in the final nanocomposites. A nanocomposite combining rGO with poly (vinylidene fluoride-co-hexafluoropropylene) was fabricated by Kumar *et al* (2019). This nanocomposite demonstrated a thermal conductivity that was significantly higher than that of the neat polymer and many common alloys. They believed that the improvement in the thermal conductivity resulted from the high conductivity of the rGO.

While high thermal conductivity is useful for many applications, it is not required for all situations. On the other hand, providing high thermal insulation properties is useful in some domains, such as in-home insulation and flame-retardants. Recently, GO had acted as an effective filler to enhance flame-retardant properties in combination with various polymeric nanocomposites (Higginbotham *et al* 2009, Yu *et al* 2014a, 2015a, Wang *et al* 2012a, Zhang *et al* 2016). For example, incorporating GO and sepiolite clay nanorods with cellulose nanofibers could develop a super-insulating flame-retardant foam that is applied as an insulating material (Wicklein *et al* 2015). This study indicates the notable potency of GO nanofillers to provide important thermal properties in polymeric systems.

10.3.3.3 Applications of graphene-2D-NLs for water treatment

As mentioned above, due to the highly tunable nature of graphene-based 2D-NLs, various mechanical, electrical, and thermal properties are achievable. These materials can be incorporated effectively into nanocomposite materials for use in a broad range of applications and present great potential for developing new and improved products. Membranes and coatings, stimuli-responsive materials, energy-storage, sensors, bioimaging, medical, and biotechnology applications are common for graphene-based 2D-NLs. Recently, it was found that incorporating graphene-based derivatives into photocatalytic composites can improve their disinfection rate (An and Jimmy 2011, Xiang *et al* 2012, Sasidharan *et al* 2021). These materials possess a long-range π-conjunction that can increase electron trapping and electron mobility in photocatalysis. In addition, GO layers have rich oxygen-containing functional groups that allow adsorbing of reactive substances and can be used for further surface modification. Finally, these materials can make a good physiochemical interaction with photocatalytic substances, contributing to increment light absorption (Xiang *et al* 2012, Ji *et al* 2016, Zeng *et al* 2017a, Ashfaq *et al* 2022).

Furthermore, the graphene-based 2D-NLs have demonstrated significant potential as a new, efficient approach to dealing with water contamination caused by microorganisms. Graphene-based derivatives have a broad spectrum of antibacterial actions against microorganisms such as viruses, bacteria, and fungi (Mohammed *et al* 2020). The antimicrobial mechanism of graphene and its derivatives has three main steps: (i) microorganism deposition on the graphene surface; (ii) causing cell membrane stress via direct contact with graphene; and (iii) producing superoxide radical anions (Mangadlao *et al* 2015). The antibacterial effects of graphite, GO, and rGO were compared on *E. coli* at the same concentration and exposure time. The results showed that GO with more oxygen-containing functional groups had the most antibacterial activity, followed by rGO and graphite (Liu *et al* 2011b). Regarding these points, graphene-based derivatives, combined with semiconductive substances, yield a multifunctional platform with several advantages, such as being a

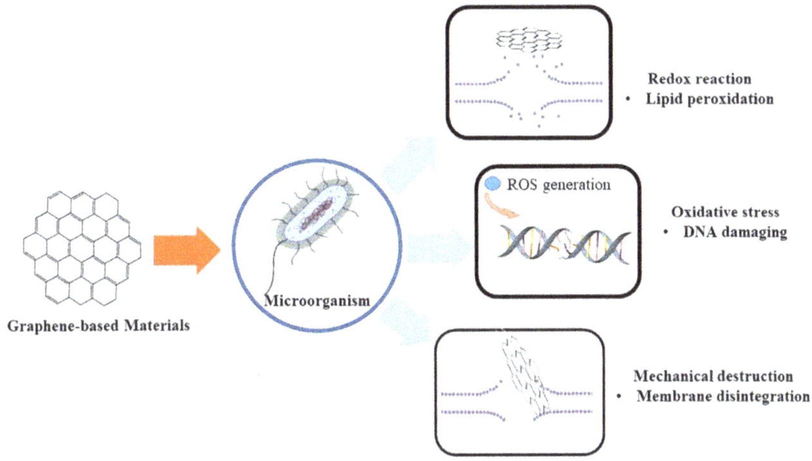

Figure 10.3. Schematic of the proposed mechanism of graphene-based materials for antimicrobial activity.

photosensitizer, electron sink, deposition substrate, conductive mediator, and absorbent (figure 10.3).

TiO_2 is a popular semiconductor with remarkable performance in photocatalytic disinfection. However, its application is limited because it only responds to UV light regions. To solve this limitation, several studies investigated incorporating GO with TiO_2. For instance, graphene nanosheets were produced from GO reduction by UV light and simultaneously deposited on the surface of an anatase TiO_2 sheet (Akhavan and Ghaderi 2009). Compared with TiO_2 film, the resultant G-TiO_2 film exhibited an enhancement in bactericidal activity in solar light. In another study, commercial TiO_2 nanoparticles were anchored to graphene through UV reduction, and the photocatalytic mechanism of the TiO_2–rGO composite was investigated (Cruz-Ortiz *et al* 2017). It was found that the main reactive species produced under UV–visible light was H_2O_2, while singlet oxygen was the main reactive species under visible light conditions. Additionally, the amount of singlet oxygen was increased in the presence of chlorine, which evolved from chloride. Liu *et al* used a two-phase hydrothermal reaction to create 1D TiO_2-nanorods on GO sheets. TiO_2 assembling on GO yielded fast charge transfer and an anti-charge recombination effect, leading to a decrease of *Escherichia coli* (an 8.2-log) under solar light after 2 h. Liu *et al* assembled TiO_2-nanorods on GO sheets by a two-phase hydrothermal reaction (Liu *et al* 2010, Liu *et al* 2011a). CdS–GO composite was prepared by using a two-phase self-assembly method for sterilization (Gao *et al* 2013a). The excellent photocatalytic performance of G-based derivatives has increased their incorporation into 2D TiO_2 nanosheets (Gao *et al* 2014). These 2D TiO_2 nanosheets can be arranged in a radiated pattern that uses incident light effectively.

The graphene-based derivatives were also applied to generate antibacterial photocatalysts in combination with ZnO as a semiconductor with a broad band gap. The ZnO acts as a photocatalyst under visible light. GO incorporation is an electron transfer booster that significantly promotes ZnO performance (Wu *et al* 2015). Gao *et al* fabricated a hierarchical SGO–ZnO–Ag composite using the synergetic advantages of sulfonated GO (SGO) and Ag (Gao *et al* 2013b). It has been reported that G incorporation forms an electronic bridge and increases light scattering related to a tree-like hierarchical structure, increasing the antibacterial rate. All these findings indicate that the G-based antibacterial photocatalyst provides increasing progress in combining with diverse semiconductors for water disinfection. More examples of the graphene-based antibacterial photocatalyst are listed in table 10.3.

GO-based membranes are also considered promising for water treatment and desalting applications thanks to their high water permeability (Hu and Mi 2013, Nair *et al* 2012, Huang *et al* 2013) and stability in water (Yeh *et al* 2015). Therefore, notable progress has been generated in developing high-permeability membranes while retaining high rejection rates over the past few years. The graphene-based derivatives with excellent electrical properties have provided a promising material for the energy-storage field, such as in lithium-ion batteries (Lin *et al* 2016, Cao *et al* 2016, Fu *et al* 2016, Huang *et al* 2015), supercapacitors (Chee *et al* 2015, Wang *et al* 2015, Lamberti *et al* 2016), and stretchable electronics (Xing *et al* 2017, Kim *et al* 2009, Jang *et al* 2016). MnO_x quantum dot–TiO_2–rGO is an example of a photocatalytic graphene-based system with an energy-storage function (Cao *et al* 2015). In this system, the MO_x could reserve electrons excited from TiO_2. The rGO can generate reactive species in the dark using storage electrons. Overall, combining ternary components in these semiconductor systems can rectify the shortcomings of binary composites and provide hybrids with more desirable features.

Despite the remarkable progress that has been made, some challenging issues still need to be resolved to incorporate graphene-based derivatives into industrial applications. Although their synthetic methods encourage large-scale production, future studies are required to further control and adjust their properties. While the various approaches can almost improve their properties, uniform properties with reproducibility are needed to develop industrial products. Most of their special functionalities have unscalable synthesis methods that are costly and time-consuming. Therefore, further study must be conducted to generate graphene-based materials using simple, common industrial techniques without losing their special characteristics. As discussed, graphene-based materials have many benefits; however, several studies have presented their negative effects on biological systems, including human health. These materials can accumulate in the environment and affect all living organisms. Therefore, their toxicity is a critical issue that needs to be evaluated before any application. It was found that their cytotoxicity depends on their dosage, size, exposure time, and functionalized compounds. Regarding these factors, it is feasible to apply graphene-based materials safely. Furthermore, it is important to gain more knowledge of their bioaccumulation, biotransformations, depuration, and exposure effects on living things.

Table 10.3. G-based 2D-NLs for water treatment and energy storage.

Materials	Field of application	Advantages	References
rGO	Photocatalytic disinfection	Preventing microbial growth.	Alayande *et al* (2019)
GO–TiO_2–Ag	Photocatalytic disinfection	Complete *E. coli* inactivation performance under solar irradiation; good reusability.	Liu *et al* (2013)
SGO–ZnO–Ag	Photocatalytic disinfection	Complete *E. coli* inactivation performance under visible light.	Gao *et al* (2013b)
TiO_2/C-dots/rGO	Photocatalytic disinfection	Increment in *E. coli* reduction in solar irradiation.	Zeng *et al* (2017b)
Porphyrin/TiO_2–G	Photocatalytic disinfection	Increment in *E. coli* reduction under visible light.	Rahimi *et al* (2015)
MnO_x quantum dots–TiO_2–rGO	Photocatalytic disinfection	Photocatalytic ability in the dark with an energy-storage function.	Cao *et al* (2015)
WO_3–rGO–TiO_2	Photocatalytic disinfection	Notable *E. coli* inactivation performance in solar irradiation; perfect reusability.	Zeng *et al* (2017c)
Ag–AgBr–G	Photocatalytic disinfection	Complete *E. coli* inactivation performance under visible light; stability for repetitive use.	Xia *et al* (2016)
Ag_3PO_4–GO	Photocatalytic disinfection	Complete bacterial cell removal; improving stability under visible light irradiation.	Liu *et al* (2012)
CdS–GO	Photocatalytic disinfection	Improving inactivation efficiency under visible light irradiation.	Deng *et al* (2016a)
Carboxylated GO-CuS	Photocatalytic disinfection	High antibacterial activity toward *E. coli* and *B. subtilis*; good recovery and reusability.	Yu *et al* (2015b)
GO–g-C_3N_4	Photocatalytic disinfection	Remarkable *E. coli* inactivation performance under visible light.	Sun *et al* (2017)
Bi_2MoO_6–GO	Photocatalytic disinfection	Enhancement in the visible-light-driven photocatalytic destruction of bacteria.	Zhang *et al* (2013)
g-C_3N_4–rGO–S_8	Photocatalytic disinfection	Complete *E. coli* inactivation performance under visible light.	Wang *et al* (2013)
TiO_2–Ag_3PO_4–GO	Photocatalytic disinfection	Increment in *E. coli* reduction under visible light.	Yang *et al* (2015)
rGO–AgNP–$Bi_2Fe_4O_9$	Photocatalytic disinfection	Complete *E. coli* inactivation performance under visible light.	Hu *et al* (2018)
Supercapacitor electrode of GO/polyurethane sponge	Energy storage	Increment in specific capacitance.	Xu *et al* (2016)

rGO/NaCl/urea film	Energy storage	Increment in specific capacitance.	Zhao *et al* (2017)
rGO electrode decorated with SnO_2	Energy storage	500 cycle life up with acceptable capacity retention.	Wang *et al* (2018)
rGO/polyamide nanocomposite films	Membranes and coatings	Increasing barrier properties, enhancing mechanical strength, thermal stability, flame resistance, and adhesive strength.	Kausar (2018)
GO/polyamide nanocomposite	Membranes and coatings	Improving the hydrophilicity of the nanocomposite membrane leads to better water desalination.	Bano *et al* (2015)

10.4 Summary and future perspective

Wastewater treatment, notably for industrial effluents and drinking water, is an important aspect of current and future water requirements. Existing studies have mostly concentrated on conventional AOP technologies for treating wastewater. The AOPs generate ROS with oxidation capabilities, such as •OH and $•O_2^-$, with strong reactivity to oxidize organic contaminants in wastewater. The AOPs are categorized as photochemical, non-photochemical, and hybrid technologies based on the ROS-generating mechanism utilized. The effective production of ROS and the safe breakdown of organic pollutants are at the core of AOPs. The AOPs are a new and effective way of treating wastewater that has the benefits of no secondary contamination and high mineralization effectiveness. The practical applicability of various AOPs is constrained in several ways. The severe reaction conditions and considerable treatment costs prevent its widespread utilization. In recent years, nanomaterials have received increased interest due to their versatility and large surface areas. There is a great deal of potential for the successful treatment of wastewater using nanomaterials as part of various AOPs. Mixed oxidation processes are one of the most effective methods for wastewater pollutant degradation in the presence of nanomaterials.

On the other hand, improved solar light use and cutting-edge interfacial architectures for quick charge separation are required for effective nanomaterial based AOPs. The many attributes of graphene have prompted scientists to consider using 2D-NLs in AOPs. 2D-NLs, such as MXenes, graphene, and its derivatives, metal oxides (such as TiO_2), and transition metal dichalcogenides (TMDs), have demonstrated strong catalytic activity against a variety of contaminants. They have remarkable contaminant selectivity thanks to their proper modification. Because of their excellent adsorption and reduction abilities, 2D-NLs and their changes may also be utilized to remove radionuclides, dyes, heavy metals, and other pollutants. Studies have also shown that 2D-NLs are highly stable and can regenerate and be reused after the adsorption of contaminates. Although 2D-NLs have many appealing qualities, they have certain disadvantages in AOPs. These materials have drawbacks since we do not fully understand how they work and because various pollutants have varying environmental effects. Furthermore, these nanomaterials have only been synthesized in laboratory settings. Thus, understanding the precise mechanism of nanomaterials used in the treatment process will be crucial for future studies to improve their effectiveness and overcome their drawbacks. The cost of production would also be reduced by developing and implementing widely a technique that obtains 2D-NLs. Despite the limitations, 2D-NLs show promise in building next-generation water treatment and energy technologies. It is envisaged that advancements in both mechanism and application research would lead to improvements in sustainable water treatment and energy technologies.

References

Abedi M, Abolmaali S S, Abedanzadeh M, Borandeh S, Samani S M and Tamaddon A M 2019 Citric acid functionalized silane coupling versus post-grafting strategy for dual pH and saline responsive delivery of cisplatin by Fe_3O_4/carboxyl functionalized mesoporous SiO_2 hybrid nanoparticles: A-synthesis, physicochemical and biological characterization *Mater. Sci. Eng.* C **104** 109922

Adeleye A S, Conway J R, Garner K, Huang Y, Su Y and Keller A A 2016 Engineered nanomaterials for water treatment and remediation: costs, benefits, and applicability *Chem. Eng. J.* **286** 640–62

Akhavan O and Ghaderi E 2009 Photocatalytic reduction of graphene oxide nanosheets on TiO_2 thin film for photoinactivation of bacteria in solar light irradiation *J. Phys. Chem.* C **113** 20214–20

Al-Hamadani Y A, Jun B M, Yoon M, Taheri-Qazvini N, Snyder S A, Jang M, Heo J and Yoon Y 2020 Applications of MXene-based membranes in water purification: a review *Chemosphere* **254** 126821

Alam S N, Sharma N and Kumar L 2017 Synthesis of graphene oxide (GO) by modified hummers method and its thermal reduction to obtain reduced graphene oxide (rGO) *Graphene* **6** 1–18

Alayande A B, Park H D, Vrouwenvelder J S and Kim I S 2019 Implications of chemical reduction using hydriodic acid on the antimicrobial properties of graphene oxide and reduced graphene oxide membranes *Small* **15** e1901023

Alimardani V, Abolmaali S S and Borandeh S 2019 Antifungal and antibacterial properties of graphene-based nanomaterials: a mini-review *J. Nanostruct.* **9** 402–13

Alimardani V, Abolmaali S S, Yousefi G, Rahiminezhad Z, Abedi M, Tamaddon A and Ahadian S 2021 Microneedle arrays combined with nanomedicine approaches for transdermal delivery of therapeutics *J. Clin. Med.* **10** 181

Ameta S and Ameta R 2018 *Advanced Oxidation Processes for Wastewater Treatment: Emerging Green Chemical Technology* (New York: Academic)

An X and Jimmy C Y 2011 Graphene-based photocatalytic composites *RSC Adv.* **1** 1426–34

Arabi Shamsabadi A, Sharifian G M, Anasori B and Soroush M 2018 Antimicrobial mode-of-action of colloidal $Ti_3C_2T_x$ MXene nanosheets *ACS Sustain. Chem. Eng.* **6** 16586–96

Arurlaj A, Mangalaraja R V and Khalid M 2022 MXene: Pioneering 2D materials ed M Khalid, A N Grace, A Arulraj and A Numan *Fundamental Aspects and Perspectives of MXenes* (Frankfrut: Springer) pp 1–16

Ashfaq M, Talreja N, Chauhan D, Rodríguez C A, Mera A C and Ramalinga Viswanathan M 2022 Synthesis of reduced graphene oxide incorporated bimetallic (Cu/Bi) nanorods based photocatalyst materials for the degradation of gallic acid and bacteria *J. Ind. Eng. Chem.* **110** 447–55

Assad H, Fatma I, Kumar A, Kaya S, Vo D V N, Al-Gheethi A and Sharma A 2022 An overview of MXene-based nanomaterials and their potential applications towards hazardous pollutant adsorption *Chemosphere* **298** 134221

Bagri A, Mattevi C, Acik M, Chabal Y J, Chhowalla M and Shenoy V B 2010 Structural evolution during the reduction of chemically derived graphene oxide *Nat. Chem.* **2** 581–7

Bano S, Mahmood A, Kim S J and Lee K H 2015 Graphene oxide modified polyamide nanofiltration membrane with improved flux and antifouling properties *J. Mater. Chem.* A **3** 2065–71

Bao C, Song L, Xing W, Yuan B, Wilkie C A, Huang J, Guo Y and Hu Y 2012 Preparation of graphene by pressurized oxidation and multiplex reduction and its polymer nanocomposites by masterbatch-based melt blending *J. Mater. Chem.* **22** 6088–96

Bartolomeu M, Neves M, Faustino M and Almeida A 2018 Wastewater chemical contaminants: remediation by advanced oxidation processes *Photochem. Photobiol. Sci.* **17** 1573–98

Bayati A, Moatar F, Karbasi A and Hasani A 2010 Role of saline water in removal of heavy elements from industrial wastewaters *Int. J. Environ. Res.* **4** 177–82

Bethi B, Sonawane S H, Bhanvase B A and Gumfekar S P 2016 Nanomaterials-based advanced oxidation processes for wastewater treatment: a review *Chem. Eng. Process.: Process Intensif.* **109** 178–89

Biswal L, Mohanty R, Nayak S and Parida K 2022 Review on MXene/TiO_2 nanohybrids for photocatalytic hydrogen production and pollutant degradations *J. Environ. Chem. Eng.* **10** 107211

Borandeh S, Alimardani V, Abolmaali S S and Seppälä J 2021 Graphene family nanomaterials in ocular applications: physicochemical properties and toxicity *Chem. Res. Toxicol.* **34** 1386–402

Thomas S C, Mishra P K and Talegaonkar S 2015 Ceramic nanoparticles: fabrication methods and applications in drug delivery *Curr. Pharm. Des.* **21** 6165–88

Cai Z, Zhao X, Wang T, Liu W and Zhao D 2017 Reusable platinum-deposited anatase/hexa-titanate nanotubes: roles of reduced and oxidized platinum on enhanced solar-light-driven photocatalytic activity *ACS Sustain. Chem. Eng.* **5** 547–55

Cao J, Chen C, Zhao Q, Zhang N, Lu Q, Wang X, Niu Z and Chen J 2016 A flexible nanostructured paper of a reduced graphene oxide–sulfur composite for high-performance lithium–sulfur batteries with unconventional configurations *Adv. Mater.* **28** 9629–36

Cao S, Chen C, Zhang J, Zhang C, Yu W, Liang B and Tsang Y 2015 MnO_x quantum dots decorated reduced graphene oxide/TiO_2 nanohybrids for enhanced activity by a UV pre-catalytic microwave method *Appl. Catal.* B **176** 500–12

Cardoso I M, Cardoso R M and da Silva J C E 2021 Advanced oxidation processes coupled with nanomaterials for water treatment *Nanomaterials* **11** 2045

Chávez A, Gimeno O, Rey A, Pliego G, Oropesa A, Álvarez P and Beltrán F 2019 Treatment of highly polluted industrial wastewater by means of sequential aerobic biological oxidation-ozone based AOPs *Chem. Eng. J.* **361** 89–98

Chee W K, Lim H N, Harrison I, Chong K F, Zainal Z, Ng C H and Huang N M 2015 Performance of flexible and binderless polypyrrole/graphene oxide/zinc oxide supercapacitor electrode in a symmetrical two-electrode configuration *Electrochim. Acta* **157** 88–94

Chen M, Tang S, Guo Z, Wang X, Mo S, Huang X, Liu G and Zheng N 2014 Core–shell Pd@ Au nanoplates as theranostic agents for *in-vivo* photoacoustic imaging, CT imaging, and photothermal therapy *Adv. Mater.* **26** 8210–6

Chen W, Wu B, Yao Q, Dong G, Zuo C, Zhang Y, Zhou Y, Liu Y and Zhang Z 2022 A MXene-based multiple catalyst for highly efficient photocatalytic removal of nitrate *Environ. Sci. Pollut. Res.* **29** 58149–160

Cheng-an T, Hao Z, Fang W, Hui Z, Xiaorong Z and Jianfang W 2017 Mechanical properties of graphene oxide/polyvinyl alcohol composite film *Polym. Polym. Compos.* **25** 11–6

Chua C K and Pumera M 2014 Chemical reduction of graphene oxide: a synthetic chemistry viewpoint *Chem. Soc. Rev.* **43** 291–312

Cruz-Ortiz B R, Hamilton J W, Pablos C, Díaz-Jiménez L, Cortés-Hernández D A, Sharma P K, Castro-Alférez M, Fernández-Ibañez P, Dunlop P S and Byrne J A 2017 Mechanism of photocatalytic disinfection using titania–graphene composites under UV and visible irradiation *Chem. Eng. J.* **316** 179–86

Cui C, Guo R, Ren E, Xiao H, Lai X, Qin Q, Jiang S, Shen H, Zhou M and Qin W 2021 Facile hydrothermal synthesis of rod-like Nb_2O_5/Nb_2CT_x composites for visible-light driven photocatalytic degradation of organic pollutants *Environ. Res.* **193** 110587

Cui Y, Kundalwal S and Kumar S 2016 Gas barrier performance of graphene/polymer nanocomposites *Carbon* **98** 313–33

Dai L, Chang D W, Baek J B and Lu W 2012 Carbon nanomaterials for advanced energy conversion and storage *Small* **8** 1130–66

De Silva K, Huang H H, Joshi R and Yoshimura M 2017 Chemical reduction of graphene oxide using green reductants *Carbon* **119** 190–9

Deng C H, Gong J L, Zeng G M, Jiang Y, Zhang C, Liu H Y and Huan S Y 2016a Graphene–CdS nanocomposite inactivation performance toward *Escherichia coli* in the presence of humic acid under visible light irradiation *Chem. Eng. J.* **284** 41–53

Deng D, Novoselov K S, Fu Q, Zheng N, Tian Z and Bao X 2016b Catalysis with two-dimensional materials and their heterostructures *Nat. Nanotechnol.* **11** 218–30

Deng H, He H, Sun S, Zhu X, Zhou D, Han F, Huang B and Pan X 2019 Photocatalytic degradation of dye by Ag/TiO_2 nanoparticles prepared with different sol–gel crystallization in the presence of effluent organic matter *Environ. Sci. Pollut. Res. Int.* **26** 35900–12

Devi L G and Kavitha R 2013 A review on non metal ion doped titania for the photocatalytic degradation of organic pollutants under UV/solar light: role of photogenerated charge carrier dynamics in enhancing the activity *Appl. Catal.* B **140** 559–87

Di J, Xiong J, Li H and Liu Z 2018 Ultrathin 2D photocatalysts: electronic-structure tailoring, hybridization, and applications *Adv. Mater.* **30** 1704548

Du M, Yi Q, Ji J, Zhu Q, Duan H, Xing M and Zhang J 2020 Sustainable activation of peroxymonosulfate by the Mo (IV) in MoS_2 for the remediation of aromatic organic pollutants *Chin. Chem. Lett.* **31** 2803–8

Eda G and Chhowalla M 2010 Chemically derived graphene oxide: towards large-area thin-film electronics and optoelectronics *Adv. Mater.* **22** 2392–415

Fang W, Xing M and Zhang J 2017 Modifications on reduced titanium dioxide photocatalysts: a review *J. Photochem. Photobiol.* C **32** 21–39

Fu K *et al* 2016 Graphene oxide-based electrode inks for 3D-printed lithium-ion batteries *Adv. Mater.* **28** 2587–94

Gao P, Li A, Sun D D and Ng W J 2014 Effects of various TiO_2 nanostructures and graphene oxide on photocatalytic activity of TiO_2 *J. Hazard. Mater.* **279** 96–104

Gao P, Liu J, Sun D D and Ng W 2013a Graphene oxide–CdS composite with high photocatalytic degradation and disinfection activities under visible light irradiation *J. Hazard. Mater.* **250** 412–20

Gao P, Ng K and Sun D D 2013b Sulfonated graphene oxide–ZnO–Ag photocatalyst for fast photodegradation and disinfection under visible light *J. Hazard. Mater.* **262** 826–35

George J S, Ramos A and Shipley H J 2015 Tanning facility wastewater treatment: analysis of physical–chemical and reverse osmosis methods *J. Environ. Chem. Eng.* **3** 969–76

Ghulam A N, Dos Santos O A, Hazeem L, Pizzorno Backx B, Bououdina M and Bellucci S 2022 Graphene oxide (GO) materials—applications and toxicity on living organisms and environment *J. Funct. Biomater.* **13** 77

Glaze W H 1987 Drinking-water treatment with ozone *Environ. Sci. Technol.* **21** 224–30

Glaze W H, Kang J W and Chapin D H 1987 The chemistry of water treatment processes involving ozone, hydrogen peroxide and ultraviolet radiation *Ozone: Science & Engineering* **9** 335–52

Gómez-Navarro C, Burghard M and Kern K 2008 Elastic properties of chemically derived single graphene sheets *Nano Lett.* **8** 2045–9

González-González R B, Rodríguez-Hernández J A, Araújo R G, Sharma P, Parra-Saldívar R, Ramirez-Mendoza R A, Bilal M and Iqbal H M N 2022 Prospecting carbon-based nanomaterials for the treatment and degradation of endocrine-disrupting pollutants *Chemosphere* **297** 134172

Gu H, Rapole S B, Huang Y, Cao D, Luo Z, Wei S and Guo Z 2013 Synergistic interactions between multi-walled carbon nanotubes and toxic hexavalent chromium *J. Mater. Chem.* A **1** 2011–21

Guo S and Dong S 2011 Graphene nanosheet: synthesis, molecular engineering, thin film, hybrids, and energy and analytical applications *Chem. Soc. Rev.* **40** 2644–72

Han R and Wu P 2019 High-performance graphene oxide nanofiltration membrane with continuous nanochannels prepared by the *in situ* oxidation of MXene *J. Mater. Chem.* A **7** 6475–81

Han X, Jing X, Yang D, Lin H, Wang Z, Ran H, Li P and Chen Y 2018 Therapeutic mesopore construction on 2D Nb_2C MXenes for targeted and enhanced chemo-photothermal cancer therapy in NIR-II biowindow *Theranostics* **8** 4491

Higginbotham A L, Lomeda J R, Morgan A B and Tour J M 2009 Graphite oxide flame-retardant polymer nanocomposites *ACS Appl. Mater. Interfaces* **1** 2256–61

Hlekelele L, Durbach S H, Chauke V P, Dziike F and Franklyn P J 2019 Resin-gel incorporation of high concentrations of W^{6+} and Zn^{2+} into TiO_2-anatase crystal to form quaternary mixed-metal oxides: effect on the a lattice parameter and photodegradation efficiency *RSC Adv.* **9** 36875–83

Horovitz I, Gitis V, Avisar D and Mamane H 2020 Ceramic-based photocatalytic membrane reactors for water treatment–where to next? *Rev. Chem. Eng.* **36** 593–622

Hou M, Xu M and Li B 2018 Enhanced electrical conductivity of cellulose nanofiber/graphene composite paper with a sandwich structure *ACS Sustain. Chem. Eng.* **6** 2983–90

Hu M and Mi B 2013 Enabling graphene oxide nanosheets as water separation membranes *Environ. Sci. Technol.* **47** 3715–23

Hu Z T, Liang Y N, Zhao J, Zhang Y, Yang E H, Chen J and Lim T T 2018 Ultra-effective integrated technologies for water disinfection with a novel 0D–2D–3D nanostructured rGO–AgNP/$Bi2Fe_4O_9$ composite *Appl. Catal.* B **227** 548–56

Huang H, Jiang R, Feng Y, Ouyang H, Zhou N, Zhang X and Wei Y 2020 Recent development and prospects of surface modification and biomedical applications of MXenes *Nanoscale* **12** 1325–38

Huang H, Song Z, Wei N, Shi L, Mao Y, Ying Y, Sun L, Xu Z and Peng X 2013 Ultrafast viscous water flow through nanostrand-channelled graphene oxide membranes *Nat. Commun.* **4** 2979

Huang J Q, Zhuang T Z, Zhang Q, Peng H J, Chen C M and Wei F 2015 Permselective graphene oxide membrane for highly stable and anti-self-discharge lithium–sulfur batteries *ACS Nano* **9** 3002–11

Huang K, Li Z, Lin J, Han G and Huang P 2018 Two-dimensional transition metal carbides and nitrides (MXenes) for biomedical applications *Chem. Soc. Rev.* **47** 5109–24

Huo W, Wang S, Zhang X, Ren K, Tan S, Fang F, Xie Z and Jiang J 2022 A strategy to improve the performance of TiO_2 nanotube array film photocatalysts by magnetron-sputtered amorphous $BiFeO_3$ *Vacuum* **202** 111135

Ihsanullah I 2022 MXenes as next-generation materials for the photocatalytic degradation of pharmaceuticals in water *J. Environ. Chem. Eng.* **10** 107381

Iqbal M A, Ali S I, Amin F, Tariq A, Iqbal M Z and Rizwan S 2019 La- and Mn-codoped bismuth ferrite/Ti_3C_2 MXene composites for efficient photocatalytic degradation of Congo red dye *ACS Omega* **4** 8661–8

Iravani S and Varma R S 2022 MXenes in photomedicine: advances and prospects *Chem. Commun.* **58** 7336–50

Jang H, Park Y J, Chen X, Das T, Kim M S and Ahn J H 2016 Graphene-based flexible and stretchable electronics *Adv. Mater.* **28** 4184–202

Jang J, Shahzad A, Woo S H and Lee D S 2020 Magnetic $Ti_3C_2T_x$ (MXene) for diclofenac degradation via the ultraviolet/chlorine advanced oxidation process *Environ. Res.* **182** 108990

Jani M, Arcos-Pareja J A and Ni M 2020 Engineered zero-dimensional fullerene/carbon dots-polymer based nanocomposite membranes for wastewater treatment *Molecules* **25** 4934

Jastrzębska A M, Karwowska E, Wojciechowski T, Ziemkowska W, Rozmysłowska A, Chlubny L and Olszyna A 2019 The atomic structure of Ti_2C and Ti_3C_2 MXenes is responsible for their antibacterial activity toward *E. coli* bacteria *J. Mater. Eng. Perform.* **28** 1272–7

Ji H, Sun H and Qu X 2016 Antibacterial applications of graphene-based nanomaterials: recent achievements and challenges *Adv. Drug Delivery Rev.* **105** 176–89

Jiang S, He Z, Li Q, Wang J, Wu G, Zhao Y and Kang M 2019 Effect of carbon fiber-graphene oxide multiscale reinforcements on the thermo-mechanical properties of polyurethane elastomer *Polym. Compos.* **40** E953–61

Jimmy J and Kandasubramanian B 2020 Mxene functionalized polymer composites: synthesis and applications *Eur. Polym. J.* **122** 109367

Kanakaraju D, Glass B D and Oelgemöller M 2018 Advanced oxidation process-mediated removal of pharmaceuticals from water: a review *J. Environ. Manage.* **219** 189–207

Kausar A 2018 Composite coatings of polyamide/graphene: microstructure, mechanical, thermal, and barrier properties *Compos. Interfaces* **25** 109–25

Khan Z U, Kausar A, Ullah H, Badshah A and Khan W U 2016 A review of graphene oxide, graphene buckypaper, and polymer/graphene composites: properties and fabrication techniques *J. Plast. Film Sheeting* **32** 336–79

Kim C B, Lee J, Cho J and Goh M 2018 Thermal conductivity enhancement of reduced graphene oxide via chemical defect healing for efficient heat dissipation *Carbon* **139** 386–92

Kim K S, Zhao Y, Jang H, Lee S Y, Kim J M, Kim K S, Ahn J H, Kim P, Choi J Y and Hong B H 2009 Large-scale pattern growth of graphene films for stretchable transparent electrodes *Nature* **457** 706–10

Kim M, Lee C and Jang J 2014 Fabrication of highly flexible, scalable, and high-performance supercapacitors using polyaniline/reduced graphene oxide film with enhanced electrical conductivity and crystallinity *Adv. Funct. Mater.* **24** 2489–99

Kim S, Gholamirad F, Yu M, Park C M, Jang A, Jang M, Taheri-Qazvini N and Yoon Y 2021 Enhanced adsorption performance for selected pharmaceutical compounds by sonicated $Ti_3C_2T_X$ MXene *Chem. Eng. J.* **406** 126789

Kim S, Yu M and Yoon Y 2020 Fouling and retention mechanisms of selected cationic and anionic dyes in a $Ti_3C_2T_x$ MXene-ultrafiltration hybrid system *ACS Appl. Mater. Interfaces* **12** 16557–65

Mahamuni N N and Adewuyi Y G 2010 Advanced oxidation processes (AOPs) involving ultrasound for waste water treatment: A review with emphasis on cost estimation *Ultrasonics Sonochemistry* **17** 990–1003

Kotagiri N, Sudlow G P, Akers W J and Achilefu S 2015 Breaking the depth dependency of phototherapy with Cerenkov radiation and low-radiance-responsive nanophotosensitizers *Nat. Nanotechnol.* **10** 370–9

Kouhbanani M A J, Beheshtkhoo N, Taghizadeh S, Amani A M and Alimardani V 2019 One-step green synthesis and characterization of iron oxide nanoparticles using aqueous leaf extract of *Teucrium polium* and their catalytic application in dye degradation *Adv. Nat. Sci.: Nanosci. Nanotechnol.* **10** 015007

Kovalevskiy N, Selishchev D, Svintsitskiy D, Selishcheva S, Berezin A and Kozlov D 2020 Synergistic effect of polychromatic radiation on visible light activity of N-doped TiO_2 photocatalyst *Catal. Commun.* **134** 105841

Kuilla T, Bhadra S, Yao D, Kim N H, Bose S and Lee J H 2010 Recent advances in graphene based polymer composites *Prog. Polym. Sci.* **35** 1350–75

Kulesza P J, Pieta I S, Rutkowska I A, Wadas A, Marks D, Klak K, Stobinski L and Cox J A 2013 Electrocatalytic oxidation of small organic molecules in acid medium: enhancement of activity of noble metal nanoparticles and their alloys by supporting or modifying them with metal oxides *Electrochim. Acta* **110** 474–83

Kumar P, Yadav M K, Panwar N, Kumar A and Singhal R 2019 Temperature dependent thermal conductivity of free-standing reduced graphene oxide/poly (vinylidene fluoride-co-hexafluoropropylene) composite thin film *Mater. Res. Express* **6** 115604

Kurian M 2021 Advanced oxidation processes and nanomaterials—a review *Cleaner Eng. Technol.* **2** 100090

Kyrylenko S, Kornienko V, Gogotsi O, Oleshko O, Kolesnyk M, Mishchenko O, Zahorodna V, Buranich V, Pogrebnjak A and Zozulia Y 2020 Bio-functionalization of electrospun polymeric nanofibers by $Ti_3C_2T_x$ MXene *2020 IEEE 10th Int. Conf. on Nanomaterials: Applications and Properties (NAP)* (Piscataway, NJ: IEEE) 02BA10-1–5

Lalliansanga , Tiwari D, Lee S M and Kim D J 2022 Photocatalytic degradation of amoxicillin and tetracycline by template synthesized nano-structured Ce^{3+}@TiO_2 thin film catalyst *Environ. Res.* **210** 112914

Lamberti A, Gigot A, Bianco S, Fontana M, Castellino M, Tresso E and Pirri C F 2016 Self-assembly of graphene aerogel on copper wire for wearable fiber-shaped supercapacitors *Carbon* **105** 649–54

Lee C, Wei X, Kysar J W and Hone J 2008 Measurement of the elastic properties and intrinsic strength of monolayer graphene *Science* **321** 385–8

Lee Y, Park J, Choe A, Cho S, Kim J and Ko H 2020 Mimicking human and biological skins for multifunctional skin electronics *Adv. Funct. Mater.* **30** 1904523

Li F, Chen J, Wang X, Xue M and Chen G 2015 Stretchable supercapacitor with adjustable volumetric capacitance based on 3D interdigital electrodes *Adv. Funct. Mater.* **25** 4601–6

Li T, Chen L, Yang X, Chen X, Zhang Z, Zhao T, Li X and Zhang J 2019a A flexible pressure sensor based on an MXene–textile network structure *J. Mater. Chem.* C **7** 1022–7

Li W, Song Z, Zhong J, Qian J, Tan Z, Wu X, Chu H, Nie W and Ran X 2019b Multilayer-structured transparent MXene/PVDF film with excellent dielectric and energy storage performance *J. Mater. Chem.* C **7** 10371–8

Li Y, Zhang P, Wan D, Xue C, Zhao J and Shao G 2020 Direct evidence of 2D/1D heterojunction enhancement on photocatalytic activity through assembling MoS_2 nanosheets onto super-long TiO_2 nanofibers *Appl. Surf. Sci.* **504** 144361

Liang H, Wang Z, Liao L, Chen L, Li Z and Feng J 2017 High performance photocatalysts: montmorillonite supported-nano TiO_2 composites *Optik* **136** 44–51

Lin D, Liu Y, Liang Z, Lee H W, Sun J, Wang H, Yan K, Xie J and Cui Y 2016 Layered reduced graphene oxide with nanoscale interlayer gaps as a stable host for lithium metal anodes *Nat. Nanotechnol.* **11** 626–32

Lin L, Li R h, Li Y, Xu J and Li X Y 2017 Recovery of organic carbon and phosphorus from wastewater by Fe-enhanced primary sedimentation and sludge fermentation *Process Biochem.* **54** 135–9

Ling Z, Ren C E, Zhao M Q, Yang J, Giammarco J M, Qiu J, Barsoum M W and Gogotsi Y 2014 Flexible and conductive MXene films and nanocomposites with high capacitance *Proc. Natl Acad. Sci.* **111** 16676–81

Liu J, Bai H, Wang Y, Liu Z, Zhang X and Sun D D 2010 Self-assembling TiO_2 nanorods on large graphene oxide sheets at a two-phase interface and their anti-recombination in photocatalytic applications *Adv. Funct. Mater.* **20** 4175–81

Liu J, Liu L, Bai H, Wang Y and Sun D D 2011a Gram-scale production of graphene oxide–TiO_2 nanorod composites: towards high-activity photocatalytic materials *Appl. Catal.* B **106** 76–82

Liu L, Bai H, Liu J and Sun D D 2013 Multifunctional graphene oxide–TiO_2–Ag nanocomposites for high performance water disinfection and decontamination under solar irradiation *J. Hazard. Mater.* **261** 214–23

Liu L, Liu J and Sun D D 2012 Graphene oxide enwrapped Ag_3PO_4 composite: towards a highly efficient and stable visible-light-induced photocatalyst for water purification *Catal. Sci. Technol.* **2** 2525–32

Liu L X, Chen W, Zhang H B, Wang Q W, Guan F and Yu Z Z 2019 Flexible and multifunctional silk textiles with biomimetic leaf-like MXene/silver nanowire nanostructures for electromagnetic interference shielding, humidity monitoring, and self-derived hydrophobicity *Adv. Funct. Mater.* **29** 1905197

Liu N, Lu N, Yu H, Chen S and Quan X 2020a Efficient day-night photocatalysis performance of 2D/2D Ti_3C_2/porous g-C_3N_4 nanolayers composite and its application in the degradation of organic pollutants *Chemosphere* **246** 125760

Liu Q, Wu D, Zhou Y, Su H, Wang R, Zhang C, Yan S, Xiao M and Zou Z 2014 Single-crystalline, ultrathin $ZnGa_2O_4$ nanosheet scaffolds to promote photocatalytic activity in CO_2 reduction into methane *ACS Appl. Mater. Interfaces* **6** 2356–61

Liu S, Zeng T H, Hofmann M, Burcombe E, Wei J, Jiang R, Kong J and Chen Y 2011b Antibacterial activity of graphite, graphite oxide, graphene oxide, and reduced graphene oxide: membrane and oxidative stress *ACS Nano* **5** 6971–80

Liu T, Li L, Geng X, Yang C, Zhang X, Lin X, Lv P, Mu Y and Huang S 2022 Heterostructured MXene-derived oxides as superior photocatalysts for MB degradation *J. Alloys Compd.* **919** 165629

Liu X, Jin X, Li L, Wang J, Yang Y, Cao Y and Wang W 2020b Air-permeable, multifunctional, dual-energy-driven MXene-decorated polymeric textile-based wearable heaters with exceptional electrothermal and photothermal conversion performance *J. Mater. Chem.* A **8** 12526–37

Long R, Yu Z, Tan Q, Feng X, Zhu X, Li X and Wang P 2021 Ti_3C_2 MXene/NH2-MIL-88B (Fe): research on the adsorption kinetics and photocatalytic performance of an efficient integrated photocatalytic adsorbent *Appl. Surf. Sci.* **570** 151244

Lou B S, Chen W T, Diyatmika W, Lu J H, Chang C T, Chen P W and Lee J W 2022 High power impulse magnetron sputtering (HiPIMS) for the fabrication of antimicrobial and transparent TiO_2 thin films *Curr. Opin. Chem. Eng.* **36** 100782

Lou M, Abdalla I, Zhu M, Wei X, Yu J, Li Z and Ding B 2020 Highly wearable, breathable, and washable sensing textile for human motion and pulse monitoring *ACS Appl. Mater. Interfaces* **12** 19965–73

Lukatskaya M R, Kota S, Lin Z, Zhao M Q, Shpigel N, Levi M D, Halim J, Taberna P L, Barsoum M W and Simon P 2017 Ultra-high-rate pseudocapacitive energy storage in two-dimensional transition metal carbides *Nat. Energy* **2** 1–6

Lv Y, Wang K, Li D, Li P, Chen X and Han W 2022 Rare Ag nanoparticles loading induced surface-enhanced pollutant adsorption and photocatalytic degradation on $Ti_3C_2T_x$ MXene-based nanosheets *Chem. Phys.* **560** 111591

Malaki M, Maleki A and Varma R S 2019 MXenes and ultrasonication *J. Mater. Chem.* A **7** 10843–57

Mangadlao J D, Santos C M, Felipe M J L, de Leon A C C, Rodrigues D F and Advincula R C 2015 On the antibacterial mechanism of graphene oxide (GO) Langmuir–Blodgett films *Chem. Commun.* **51** 2886–9

Marcano D C, Kosynkin D V, Berlin J M, Sinitskii A, Sun Z, Slesarev A, Alemany L B, Lu W and Tour J M 2010 Improved synthesis of graphene oxide *ACS Nano* **4** 4806–14

Mashtalir O, Naguib M, Mochalin V N, Dall'Agnese Y, Heon M, Barsoum M W and Gogotsi Y 2013 Intercalation and delamination of layered carbides and carbonitrides *Nat. Commun.* **4** 1–7

Mayerberger E A, Street R M, McDaniel R M, Barsoum M W and Schauer C L 2018 Antibacterial properties of electrospun $Ti_3C_2T_z$ (MXene)/chitosan nanofibers *RSC Adv.* **8** 35386–94

Mohammed H, Kumar A, Bekyarova E, Al-Hadeethi Y, Zhang X, Chen M, Ansari M S, Cochis A and Rimondini L 2020 Antimicrobial mechanisms and effectiveness of graphene and graphene-functionalized biomaterials. a scope review *Front. Bioeng. Biotechnol.* **8** 465

Monga D, Sharma S, Shetti N P, Basu S, Reddy K R and Aminabhavi T M 2021 Advances in transition metal dichalcogenide-based two-dimensional nanomaterials *Mater. Today Chem.* **19** 100399

Mostafavi E and Iravani S 2022 Mxene-graphene composites: a perspective on biomedical potentials *Nano-Micro Lett.* **14** 1–12

Musolff A, Leschik S, Reinstorf F, Strauch G and Schirmer M 2010 Micropollutant loads in the urban water cycle *Environ. Sci. Technol.* **44** 4877–83

Mustafa R and Asmatulu E 2020 Preparation of activated carbon using fruit, paper and clothing wastes for wastewater treatment *J. Water Process Eng.* **35** 101239

Naguib M, Kurtoglu M, Presser V, Lu J, Niu J, Heon M, Hultman L, Gogotsi Y and Barsoum M W 2011 Two-dimensional nanocrystals produced by exfoliation of Ti_3AlC_2 *Adv. Mater.* **23** 4248–53

Naguib M, Mashtalir O, Carle J, Presser V, Lu J, Hultman L, Gogotsi Y and Barsoum M W 2012 Two-dimensional transition metal carbides *ACS Nano* **6** 1322–31

Naguib M, Mochalin V N, Barsoum M W and Gogotsi Y 2014 25th anniversary article: MXenes: a new family of two-dimensional materials *Adv. Mater.* **26** 992–1005

Nair R, Wu H, Jayaram P N, Grigorieva I V and Geim A 2012 Unimpeded permeation of water through helium-leak–tight graphene-based membranes *Science* **335** 442–4

Naseem T and Durrani T 2021 The role of some important metal oxide nanoparticles for wastewater and antibacterial applications: a review *Environ. Chem. Ecotoxicol.* **3** 59–75

Negishi N, Takeuchi K and Ibusuki T 1998 Surface structure of the TiO_2 thin film photocatalyst *J. Mater. Sci.* **33** 5789–94

Ni Y, Yan K, Xu F, Zhong W, Zhao Q, Liu K, Yan K and Wang D 2019 Synergistic effect on TiO_2 doped poly (vinyl alcohol-co-ethylene) nanofibrous film for filtration and photocatalytic degradation of methylene blue *Compos. Commun.* **12** 112–6

Niyogi S, Bekyarova E, Itkis M E, McWilliams J L, Hamon M A and Haddon R C 2006 Solution properties of graphite and graphene *J. Am. Chem. Soc.* **128** 7720–1

Nobuaki N, Tomokazu I, Kazuhito H and Akira F 1995 Preparation of transparent TiO_2 thin film photocatalyst and its photocatalytic activity *Chem. Lett.* **24** 841–2

Novoselov K S, Colombo L, Gellert P, Schwab M and Kim K 2012 A roadmap for graphene *Nature* **490** 192–200

Nzaba S K M, Nyoni H H, Mamba B B and Kuvarega A T 2019 Comparative study of dendrimer-templated nitrogen–platinum co-doped TiO_2 for the photocatalytic degradation of azo dyes in contaminated water *ChemistrySelect* **4** 12156–63

Park S and Ruoff R S 2009 Chemical methods for the production of graphenes *Nat. Nanotechnol.* **4** 217–24

Pei S and Cheng H M 2012 The reduction of graphene oxide *Carbon* **50** 3210–28

Pei S, Zhao J, Du J, Ren W and Cheng H M 2010 Direct reduction of graphene oxide films into highly conductive and flexible graphene films by hydrohalic acids *Carbon* **48** 4466–74

Pogorielov M, Smyrnova K, Kyrylenko S, Gogotsi O, Zahorodna V and Pogrebnjak A 2021 MXenes—a new class of two-dimensional materials: structure, properties and potential applications *Nanomaterials* **11** 3412

Priyadarsini S, Mohanty S, Mukherjee S, Basu S and Mishra M 2018 Graphene and graphene oxide as nanomaterials for medicine and biology application *J. Nanostruct. Chem.* **8** 123–37

Radha E, Komaraiah D, Sayanna R and Sivakumar J 2022 Photoluminescence and photocatalytic activity of rare Earth ions doped anatase TiO_2 thin films *J. Lumin.* **244** 118727

Rahimi R, Zargari S, Yousefi A, Berijani M Y, Ghaffarinejad A and Morsali A 2015 Visible light photocatalytic disinfection of *E. coli* with TiO_2–graphene nanocomposite sensitized with tetrakis (4-carboxyphenyl) porphyrin *Appl. Surf. Sci.* **355** 1098–106

Rajib D, Vibhav A and Partha Pratim B 2021 Titanium dioxide and its applications in mechanical, electrical, optical, and biomedical fields *Titanium Dioxide* ed A Hafiz Muhammad (Rijeka: IntechOpen)

Rasool K, Helal M, Ali A, Ren C E, Gogotsi Y and Mahmoud K A 2016 Antibacterial activity of $Ti_3C_2T_x$ MXene *ACS Nano* **10** 3674–84

Rasool K, Mahmoud K A, Johnson D J, Helal M, Berdiyorov G R and Gogotsi Y 2017 Efficient antibacterial membrane based on two-dimensional $Ti_3C_2T_x$ (MXene) nanosheets *Sci. Rep.* **7** 1–11

Razaq A, Bibi F, Zheng X, Papadakis R, Jafri S H M and Li H 2022 Review on graphene-, graphene oxide-, reduced graphene oxide-based flexible composites: from fabrication to applications *Materials* **15** 1012

Renteria J D, Ramirez S, Malekpour H, Alonso B, Centeno A, Zurutuza A, Cocemasov A I, Nika D L and Balandin A A 2015 Strongly anisotropic thermal conductivity of free-standing reduced graphene oxide films annealed at high temperature *Adv. Funct. Mater.* **25** 4664–72

Rueda-Márquez J J, Palacios-Villarreal C, Manzano M, Blanco E, Ramírez del Solar M and Levchuk I 2020 Photocatalytic degradation of pharmaceutically active compounds (PhACs) in urban wastewater treatment plants effluents under controlled and natural solar irradiation using immobilized TiO_2 *Sol. Energy* **208** 480–92

Sasidharan V, Sachan D, Chauhan D, Talreja N and Ashfaq M 2021 Three-dimensional (3D) polymer–metal–carbon framework for efficient removal of chemical and biological contaminants *Sci. Rep.* **11** 7708

Sgroi M, Snyder S A and Roccaro P 2021 Comparison of AOPs at pilot scale: energy costs for micro-pollutants oxidation, disinfection by-products formation and pathogens inactivation *Chemosphere* **273** 128527

Shahzad A, Rasool K, Nawaz M, Miran W, Jang J, Moztahida M, Mahmoud K A and Lee D S 2018 Heterostructural $TiO_2/Ti_3C_2T_x$ (MXene) for photocatalytic degradation of antiepileptic drug carbamazepine *Chem. Eng. J.* **349** 748–55

Shahzad F, Alhabeb M, Hatter C B, Anasori B, Man Hong S, Koo C M and Gogotsi Y 2016 Electromagnetic interference shielding with 2D transition metal carbides (MXenes) *Science* **353** 1137–40

Shaker Ardakani L, Alimardani V, Tamaddon A M, Amani A M and Taghizadeh S 2021 Green synthesis of iron-based nanoparticles using *Chlorophytum comosum* leaf extract: methyl orange dye degradation and antimicrobial properties *Heliyon* **7** e06159

Singh G, Sharma M and Vaish R 2020 Transparent ferroelectric glass–ceramics for wastewater treatment by piezocatalysis *Commun. Mater.* **1** 100

Siwal S S, Sheoran K, Mishra K, Kaur H, Saini A K, Saini V, Vo D V N, Nezhad H Y and Thakur V K 2022 Novel synthesis methods and applications of MXene-based nanomaterials (MBNs) for hazardous pollutants degradation: future perspectives *Chemosphere* **293** 133542

Somanathan T, Prasad K, Ostrikov K, Saravanan A and Mohana Krishna V 2015 Graphene oxide synthesis from agro waste *Nanomaterials* **5** 826–34

Song S, Zhai Y and Zhang Y 2016 Bioinspired graphene oxide/polymer nanocomposite paper with high strength, toughness, and dielectric constant *ACS Appl. Mater. Interfaces* **8** 31264–72

Sopyan I, Watanabe M, Murasawa S, Hashimoto K and Fujishima A 1996 An efficient TiO_2 thin-film photocatalyst: photocatalytic properties in gas-phase acetaldehyde degradation *J. Photochem. Photobiol.* A **98** 79–86

Soundiraraju B and George B K 2017 Two-dimensional titanium nitride (Ti_2N) MXene: synthesis, characterization, and potential application as surface-enhanced Raman scattering substrate *ACS Nano* **11** 8892–900

Sreedhar A and Noh J S 2021 Recent advances in partially and completely derived 2D Ti_3C_2 MXene based TiO_2 nanocomposites towards photocatalytic applications: a review *Sol. Energy* **222** 48–73

Stankovich S, Dikin D A, Dommett G H, Kohlhaas K M, Zimney E J, Stach E A, Piner R D, Nguyen S T and Ruoff R S 2006 Graphene-based composite materials *Nature* **442** 282–6

Stankovich S, Dikin D A, Piner R D, Kohlhaas K A, Kleinhammes A, Jia Y, Wu Y, Nguyen S T and Ruoff R S 2007 Synthesis of graphene-based nanosheets via chemical reduction of exfoliated graphite oxide *Carbon* **45** 1558–65

Suárez S, Jansson I, Ohtani B and Sánchez B 2019 From titania nanoparticles to decahedral anatase particles: photocatalytic activity of TiO_2/zeolite hybrids for VOCs oxidation *Catal. Today* **326** 2–7

Suk J W, Piner R D, An J and Ruoff R S 2010 Mechanical properties of monolayer graphene oxide *ACS Nano* **4** 6557–64

Sun B, Tao F, Huang Z, Yan W, Zhang Y, Dong X, Wu Y and Zhou G 2021 Ti_3C_2 MXene-bridged Ag/Ag_3PO_4 hybrids toward enhanced visible-light-driven photocatalytic activity *Appl. Surf. Sci.* **535** 147354

Sun L, Du T, Hu C, Chen J, Lu J, Lu Z and Han H 2017 Antibacterial activity of graphene oxide/ g-C_3N_4 composite through photocatalytic disinfection under visible light *ACS Sustain. Chem. Eng.* **5** 8693–701

Sun Y, Gao S, Lei F and Xie Y 2015 Atomically-thin two-dimensional sheets for understanding active sites in catalysis *Chem. Soc. Rev.* **44** 623–36

Szuplewska A, Kulpińska D, Dybko A, Chudy M, Jastrzębska A M, Olszyna A and Brzózka Z 2020 Future applications of MXenes in biotechnology, nanomedicine, and sensors *Trends Biotechnol.* **38** 264–79

Tajeddine L, Nemmaoui M, Mountacer H, Dahchour A and Sarakha M 2010 Photodegradation of fenamiphos on the surface of clays and soils *Environ. Chem. Lett.* **8** 123–8

Talreja N, Ashfaq M, Chauhan D and Mangalaraja R V 2023 Cu-MXene: a potential biocide for the next-generation biomedical application *Mater. Chem. Phys.* **294** 127029

Tang L, Li X, Ji R, Teng K S, Tai G, Ye J, Wei C and Lau S P 2012 Bottom-up synthesis of large-scale graphene oxide nanosheets *J. Mater. Chem.* **22** 5676–83

Tarcan R, Todor-Boer O, Petrovai I, Leordean C, Astilean S and Botiz I 2020 Reduced graphene oxide today *J. Mater. Chem.* C **8** 1198–224

Thirumal V, Yuvakkumar R, Kumar P S, Ravi G, Keerthana S and Velauthapillai D 2022 Facile single-step synthesis of MXene@ CNTs hybrid nanocomposite by CVD method to remove hazardous pollutants *Chemosphere* **286** 131733

Tian J, Zhao Z, Kumar A, Boughton R I and Liu H 2014 Recent progress in design, synthesis, and applications of one-dimensional TiO_2 nanostructured surface heterostructures: a review *Chem. Soc. Rev.* **43** 6920–37

Tobiszewski M and Namieśnik J 2012 Abiotic degradation of chlorinated ethanes and ethenes in water *Environ. Sci. Pollut. Res.* **19** 1994–2006

Velempini T, Prabakaran E and Pillay K 2021 Recent developments in the use of metal oxides for photocatalytic degradation of pharmaceutical pollutants in water—a review *Mater. Today Chem.* **19** 100380

Wang B *et al* 2018 Folding graphene film yields high areal energy storage in lithium-ion batteries *ACS Nano* **12** 1739–46

Wang M *et al* 2015 All-solid-state reduced graphene oxide supercapacitor with large volumetric capacitance and ultralong stability prepared by electrophoretic deposition method *ACS Appl. Mater. Interfaces* **7** 1348–54

Wang Q, Mei Y, Zhou R, Komarneni S and Ma J 2022 Persulfate activation of CuS@ Ti_3C_2-based MXene with bi-active centers toward orange II removal under visible light *Colloids Surf.* A **648** 129315

Wang W, Yu J C, Xia D, Wong P K and Li Y 2013 Graphene and g-C_3N_4 nanosheets cowrapped elemental α-sulfur as a novel metal-free heterojunction photocatalyst for bacterial inactivation under visible-light *Environ. Sci. Technol.* **47** 8724–32

Wang X, Song L, Yang H, Xing W, Kandola B and Hu Y 2012a Simultaneous reduction and surface functionalization of graphene oxide with POSS for reducing fire hazards in epoxy composites *J. Mater. Chem.* **22** 22037–43

Wang Z, Nelson J K, Hillborg H, Zhao S and Schadler L S 2012b Graphene oxide filled nanocomposite with novel electrical and dielectric properties *Adv. Mater.* **24** 3134–7

Wicklein B, Kocjan A, Salazar-Alvarez G, Carosio F, Camino G, Antonietti M and Bergström L 2015 Thermally insulating and fire-retardant lightweight anisotropic foams based on nanocellulose and graphene oxide *Nat. Nanotechnol.* **10** 277–83

Williams G, Seger B and Kamat P V 2008 TiO_2–graphene nanocomposites. UV-assisted photocatalytic reduction of graphene oxide *ACS Nano* **2** 1487–91

Wu D, An T, Li G, Wang W, Cai Y, Yip H Y, Zhao H and Wong P K 2015 Mechanistic study of the visible-light-driven photocatalytic inactivation of bacteria by graphene oxide–zinc oxide composite *Appl. Surf. Sci.* **358** 137–45

Wu S, Su Y, Zhu Y, Zhang Y and Zhu M 2020a *In-situ* growing Bi/BiOCl microspheres on Ti_3C_2 nanosheets for upgrading visible-light-driven photocatalytic activity *Appl. Surf. Sci.* **520** 146339

Wu W, Fang H, Ma H, Wu L, Zhang W and Wang H 2021 Boosting transport kinetics of ions and electrons simultaneously by $Ti_3C_2T_x$ (MXene) addition for enhanced electrochromic performance *Nano-Micro Lett.* **13** 1–13

Wu Z, Liang Y, Yuan X, Zou D, Fang J, Jiang L, Yang H and Xiao Z 2020b MXene Ti_3C_2 derived Z-scheme photocatalyst of graphene layers anchored TiO_2/g-C_3N_4 for visible light photocatalytic degradation of refractory organic pollutants *Chem. Eng. J.* **394** 124921

Xia D, An T, Li G, Wang W, Zhao H and Wong P K 2016 Synergistic photocatalytic inactivation mechanisms of bacteria by graphene sheets grafted plasmonic AgAgX (X = Cl, Br, I) composite photocatalyst under visible light irradiation *Water Res.* **99** 149–61

Xiang Q, Yu J and Jaroniec M 2012 Graphene-based semiconductor photocatalysts *Chem. Soc. Rev.* **41** 782–96

Xiao Z, Ruan S, Kong L B, Que W, Zhou K, Liu Y and Zhang T 2020 Synthesis and properties of MXenes *MXenes and MXenes-Based Composites* (Berlin: Springer)

Xing F, Li T, Li J, Zhu H, Wang N and Cao X 2017 Chemically exfoliated MoS_2 for capacitive deionization of saline water *Nano Energy* **31** 590–5

Xu C, Wang L, Liu Z, Chen L, Guo J, Kang N, Ma X L, Cheng H M and Ren W 2015 Large-area high-quality 2D ultrathin Mo_2C superconducting crystals *Nat. Mater.* **14** 1135–41

Xu J, Tan Z, Zeng W, Chen G, Wu S, Zhao Y, Ni K, Tao Z, Ikram M and Ji H 2016 A hierarchical carbon derived from sponge-templated activation of graphene oxide for high-performance supercapacitor electrodes *Adv. Mater.* **28** 5222–8

Yang X, Qin J, Jiang Y, Chen K, Yan X, Zhang D, Li R and Tang H 2015 Fabrication of P25/Ag_3PO_4/graphene oxide heterostructures for enhanced solar photocatalytic degradation of organic pollutants and bacteria *Appl. Catal.* B **166** 231–40

Yang X, Wang Z and Shao L 2018 Construction of oil-unidirectional membrane for integrated oil collection with lossless transportation and oil-in-water emulsion purification *J. Membr. Sci.* **549** 67–74

Yang Y, Cao Z, He P, Shi L, Ding G, Wang R and Sun J 2019 $Ti_3C_2T_x$ MXene–graphene composite films for wearable strain sensors featured with high sensitivity and large range of linear response *Nano Energy* **66** 104134

Yao C, Zhang W, Xu L, Cheng M, Su Y, Xue J, Liu J and Hou S 2021 A facile synthesis of porous MXene-based freestanding film and its spectacular electrosorption performance for organic dyes *Sep. Purif. Technol.* **263** 118365

Yao Z, Sun H, Sui H and Liu X 2020 2D/2D heterojunction of R-scheme Ti_3C_2 MXene/MoS_2 nanosheets for enhanced photocatalytic performance *Nanoscale Res. Lett.* **15** 1–12

Yeh C N, Raidongia K, Shao J, Yang Q H and Huang J 2015 On the origin of the stability of graphene oxide membranes in water *Nat. Chem.* **7** 166–70

Yildiz A, Lisesivdin S B, Kasap M and Mardare D 2008 Electrical properties of TiO_2 thin films *J. Non-Cryst. Solids* **354** 4944–7

Yildiz A, Serin N, Kasap M, Serin T and Mardare D 2010 The thickness effect on the electrical conduction mechanism in titanium oxide thin films *J. Alloys Compd.* **493** 227–32

Yoshida T, Misu Y, Yamamoto M, Tanabe T, Kumagai J, Ogawa S and Yagi S 2020 Effects of the amount of Au nanoparticles on the visible light response of TiO_2 photocatalysts *Catal. Today* **352** 34–8

Yu B, Shi Y, Yuan B, Qiu S, Xing W, Hu W, Song L, Lo S and Hu Y 2015a Enhanced thermal and flame retardant properties of flame-retardant-wrapped graphene/epoxy resin nanocomposites *J. Mater. Chem.* A **3** 8034–44

Yu B *et al* 2014a Functionalized graphene oxide/phosphoramide oligomer hybrids flame retardant prepared via *in situ* polymerization for improving the fire safety of polypropylene *RSC Adv.* **4** 31782–94

Yu H, Zhang B, Bulin C, Li R and Xing R 2016 High-efficient synthesis of graphene oxide based on improved hummers method *Sci. Rep.* **6** 1–7

Yu J, Low J, Xiao W, Zhou P and Jaroniec M 2014b Enhanced photocatalytic CO_2-reduction activity of anatase TiO_2 by coexposed {001} and {101} facets *J. Am. Chem. Soc.* **136** 8839–42

Yu S, Liu J, Zhu W, Hu Z T, Lim T T and Yan X 2015b Facile room-temperature synthesis of carboxylated graphene oxide-copper sulfide nanocomposite with high photodegradation and disinfection activities under solar light irradiation *Sci. Rep.* **5** 1–12

Zeng X, Wang G, Liu Y and Zhang X 2017a Graphene-based antimicrobial nanomaterials: rational design and applications for water disinfection and microbial control *Environ. Sci. Nano* **4** 2248–66

Zeng X, Wang Z, Meng N, McCarthy D T, Deletic A, Pan J H and Zhang X 2017b Highly dispersed TiO_2 nanocrystals and carbon dots on reduced graphene oxide: ternary nanocomposites for accelerated photocatalytic water disinfection *Appl. Catal.* B **202** 33–41

Zeng X, Wang Z, Wang G, Gengenbach T R, McCarthy D T, Deletic A, Yu J and Zhang X 2017c Highly dispersed TiO_2 nanocrystals and WO_3 nanorods on reduced graphene oxide: Z-scheme photocatalysis system for accelerated photocatalytic water disinfection *Appl. Catal.* B **218** 163–73

Zeribi F, Attaf A, Derbali A, Saidi H, Benmebrouk L, Aida M S, Dahnoun M, Nouadji R and Ezzaouia H 2022 Dependence of the physical properties of titanium dioxide (TiO_2) thin films grown by sol–gel (spin-coating) process on thickness *ECS J. Solid State Sci. Technol.* **11** 023003

Zhang H B, Zheng W G, Yan Q, Yang Y, Wang J W, Lu Z H, Ji G Y and Yu Z Z 2010 Electrically conductive polyethylene terephthalate/graphene nanocomposites prepared by melt compounding *Polymer* **51** 1191–6

Zhang H 2015 Ultrathin two-dimensional nanomaterials *ACS Nano* **9** 9451–69

Zhang H, Li M, Zhu C, Tang Q, Kang P and Cao J 2020a Preparation of magnetic α-Fe_2O_3/$ZnFe_2O_4$@ Ti_3C_2 MXene with excellent photocatalytic performance *Ceram. Int.* **46** 81–8

Zhang L, Guo J, Hao B and Ma H 2022a WO_3/TiO_2 heterojunction photocatalyst prepared by reactive magnetron sputtering for Rhodamine B dye degradation *Opt. Mater.* **133** 113035

Zhang L, Ma P, Dai L, Bu Z, Li X, Yu W, Cao Y and Guan J 2022b Removal of pollutants via synergy of adsorption and photocatalysis over MXene-based nanocomposites *Chem. Eng. J. Adv.* **10** 100285

Zhang S and Lu X 2018 Treatment of wastewater containing reactive brilliant blue KN-R using TiO_2/BC composite as heterogeneous photocatalyst and adsorbent *Chemosphere* **206** 777–83

Zhang T, Liu Y, Zhong S and Zhang L 2020b AOPs-based remediation of petroleum hydrocarbons-contaminated soils: efficiency, influencing factors and environmental impacts *Chemosphere* **246** 125726

Zhang X, Shen Q, Zhang X, Pan H and Lu Y 2016 Graphene oxide-filled multilayer coating to improve flame-retardant and smoke suppression properties of flexible polyurethane foam *J. Mater. Sci.* **51** 10361–74

Zhang X, Zhang M, Liu H, Gu J and Liu Y 2019 Environmental sustainability: a pressing challenge to biological sewage treatment processes *Curr. Opin. Environ. Sci. Health* **12** 1–5

Zhang Y Z, El-Demellawi J K, Jiang Q, Ge G, Liang H, Lee K, Dong X and Alshareef H N 2020c MXene hydrogels: fundamentals and applications *Chem. Soc. Rev.* **49** 7229–51

Zhang Y, Zhu Y, Yu J, Yang D, Ng T W, Wong P K and Jimmy C Y 2013 Enhanced photocatalytic water disinfection properties of Bi_2MoO_6–RGO nanocomposites under visible light irradiation *Nanoscale* **5** 6307–10

Zhang Z, Cai Z, Zhang Y, Peng Y, Wang Z, Xia L, Ma S, Yin Z, Wang R and Cao Y 2021 The recent progress of MXene-based microwave absorption materials *Carbon* **174** 484–99

Zhao Y, Liu J, Wang B, Sha J, Li Y, Zheng D, Amjadipour M, MacLeod J and Motta N 2017 Supercapacitor electrodes with remarkable specific capacitance converted from hybrid graphene oxide/NaCl/urea films *ACS Appl. Mater. Interfaces* **9** 22588–96

Zheng X, Shen J, Hu Q, Nie W, Wang Z, Zou L and Li C 2021 Vapor phase polymerized conducting polymer/MXene textiles for wearable electronics *Nanoscale* **13** 1832–41

Zhong Q, Liu J, Wang Z, Ghasemi J B and Zhang G 2021 Ti_3C_2 MXene/Ag_2ZnGeO_4 Schottky heterojunctions with enhanced photocatalytic performances: efficient charge separation and mechanism studies *Sep. Purif. Technol.* **278** 119560

Zhou W, Yu B, Zhu J, Li K and Tian S 2022 Hierarchical ZnO/MXene (Nb_2C and V_2C) heterostructure with efficient electron transfer for enhanced photocatalytic activity *Appl. Surf. Sci.* **590** 153095

Zhou Y, Meng F and Zhang J 2008 New MAX-phase compounds in the V–Cr–Al–C system *J. Am. Ceram. Soc.* **91** 1357–60

Zhu X, Liu P, Xue T, Ge Y, Ai S, Sheng Y, Wu R, Xu L, Tang K and Wen Y 2021 A novel graphene-like titanium carbide MXene/Au–Ag nanoshuttles bifunctional nanosensor for electrochemical and SERS intelligent analysis of ultra-trace carbendazim coupled with machine learning *Ceram. Int.* **47** 173–84

Zhu Y, Murali S, Cai W, Li X, Suk J W, Potts J R and Ruoff R S 2010 Graphene and graphene oxide: synthesis, properties, and applications *Adv. Mater.* **22** 3906–24

Zou X, Zhang L, Wang Z and Luo Y 2016 Mechanisms of the antimicrobial activities of graphene materials *J. Am. Chem. Soc.* **138** 2064–77

IOP Publishing

Materials Technology for the Energy and Environmental Nexus, Volume 1

R V Mangalaraja, Arunachalam Arulraj, Nalandhiran Pugazhenthiran and Chan Siew Hwa

Chapter 11

Nanostructured graphitic carbon nitride for remediation of environmental contaminants

Pramila Murugesan and J A Moses

Graphitic carbon nitride (g-C_3N_4) is a widely studied carbon-based (non-metal oxide) photocatalytic material for both energy and environmental applications such as gas conversion, water contaminant removal (organic pollutants and microbial species), and hydrogen generation. Furthermore, g-C_3N_4 has unique properties such as biocompatibility and improved chemical, mechanical, and thermal stability. This chapter discusses the fundamental physicochemical properties of nanostructured g-C_3N_4, strategies for band gap tuning (surface modification, doping, sensitization, and hybridization), mechanisms behind graphitic carbon nitride in photocatalytic removal of environmental contaminants (gas, toxic chemicals, and microbial species), water splitting towards hydrogen production applications, and their potential and prospects for the future.

11.1 Introduction

The distinctive physicochemical features of two-dimensional layered materials have attracted interest in photocatalytic applications. Many two-dimensional structures have been developed, such as black phosphorus, metal carbides, hexagonal boron nitride, dichalcogenides, and transition metal oxides. In addition, the high crystallinity and small thickness of the two-dimensional structured materials offer a short distance for the photogenerated charge carrier transfer on the surface of the catalyst and facilitate in-plane charge transfer, bringing about improved photogenerated charge carrier pair separation efficiency in various photocatalytic reactions. To date, two-dimensional polymeric graphitic carbon nitride (g-C_3N_4) has received substantial research interest in energy and environmental applications (gas conversion, disinfection, chemical pollutant degradation, and hydrogen generation), due to its

doi:10.1088/978-0-7503-5724-1ch11

distinctive features of non-toxicity, biocompatibility, appealing electronic structure (energy band gap—2.7 eV), and strong thermal, chemical, and mechanical stability [1]. However, it has some drawbacks such as the specific surface area being low, the electron and hole separation still having poor efficiency, and the utilization of visible light being limited, which could reduce the photocatalytic efficiency. Therefore, modification of g-C_3N_4 has become essential to enhancing its photocatalytic activity and advancing practical applications. Four different kinds of strategies, namely textural modification, metal or non-metal doping, surface sensitization, and surface hybridization, have been adapted to enhance the photo-response characteristics, surface/interface properties, and separation efficiency of photogenerated charge carriers in different perspectives [2–4].

This chapter highlights the emerging applications of g-C_3N_4-based photocatalysts for energy and environmental applications. A summary of the fundamental aspects of g-C_3N_4-based photocatalysts and the basic mechanisms involved in various photocatalytic applications are also discussed.

11.2 Two-dimensional graphitic carbon nitride

11.2.1 Structure and properties

The n-type π-conjugative polymeric g-C_3N_4 comprises mainly carbon and nitrogen [5, 6]. There are many surface properties of g-C_3N_4 that are crucial to catalysis, including fundamental surface functions, electron-rich characteristics, H-bonding motifs, etc, as shown in figure 11.1. Berzelius first synthesized polymeric g-C_3N_4 in 1834 and Liebig named it [7]. g-C_3N_4 contains C–N layers that exhibit a laminar form, which is comparable to the graphite structure [8]. It usually has a long pair of nitrogen-based tri-s-triazine structural units coupled with planar amino groups in each layer and weak van der Waals interaction between layers. It can be synthesized easily from nitrogen-rich precursors via a thermal polycondensation route. Numerous benefits of g-C_{34} synthesis include the production of tiny, homogeneous particles, reduced energy consumption, and economic viability. Additionally, there are several allotropes of carbon nitride, including g-o-triazine (0.93 eV), pseudo-cubic-C_3N_4 (4.13 eV), cubic-C_3N_4 (4.3 eV), g-h-heptazine (2.87 eV), g-h-triazine (2.97 eV), β-g-C_3N_4 (4.85 eV), and α-g-C_3N_4 (5.5 eV). Among the all allotropes of g-C_3N_4, the tri-s-triazine-based g-C_3N_4 is energetically preferred and most stable structure [8–10]. The advantages of g-C_3N_4 for energy and environmental applications are as follows:

1. The band gap level of g-C_3N_4 is suitable for a redox reaction to adequate hydrocarbons.
2. The two-dimensional layered structure of g-C_3N_4 could enable the transfer rate of electrons to the active sites of the catalyst surface.
3. g-C_3N_4 has a tunable structure, which could also be readily utilized to form highly tailorable hybrid photocatalysts.
4. g-C_3N_4 absorbs visible light when compared to wide-band gap semiconductors (often TiO_2 and ZnO).

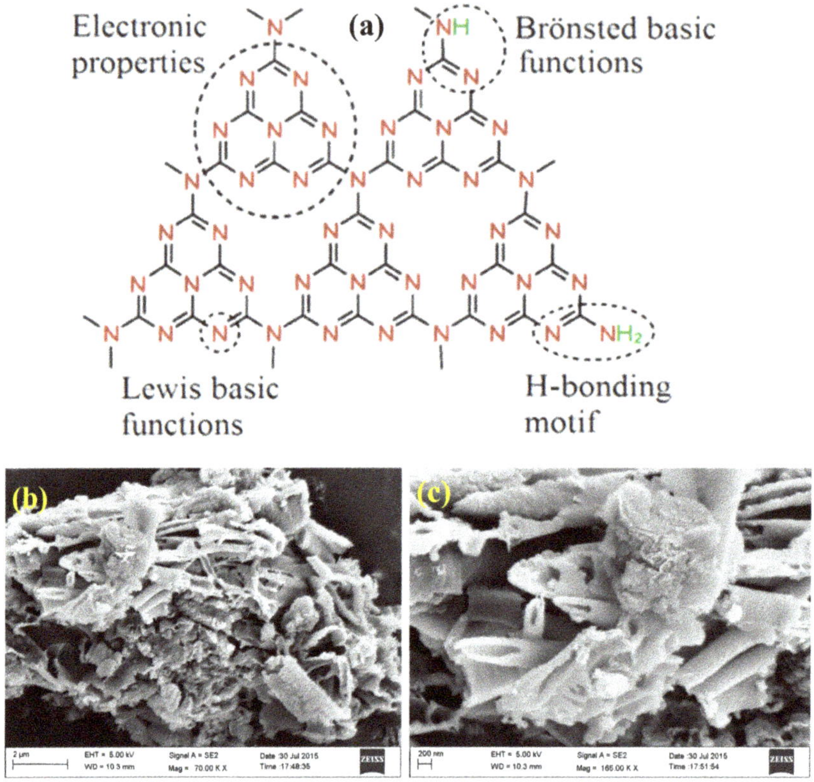

Figure 11.1. (a) The basic structure of g-C_3N_4. (Reproduced with permission from [8]. Copyright 2008 the Royal Society of Chemistry.) (b) and (c) Morphological structure of g-C_3N_4. (Reproduced with permission from [2]. Copyright 2017 Elsevier.)

5. Unlike narrow band gap structures, it does not exhibit the photo-etching effect.
6. It is highly stable in nature.
7. It has low cost of synthesis and is non-toxic.

11.2.2 Synthesis of g-C_3N_4

Bulk g-C_3N_4 is produced by thermal polycondensation while being heated from organic molecules rich in amino groups, such as melamine, dicyandiamide, guanidine thiocyanate, guanidinium chloride, cyanamide, urea, and thiourea [1, 11–16]. The reaction environment may be air or an inert gas such as nitrogen or argon, and the reaction temperature is typically between 500 °C and 600 °C. The properties of bulk g-C_3N_4 are affected by factors such as heating rate, inert gas flow rate, temperature, and time. In addition, the type of precursor also influences the optical and textural characteristics of g-C_3N_4. For example, g-C_3N_4 generated from urea has a somewhat higher energy band gap in comparison to g-C_3N_4 formed from thiourea [17]. The lower condensation temperature helps to tune the g-C_3N_4

properties such as surface area owing to its lower aggregation. Because of its graphitic layered structure, the surface area of g-C_3N_4 produced by direct polycondensation of feedstock containing carbon, nitrogen, and hydrogen is typically minimal, which leads poor performance in heterogeneous catalysis. So as to enhance the surface area of the g-C_3N_4, soft template, ultrasonic dispersion, acidic solution impregnation, and chemical functionalization routes were discovered. Some thermal treatments such as solid-state reaction, solvothermal, and chemical vapor deposition are also adapted to polymerizing the oxygen-free and nitrogen-rich compound precursors with pre-bonded C–N core structures into g-C_3N_4 [18–20].

11.2.3 Photocatalytic mechanism of g-C_3N_4

The photocatalytic activity of g-C_3N_4 is described by the creation of photogenerated charge carriers. Upon irradiation, the g-C_3N_4 material absorbs the energy and generates charge carriers. However, this is only possible if the energy of the light is greater than or equal to the energy band (E_g) of the semiconductor ($h\nu \geqslant E_g$):

$$g - C_3N_4 + h\upsilon \rightarrow g - C_3N_4(e^- + h^+). \tag{11.1}$$

Then, the electron is excited from the valance band to the conduction band. The following reactions may occur in these exciting species:

$$g - C_3N_4(h^+) + H_2O \rightarrow g - C_3N_4 + OH^- + H^+ \tag{11.2}$$

$$g - C_3N_4(h^+) + OH^- \rightarrow g - C_3N_4 + OH^{\cdot} \tag{11.3}$$

$$g - C_3N_4(e^-) + O_2 \rightarrow g - C_3N_4 + O_2^{\cdot -} \tag{11.4}$$

$$O_2^{\cdot -} + HO_2^{\cdot} \rightarrow OH^{\cdot} + O_2 + H_2O_2 \tag{11.5}$$

$$2HO_2^{\cdot -} \rightarrow O_2 + H_2O_2 \tag{11.6}$$

$$g - C_3N_4(e^-) + H_2O_2 \rightarrow g - C_3N_4 + OH^- + OH^{\cdot}. \tag{11.7}$$

In the recombination process there is no photocatalytic activity and it generates only heat:

$$e^- + h^+ \rightarrow \text{heat}. \tag{11.8}$$

11.2.4 Design strategies for g-C_3N_4 based nanocomposites

To enhance the photocatalytic efficiency of g-C_3N_4 various strategies have been adapted such as textural and morphological modifications, doping (cationic or anionic), sensitization, and hybridization of g-C_3N_4, as listed in table 11.1.

Table 11.1. Different surface modification strategies for g-C_3N_4.

Strategies	Description	Advantages	Disadvantages	Example
Textural and morphological modifications	Enhancing the surface area via altering the fabrication routes.	• A large number of surface active sites. • Low band gap. • Better electron transport ability. • Improves the photocatalytic activity.	• High recombination rate.	Mesoporous g-C_3N_4, g-C_3N_4 sheets, films, nanotubes, and edge-functionalized nanosheets.
Cation and anion doping	Incorporation of metals and non-metals on the surface of g-C_3N_4, which can act as sinks for charge particles.	• Narrows the energy band gap. • Improves the excitation of surface electrons. • Reduces the recombination rate.	• Photocorrosion of doped metals. • The incorporated metal itself may play a charge carrier recombination center.	Ag/g-C_3N_4, Au/g-C_3N_4, Pt/g-C_3N_4, Ni/g-C_3N_4 B/g-C_3N_4, P/g-C_3N_4, S/g-C_3N_4, and F/g-C_3N_4
Hybridization with another semiconducting material	Introduction of another semiconductor material on the g-C_3N_4 surface.	• Enhanced charge separation efficiency via Z-scheme mechanisms or double charge transfer. • Lowers the energy band gap. • Enhances the stability.		TiO_2/g-C_3N_4, ZnO/ g-C_3N_4, AgX (Br, Cl or I)/ g-C_3N_4, SnO_2/g-C_3N_4, Bi_2MoO_6/g-C_3N_4
Sensitization	Introduction of visible light-driven compounds (or semiconductor materials or narrow band gap dye compounds) with a large band gap semiconductor.	• Increases the visible light utilization efficiency. • Improves the charge separation efficiency. • Enhances the charge carrier mobility.	• Issues with dye solubility in aqueous solutions. • Photocorrosion reaction occurring in coupled semiconductors.	Rhodamine blue/g-C_3N_4, erythrosin B dye/ erythrosin B dye

11.3 Photocatalytic applications of g-C_3N_4

The g-C_3N_4-based catalyst is appropriate for many photocatalytic applications, such as gas conversion (CO_2 activation and nitrogen fixation), disinfection, organic pollutant degradation, and water splitting applications.

11.3.1 Gas conversion

11.3.1.1 CO_2 conversion

Among the many technologies mentioned above, CO_2 photoreduction is a promising approach promoting energy-bearing products (hydrocarbons) and mitigating the increased concentration of atmospheric CO_2 levels [21]. Traditionally, several photocatalytic materials have been explored for CO_2 reduction. Some of the vital photocatalytic materials are mentioned in figure 11.2. Remarkably, the narrow band gap semiconductor materials (e.g. Cu_2O, MoS_2, CdS, Fe_2O_3, WO_3, and Ag_3PO_4) absorb the greater portion of visible light. Only a few of them are photoactive because the band gap of the conduction and valance bands is unsuitable for redox reactions and has issues related to stability. Also, it exhibits poor charge separation and a high recombination rate on its surface thus suppressing the photocatalytic performance. While large energy band gap (E_g >3.1 eV) materials (TiO_2, ZnO, $SrTiO_3$, and $NaTaO_3$) are appropriate for the oxidation and reduction reactions due to their sufficient valance and conduction band potential, they have a significant energy input and only absorb a small amount of visible light [22]. Smaller band gap materials are more beneficial for the CO_2 photoreduction process. Among all the photocatalysts, two-dimensional structured g-C_3N_4 has an intermediate band gap with a suitable conduction band (−1.12 eV) and valance band position (1.58 eV) and makes it an appropriate photocatalyst in CO_2 reduction applications. Moreover, it is predicted to have excellent potential in CO_2 fixation and activation since it has Lewis

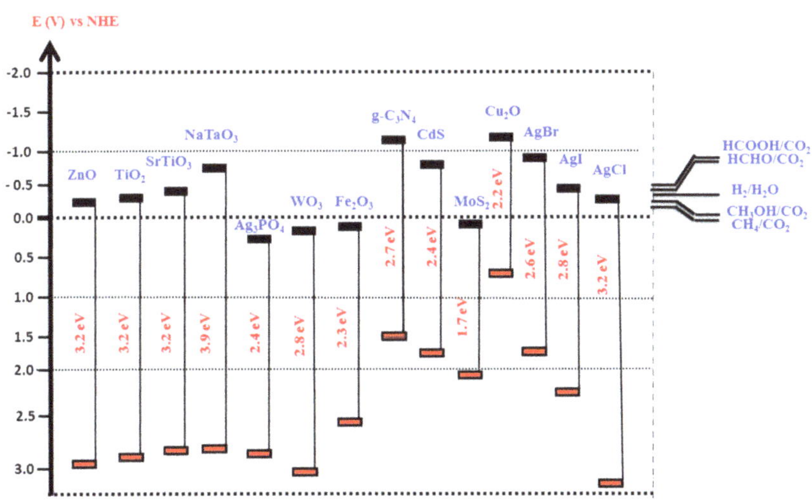

Figure 11.2. Band gap potential of different photocatalytic materials.

and Bronsted basic sites that are nitrogen-rich. The reactions that occur during CO_2 photoreduction are as follows:

$$g - C_3N_4 + h\upsilon \rightarrow e^- + h^+ \tag{11.9}$$

$$H_2O \rightarrow H^+ + OH^- \tag{11.10}$$

$$H^+ + e^- \rightarrow H^{\cdot} \tag{11.11}$$

$$OH^- + h^+ \rightarrow OH^{\cdot} \tag{11.12}$$

$$CO_2 + e^- \rightarrow CO_2^{\cdot -} \tag{11.13}$$

$$CO_2^{\cdot -} + H^{\cdot} \rightarrow CO + OH^- \tag{11.14}$$

$$CO + e^- \rightarrow CO^{\cdot -} \tag{11.15}$$

$$CO^{\cdot -} + H^{\cdot} \rightarrow C^{\cdot} + OH^- \tag{11.16}$$

$$C^{\cdot} + H^+ + e^- \rightarrow CH^{\cdot} \rightarrow CH_2^{\cdot} \rightarrow CH_3^{\cdot} \tag{11.17}$$

$$CH_3 + H^+ + e^- \rightarrow CH_4(\text{methane}) \tag{11.18}$$

$$2CH_3 + CO^{\cdot -} \rightarrow CH_3COCH_3(\text{acetone}) \tag{11.19}$$

$$H^+ + 2e^- \rightarrow H_2(\text{hydrogen}). \tag{11.20}$$

The porous g-C_3N_4 material shows poorer photocatalytic activity towards CO_2 reduction than bulk g-C_3N_4, because porousification improves the band gap and the presence of numerous defects could influence the efficiency of photoreduction [23]. However, in 2014 Wang *et al* used a melamine precursor via the chemical vapor deposition (CVD) method to prepare mesoporous g-C_3N_4 which displayed an outstanding photocatalytic effect in the reduction of CO_2 into CO [24]. The differences in the crystallinity and microstructure of g-C_3N_4 result in a variety of products. For example, C_2H_5OH is formed from CO_2 when melamine-derived g-C_3N_4 is employed as a photocatalyst, but urea-derived g-C_3N_4 produces CH_3OH and C_2H_5OH [16]. Thus far, different structures of g-C_3N_4 nanostructures, including nanowires, nanosheets, and mesoporous structures, have been synthesized successfully for improved CO_2 photoreduction. The g-C_3N_4 nanosheets would assist in shifting the electron to the active sites on the catalyst surface and alter the product selectivity from CO_2 photoreduction. The main products in g-C_3N_4 with band gap values of 2.97 eV and 2.77 eV are CH_4 and CH_3CHO, respectively [25]. Furthermore, the g-C_3N_4 containing amino groups are thought to play an important role in the adsorption and activation of CO_2 on its surface [26]. Thiourea-derived g-C_3N_4 exhibited an excellent CO_2 reduction performance owing to the appearance of a number of NH_x groups as well as a suitable conduction band potential and high charge separation efficiency [27]. In 2017 Xia *et al* developed ultrathin g-C_3N_4 nanosheets functionalized with amine via

NH_3-mediated thermal exfoliation of dicyandiamide-derived g-C_3N_4, which exhibited improved CO_2 adsorption, light-harvesting ability, and charge carrier separation, thus leading to remarkable enhancement in the production of CH_4 and CH_3OH from CO_2 reduction [28]. In 2017 He *et al* synthesized quasi-hexagonal prism-shaped carbon nitride using urea-derived g-C_3N_4 via the alkaline hydrothermal process and suggested that –NH_2 and –NH-groups are the major contributors for the improved photocatalytic performance of quasi-hexagonal prism-shaped carbon nitride compared to urea-derived g-C_3N_4 [29]. However, due to the rapid electron–hole pair recombination, the photocatalytic processes over g-C_3N_4 still had poor efficiencies [17, 30–32]. Therefore, a new strategy must be designed and fabricated to solve this problem. Alteration of and morphological properties and band structure can improve the activity of g-C_3N_4 photocatalysts.

Generally, large band gap materials limit their photocatalytic activity under visible light. Doping of anions and cations seems to be a suitable approach for narrowing the band gap and improving the absorbance region into visible light. Usually, anionic doping can shift the g-C_3N_4 valance band energy level via the creation of impurity energy levels above the valance band maximum, thus, narrowing the band gap and promoting a broader range of visible light utilization efficiency. Different types of anions such as nitrogen (N), sulfur (S), boron (B), phosphorous (P), fluorine (F), and iodine (I) have been adapted extensively as dopants for many semiconductor materials. The S-doped g-C_3N_4 for CO_2 conversion was examined and it was established that the sulfur-doped g-C_3N_4 provides more CH_3OH (1.12 mol g^{-1}) formation than un-doped g-C_3N_4 [33]. The synergistic interaction of surface phosphate groups and mesoporous structure and improved separation of photogenerated electron–hole pairs of phosphorylated g-C_3N_4 brings about superior performance for the conversion of CO_2 to fuel (figure 11.3(a)) [34]. The rate of fuel gas evolution using phosphorylated g-C_3N_4 reached 7.7 and 14.6 mol h^{-1} for CO and CH_4, respectively. In addition, the introduction of g-C_3N_4 to the red phosphorous (r-P) surface improved photocatalytic performance for CO_2 conversion and H_2 generation in the presence of H_2O vapor [35]. The CH_4 yield for r-P/g-C_3N_4–30, pure r-P, and g-C_3N_4 were 295 μmol h^{-1} g^{-1}, 145 μmol h^{-1} g^{-1}, and 107 μmol h^{-1} g^{-1}, respectively. Furthermore, P doping on the g-C_3N_4 surface improves light absorption, charge separation and migration, adsorption capacity, and surface area for enhanced activity. In particular, the surface amino-rich groups and more negative surface of P-g-C_3N_4 enhance the CO_2 adsorption and H^+ and the distinctive nanostructured structure promotes gas transport by enabling the production of CH_4 (1.81 μmol h^{-1} g^{-1}) and CO (2.37 μmol h^{-1} g^{-1}) formation [36]. Transition metals (cerium (Ce), cobalt (Co), iron (Fe), nickel (Ni), and copper (Cu)) and noble metals (silver (Ag), gold (Au), platinum (Pt), and palladium (Pd)) are commonly adopted as cationic co-catalysts in CO_2 conversion applications. Cu-modified polymeric g-C_3N_4 nanorods could efficiently convert the CO_2 into CH_4 and CH_3OH with a trace amount of C_2H_4 and C_2H_6 hydrocarbons in the CO_2–H_2O system and CO, H_2, C_2H_6, and C_3H_8 from the CO_2–CH_4 system (figure 11.3(b)) [37]. The addition of platinum to g-C_3N_4 may improve visible light absorption. The Pt co-catalyst deposition affects not only the selectivity of the product but also the activity of the reaction. Moreover, the Pt co-catalyst not only improves interfacial electron

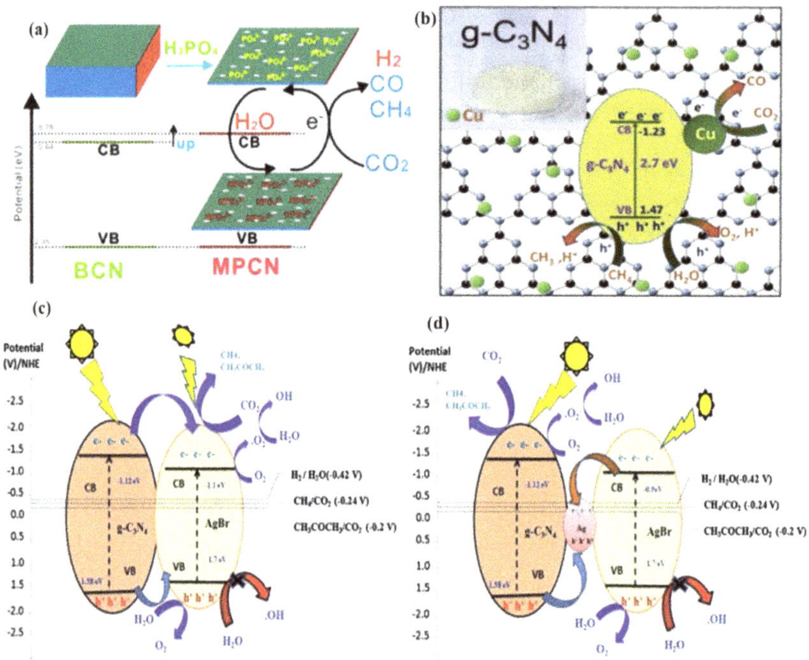

Figure 11.3. Photocatalytic reduction of CO_2 using a g-C_3N_4 based photocatalyst: (a) P/g-C_3N_4, (b) Cu/g-C_3N_4, (c) AgBr/g-C_3N_4 (double charge transfer), and (d) AgBr/g-C_3N_4 (Z-scheme charge transfer). ((a) Reproduced from [34]. Copyright 2016 Elsevier. (b) Reproduced with permission from [37]. Copyright 2017 Elsevier. (c) and (d) Reproduced with permission from [2]. Copyright 2017 Elsevier.)

transfer from g-C_3N_4 to Pt nanoparticles but also lowers the excess potential for CO_2 reduction [38].

Light absorbing compounds (eosin Y dye, magnesium phthalocyanine (MgPc) dye, cobalt-phthalocyanine (CoPc) dye, erythrosin B dye, and Tris(bipyridine) ruthenium (II) chloride ($[Ru(Bpy)_3]^{2+}$ dye)) are often employed as sensitizing materials to improve the visible light usage efficiency of g-C_3N_4. However, the modification of the photocatalyst will change the reaction pathway and selectivity of product formation [39]. For example, the ruthenium complex (cis,trans-[Ru{4,40-$(CH_2PO_3H_2)_2$-2,20-bipyridine}$(CO)_2Cl_2$]) (abbreviated as Ru) based polymeric g-C_3N_4 hybrid structure could effectively reduce the CO_2 into HCOOH (major product), and CO and H_2 (minor products) with the existence of triethanolamine under visible light irradiation. Because the Ru based complex acts as a light-harvesting unit and catalytic center for CO_2 reduction reaction [40, 41]. In 2014 Lin *et al* developed polymeric g-C_3N_4 with Co species for the production of CO from the CO_2 reduction reaction with visible light and suggested g-C_3N_4 as both CO_2 activator and photocatalyst and Co species as redox promoters [42].

Generally, the g-C_3N_4-based heterojunction displays two different types of charge separation mechanisms, namely double charge transfer (conventional) and Z-scheme mechanism (figures 11.3(c) and (d)). The brookite TiO_2/g-C_3N_4

nanocomposite via a facile calcination process, shows enhanced visible-light-driven CO_2 reduction activity and selectivity for CH_4 generation as compared to g-C_3N_4 [43]. The enhanced photocatalytic activity of TiO_2/g-C_3N_4 nanocomposite was attributed to reasons such as (i) uniform dispersion of g-C_3N_4 nanodots on brookite TiO_2 quasi nanocube facilitates better adsorption capacity of reactants for CO_2 reduction, (ii) well-aligned band structures, and (iii) better carbonate and OH adsorption which creates additional protons for CH_4 formation. Likewise, $LaPO_4$/g-C_3N_4 core–shell nanocomposites enhance light sensitivity greatly and accelerate the separation of charge carriers across the bonding interaction between g-C_3N_4 and $LaPO_4$ effectively, and achieved a CO production rate of 14.4 μmol h^{-1} g^{-1} under Xe lamp irradiation [44]. The generation of the internal electrical field between the g-C_3N_4 and B_4C increases the light absorbance range significantly and splits the charge carriers [45]. The Pt loaded B_4C/g-C_3N_4 showed a higher CH_4 (0.84 μmol h^{-1} g^{-1}) production rate than g-C_3N_4 (0.14 μmol h^{-1} g^{-1}). The well-dispersed In_2O_3 nanocrystals on sheet-like g-C_3N_4 surfaces form intimate contact, which allows for effective interfacial charge transfer across the In_2O_3/g-C_3N_4 heterojunction. The yield of CH_4 from CO_2 reduction is around 76.7 ppm [15].

g-C_3N_4/$NaNbO_3$ nanowire was fabricated by a facile hydrothermal route [46]. The photocatalytic of the as-prepared g-C_3N_4/$NaNbO_3$ nanowire was tested toward CO_2 conversion under visible light irradiation. The CH_4 (6.4 μmol^{-1} h^{-1} g^{-1}) production from CO_2 conversion was almost eight times greater compared to pristine g-C_3N_4. In addition, the photocatalytic activity of g-C_3N_4/$KNbO_3$ for CO_2 reduction was almost four times higher than that of pristine g-C_3N_4 under visible light irradiation [47]. Such outstanding improvement of performance was attributed to the separation and movement of charge carrier g-C_3N_4/$NaNbO_3$ or g-C_3N_4/$KNbO_3$ composites.

Recently, it has been found that the introduction of co-catalyst on hybrid structure composites promotes photocatalytic efficiency. The photocatalytic CO_2 reduction by H_2O was investigated over g-C_3N_4/Ag-TiO_2 hybrid [48]. There is a synergistic impact of combining g-C_3N_4 with Ag NP coated TiO_2 (CN/AgTi) in improving CO_2 conversion. The better efficiency is due to the slower recombination rate of charge carriers and superior light-harvesting capacity. In 2018 Yang *et al* reported Sb/SnO_2/g-C_3N_4 nanosheets that provided an superior performance towards CO_2 reduction under visible light irradiation. The improved photocatalytic activity is observed at an optimal mass ratio of 5wt% of Sb/SnO_2/g-C_3N_4 the nanosheet [49]. Pt incorporated g-C_3N_4/kNbO_3 photocatalyst has been investigated for enhanced CO_2 photoreduction in visible light irradiation [47]. The composites of g-C_3N_4 and N-TiO_2 exhibit good visible light sensitization and high efficiency of charge separation during light irradiation. Thus, the optimized samples produced the greatest CO evolution quantity of 14.73 mol under 12 h light irradiation, resulting in strong photocatalytic performance and photostability in CO_2 conversion [50]. g-C_3N_4 with oxidative polyoxometalates composites exhibited an outstanding absorption capacity of visible light and better capability for the CO_2 capture and conversion process [30]. Moreover, ternary chalcogenide ($CdIn_2S_4$) deposited g-C_3N_4 composite was employed for conversion of CO_2 into CH_3OH. It was found that $CdIn_2S_4$/g-C_3N_4 composite not only increases the CO_2 adsorption capacity, it also increases the separation of photogenerated electron–hole pairs [51].

Carbon nanomaterials such as graphene nanosheets, carbon nanotubes, and reduced graphene oxide (rGO) have been used widely for enhancing the photocatalytic performance of g-C_3N_4 due to the availability, low cost, and good resistance to corrosion in the long term. Two-dimensional-RGO hybridized protonated g-C_3N_4 nanostructures are fabricated by Ong *et al* who reported that 15rGO/g-C_3N_4 produced a remarkable 5.4-fold enhancement of CH_4 evolution over g-C_3N_4 due to the effective separation of photogenerated charge carriers [52]. The photocatalytic performance of ultrathin g-C_3N_4 nanosheets and a series of rGO/g-C_3N_4 nanocomposites in a gaseous system under visible light illumination were then evaluated. The CH_4 evolution from ultrathin g-C_3N_4 and 3rGO/g-C_3N_4 is about 7.49 μmol/10 g catalyst and 17.36 μmol/10 g catalyst, respectively, for a period of 3 h [53]. Under visible light irradiation, the 2D/2D interface heterojunction of NiAl-LDH with g-C_3N_4 showed remarkable performance for photocatalytic conversion of CO_2 to CO and H_2, because of the effective separation of charge carrier, suppressed recombination rate, and efficient transfer of photoexcited charge carriers. The Z-scheme Ag_3PO_4/g-C_3N_4 composite promoted effectively the separation of electron–hole pairs through the Z-scheme mechanism, and eventually raised the CO_2 conversion efficiency of Ag_3PO_4/g-C_3N_4. The main product is CO and a certain amount of CH_4, CH_3OH, and C_2H_5OH was formed [32]. Another Z-scheme structure of AgBr/g-C_3N_4/N-graphene composite prepared by Li *et al* in 2015 displayed admirable performance for the reduction of CO_2 into CH_3OH (105.89 μmol g^{-1}) and C_2H_5OH (256.45 μmol g^{-1}) formation. Such excellent performance of the AgBr/g-C_3N_4/N-graphene composite is mainly due to the improvement in the photoinduced charge carrier separation efficiency and suppression of recombination of the electron–hole pair [54]. In addition, some of the metal oxides were also found to form a Z-scheme heterostructure with g-C_3N_4, including SnO_2 [55] and ZnO [56].

Moreover g-C_3N_4/Bi_2WO_6 nanocomposites showed the generation of CO from the CO_2 photoreduction process using H_2O under visible light irradiation. The as-prepared composite produced 5.19 mol h^{-1} g^{-1} of CO, which is 22 times higher than pristine g-C_3N_4. This improvement may be attributed mostly to band control and light sensitization, which resulted in more effective separation of excited electrons at the conduction band and holes at the valance band [57]. Bai *et al* and Wang *et al* reported the realignment of the band structures of two semiconductor materials to form a Z-scheme structure. For example, Z-scheme g-C_3N_4/$Bi_4O_5I_2$ and g-C_3N_4/Bi_4O could enrich the CO production yield of 17.23 μmol g^{-1} and 45.6 μmol g^{-1}, respectively [58, 59].

11.3.1.2 Nitrogen fixation

g-C_3N_4 with suitable band gap positions could strongly oxidize the nitrous oxide (NO) to nitrite (NO_2^-) or nitrate (NO_3^-) under exposure to visible light [19]. The reaction is as follows:

$$g - C_3N_4 + h\upsilon \rightarrow e^- + h^+ \tag{11.21}$$

$$e^- + O_2 \rightarrow .O_2^- \tag{11.22}$$

$$O_2^- + NO \rightarrow NO_3^- \tag{11.23}$$

$$h^+ + OH^- \rightarrow OH^{\cdot} \tag{11.24}$$

$$NO + 2OH^{\cdot} \rightarrow NO_2 + H_2O \tag{11.25}$$

$$NO_2 + OH^{\cdot} \rightarrow NO_3^- + H^+. \tag{11.26}$$

The co-functionalization of g-C_3N_4 with other metals or metal oxides could enhance the NO conversion efficiency due to the generation of localized electrons on the surface of g-C_3N_4. For instance, the NO conversion efficiency improved when SrO, O/Ba, and O/La were incorporated on the surface of the g-C_3N_4 [60–62]. Additionally, Li *et al* showed that the functioning of the localized excess electrons surrounding Ca improved the photocatalytic performance and selectivity of Ca/g-C_3N_4 for NO elimination significantly [63]. In a different instance, the insertion of Sr results in an uneven electron distribution on the surface of g-C_3N_4, which speeds up the separation and shifting of photogenerated electron–hole pairs as well as the capacity to absorb light. Because of this, active species may effectively oxidize NO_x, turning it into the intended products NO_2 and NO_3 instead of other harmful by-products [64]. When related to g-C_3N_4 and K-g-C_3N_4, the addition of Cl and K ions in the g-C_3N_4 could decrease the charge transfer barrier and work as a dual channel for charge carrier transfer, extending their lifespan. This may improve the NO_x removal efficiency under visible light exposure [65]. The enhanced N_2 fixation efficiency of nitrogen vacancy endowed g-C_3N_4 photocatalysts is mainly attributed to the three different reasons: (i) selective absorption and activation of nitrogen owing to the same size and shape; (ii) the nitrogen vacancy improves the photo-generated charge separation efficiency and produces more electrons; and (iii) it stimulate the transfer of photogenerated electrons from the g-C_3N_4 to the adsorbed nitrogen. Another study demonstrated that g-C_3N_4/$BiVO_4$ composites convert four times more effectively than g-C_3N_4 alone due to the effective separation of photo-generated charge carriers.

11.3.2 Disinfection of microbial species

In general, the photocatalytic disinfection species are the superoxide (O_2.), hydrogen peroxide (H_2O_2), and hydroxyl (OH.) radicals [66–68]. It is generally recognized that holes in the valance band may breakdown (oxidize) water, resulting in the inactivation of microbiological organisms, particularly bacteria. Furthermore, additional reactive active species (H_2O_2 and O_2.) are mostly produced as a result of O_2 reduction ($O_2 \rightarrow O_2^- \leftrightarrow HO_2^{\cdot} \rightarrow H_2O_2 \leftrightarrow OH^{\cdot}$), because they efficiently react with biological components, promote the cleavage of microbial cell walls, and, eventually, cause cell death (figure 11.4) [69, 70]:

$$g - C_3N_4 + h\upsilon \rightarrow g - C_3N_4\,(e^- + h^+) \tag{11.27}$$

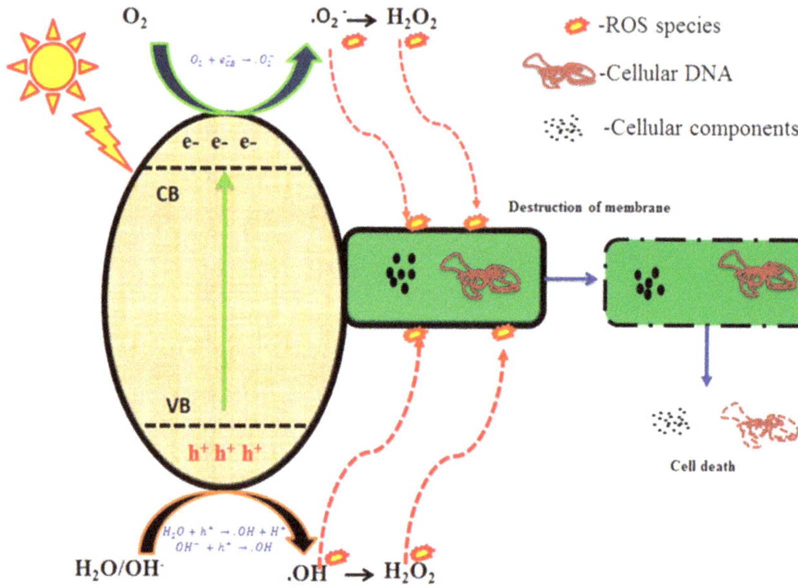

Figure 11.4. Photocatalytic disinfection mechanism over g-C_3N_4 photocatalyst. (Reproduced with permission from [4]. Copyright 2019 Springer Nature.)

$$g-C_3N_4(h^+) + H_2O \rightarrow g-C_3N_4 + OH^- + H^+ \tag{11.28}$$

$$g-C_3N_4(h^+) + OH^- \rightarrow g-C_3N_4 + OH^{\cdot} \tag{11.29}$$

$$g-C_3N_4(e^-) + O_2 \rightarrow g-C_3N_4 + O_2^{\cdot-} \tag{11.30}$$

$$O_2^{\cdot-} + HO_2^{\cdot} \rightarrow OH^{\cdot} + O_2 + H_2O_2 \tag{11.31}$$

$$2HO_2^{\cdot} \rightarrow O_2 + H_2O_2 \tag{11.32}$$

$$g-C_3N_4(e^-) + H_2O_2 \rightarrow g-C_3N_4 + OH^- + OH^{\cdot}. \tag{11.33}$$

Mesoporous g-C_3N_4, porous g-C_3N_4 nanosheets, and g-C_3N_4 films exhibit improved photodisinfection activity against different microbial species. The improved disinfection activity is mainly attributed to the high surface area, increased electron transport facility, and reduced band gap energy. Notably, the improved surface area has high reactive sites on its surface, effective separation of photogenerated charge carriers, and charge transfer efficiency. For example, urea-derived g-C_3N_4 film shows enhanced disinfection activity toward gram positive and gram negative bacterial species with comparison to bulk g-C_3N_4 [69, 70]. Visible light active porous g-C_3N_4 nanosheets with a large surface area (190.1 $m^2\ g^{-1}$) were able to kill all *E. coli* within 4 h [71]. The metal loading on the g-C_3N_4 surface may also considerably promote bactericidal action. The coupling of silver nanoparticles into g-C_3N_4 might increase the visible light utilization efficiency, reducing the charge carrier recombination, stimulating the electron–hole pair separation efficiency, and promote the charge carrier to have a longer lifetime [72]. The disinfection efficiency of $g\text{-}C_3N_4$ and $Ag/g\text{-}C_3N_4$ are 29.6% and 99.4%, respectively [73]. In another study, Ag NP hybridized $g\text{-}C_3N_4$ NP nanocomposite shown more efficient bactericidal action against *B. subtilis*, *P. aeruginosa*, *E. coli*, and *S. aureus*, in the absence of light as compared to pure Ag NPs. Many factors can cause this: (i) Ag NPs on the $g\text{-}C_3N_4$ surface causing the inactivation of some bacterial species; (ii) the formation of highly stable Ag NPs due to interactions between the surface of silver containing ligand molecules from the grape extract and the $g\text{-}C_3N_4$ that have been integrated; and (iii) $g\text{-}C_3N_4/Ag$ NP nanocomposites suitable for gram negative rather than gram positive bacterial species [74]. Ma *et al* reported that the main species in the destruction of bacterial cell wall structure are O_2^-, h^+, and e^- when $Ag/g\text{-}C_3N_4$ nanocomposite was used [72].

The performance of the photocatalytic system can be improved by $g\text{-}C_3N_4$ being hybridized with different semiconductor materials. It achieves effective electron–hole pair separation via transferring the holes from a higher valance band to a lower valance band and electrons from the high conduction band gap level to the lower conduction band gap and enhances the stability of low band gap semiconducting material. Different $g\text{-}C_3N_4$ based hybrid nanocomposites have been fabricated for bacteriacidal activity under visible light irradiation such as TiO_2 based hybrid structure ($g\text{-}C_3N_4$/microsized TiO_2, $g\text{-}C_3N_4/TiO_2$ nanotube arrays, and $g\text{-}C_3N_4$ quantum dots/TiO_2 nanotube arrays), bismuth based hybrids ($g\text{-}C_3N_4/BiVO_4$, $Bi_2MoO_6/g\text{-}C_3N_4$ nanosheets, and $g\text{-}C_3N_4/Bi_2O_4$), plasmonic catalyst ($g\text{-}C_3N_4/AgBr$, $Ag_2WO_4/g\text{-}C_3N_4$, and $g\text{-}C_3N_4/AgVO_3$), carbon-based composites (graphene oxide/$g\text{-}C_3N_4$, $g\text{-}C_3N_4$/reduced graphene oxide), fullerene/$g\text{-}C_3N_4$, $g\text{-}C_3N_4$/EP (porous expanded perlite) $Ni_2P/g\text{-}C_3N_4$, $g\text{-}C_3N_4/CdCO_3$, and $ZnO/g\text{-}C_3N_4$ [75–85]. In addition, the photocatalytic disinfection performance of $g\text{-}C_3N_4$ hybrid composites or ternary composites may be increased significantly by adding noble metals to it. Ag-decorated TiO_2 NFs on $g\text{-}C_3N_4$ sheets, $Ag/ZnO/g\text{-}C_3N_4$, $Ag/AgO/g\text{-}C_3N_4$ microspheres, and $BiVO_4/Ag/g\text{-}C_3N_4$ have shown improved antimicrobial and antifouling activity. Moreover, B/Phenyl/$g\text{-}C_3N_4$ showed a wide spectrum of absorption from ultraviolet to near-infrared light. B/phenyl/$g\text{-}C_3N_4$ composites inhibited over 99% of *E. coli* for 180 min while exposed to visible light [86]. The vacuum filtration approach of building $g\text{-}C_3N_4$ nanosheet/rGO photocatalysts (specifically $g\text{-}C_3N_4NS$/rGO/CA) may kill all bacteria (6.5 log) under visible light in 2 h [87]. In line with this, modifying TiO_2 nanotubes and Ti plates with $g\text{-}C_3N_4\text{-}SnO_2$ composites may enhance photocatalytic activity, and because of the efficient charge carrier separation between $g\text{-}C_3N_4\text{-}SnO_2$ and TiO_2 nanotubes/Ti in the ternary composite, the $g\text{-}C_3N_4\text{-}SnO_2/TiO_2$ nanotubes/Ti plate's photocatalytic and bactericidal activity was improved [88]. Moreover, $g\text{-}C_3N_4$ based composites could eliminate the algal species via the photooxidation process. For example, the $g\text{-}C_3N_4$ and TiO_2 immobilized Al_2O_3/expanded perlite nanocomposites removed 88.1% of 2.7×10^6 CFU ml^{-1} algal species after 6 h of reaction. After two hours of radiation, functional expanded graphite covered carbon layer composites hybridized with nitrogen-phosphorus co-doped TiO_2 could eliminate 98.2% of algal cells (*M. aeruginosa*) [89].

11.3.3 Hydrogen generation

Among the other hydrocarbon fuels, hydrogen is considered a sustainable from of solar energy with a heat value of 120–142 MJ kg^{-1}. Despite the fact that there are other ways to create hydrogen, the photocatalytic hydrogen generation approach has received substantial consideration for its variety of potential applications for the environment and energy. Two different reactions occur during water splitting, namely oxygen evaluation and hydrogen evaluation. The equation is as follows:

Oxygen evaluation reaction:

$$2H_2O \rightarrow 4H^+ + 4e^- + O_2, \quad E^\circ = 1.23\ \text{eV}. \tag{11.34}$$

Hydrogen evaluation reaction:

$$4H^+ + 4e^- \rightarrow 2H_2, \quad E^\circ = 0.00\ \text{eV}. \tag{11.35}$$

Overall reaction:

$$2H_2O + h\vartheta \rightarrow 2H_2 + O_2, \quad E^\circ = -1.23\ \text{eV}. \tag{11.36}$$

Typically, there are three phases involved in photocatalytic hydrogen generation: (i) migration of valance band electrons to the photocatalyst's conduction band upon light irradiation to create an electron–hole pair; (ii) movement of photogenerated charges to the surface of the photocatalyst; and (iii) redox reactions occurring in the valance band and conduction band positions. The proposed mechanism of hydrogen evaluation over bulk g-C_3N_4 and g-C_3N_4 nanosheets is represented in figure 11.5. For hydrogen evaluation, the conduction band potential of semiconductor material must have the valance band and conduction band potential of above 1.23 eV and below 0 eV, respectively. According to reports, the energy band gap level of g-C_3N_4 is suitable for hydrogen evaluation applications. At first, g-C_3N_4 was synthesized by Wang and colleagues for the hydrogen evaluation reaction [1]. In comparison to bulk g-C_3N_4, the H_2 production activity of g-C_3N_4 nanosheets, nanorods, and quantum dots has been shown to be improved. Moreover, the absorption range of

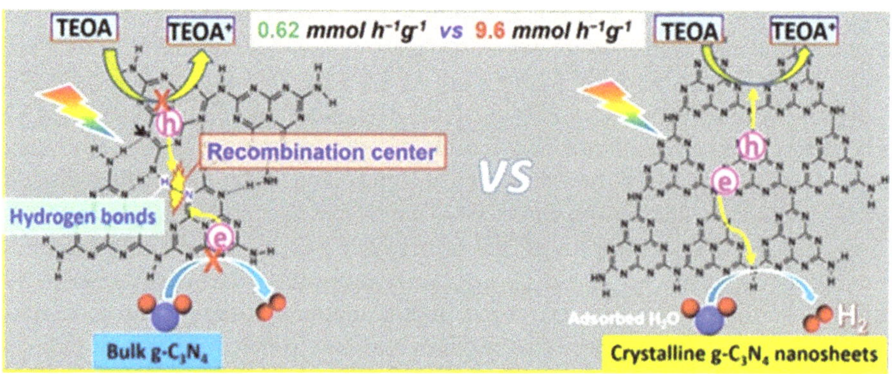

Figure 11.5. Schematic representation of the proposed mechanism of hydrogen evaluation over bulk g-C_3N_4 and crystalline g-C_3N_4 nanosheets. (Reproduced from [91]. Copyright 2018 Elsevier.)

$g\text{-}C_3N_4$ is extended by incorporating it with metal/non-metal nanoparticles and another semiconductor material that exhibits visible light surface plasmon resonances. Different metal (Sn, Zn, Au, Ag, Pt, and Rh), non-metal (P, S, C, F, and I), hybridization (metal oxide (NiO, CeO_2, InO_3, MoS_2, TiO_2, Al_2O_3, and $Cu(OH)_2$)), and metal chalcogenides ($CdS/g\text{-}C_3N_4$) have shown superior photocatalytic performance toward hydrogen generation than pristine $g\text{-}C_3N_4$. For example, the addition of metallic co-catalyst on the $g\text{-}C_3N_4$ surface could develop the Schottky effect between the metal and $g\text{-}C_3N_4$, which facilitates the separation of electron–hole pairs. In addition, the metallic co-catalyst serves as an electron sink for the photoelectrons generated on the surface of $g\text{-}C_3N_4$, thus increasing the lifespan of the photo-generated charge carriers. Moreover, a metallic co-catalyst addition also supplied more active sites for hydrogen evolution [90]. Likewise, the addition of a semiconducting co-catalyst could enhance the hydrogen generation efficiency by aligning their band gap with $g\text{-}C_3N_4$ and limiting the backflow of electrons. However, with semiconductor co-catalysts, particularly when the band gap is smaller than that of pure $g\text{-}C_3N_4$, building van der Waals heterojunction structures on $g\text{-}C_3N_4$ was suggested as a way to improve electron transport capabilities. More H_2 production was accomplished in network-like $MoSe_2$ than over flower-like $MoSe_2$ nanostructures, which is ascribed to the very effective charge carrier migration and separation based on the synergistic effect originating from the special sheet-on-sheet hetero-interface in porous $g\text{-}C_3N_4$ nanosheets [91].

Sacrificial substances such as lactic acid, methanol, and triethanolamine are the key requirements for $g\text{-}C_3N_4$-based hydrogen evaluation. Among the all sacrificial substances, triethanolamine shows superior performance towards hydrogen generation due to the enhanced electron donation behavior. This leads to the quenching effect towards the oxidation reaction, which limits the number of photogenerated holes. The quenching mechanism for triethanolamine is shown by the following:

$$g - C_3N_4 + h\vartheta \rightarrow e^- + h^+ \tag{11.37}$$

$$C_6H_{15}NO_3 + 2h^+ \rightarrow C_6H_{15}NO_3{}^+ \tag{11.38}$$

$$C_6H_{15}NO_3{}^+ \rightarrow C_6H_{15}NO_3{}^{\cdot} \tag{11.39}$$

$$C_6H_{15}NO_3{}^{\cdot} \rightarrow C_6H_{15}NO_3{}^+ + e^- \tag{11.40}$$

$$C_6H_{15}NO_3{}^+ + H_2O \rightarrow C_4H_{11}NO_3 + CH_3CHO + H^+. \tag{11.41}$$

Moreover, after triethanolamine is adsorbed onto the surface of $g\text{-}C_3N_4$, the Fermi level can be increased dramatically.

11.3.4 Decomposition of organic pollutants

Enormous quantities of persistent inorganic and organic contaminants have been released inexorably into the environment due to the rapid development of industry and urbanization. Owing to its high stability, visible light absorption capability, and improved adsorption ability for binding of organic and inorganic molecules, $g\text{-}C_3N_4$

based catalysts have been utilized as photocatalysts in the elimination of water contaminants. Generally, h^+ and O_2^- often play a substantial role in the degradation of various organic contaminants into H_2O, CO_2, and other negligible amounts of organic acids [4] as shown in equations (11.42) and (11.43) below. Organic pollutants are initially adsorbed on the surface of g-C_3N_4. The photogeneration of reactive species then oxidizes or disintegrates these adsorbed particles. Following that, the g-C_3N_4 surface is desorbed of intermediate degradation products to its liquid solution:

$$\text{organic pollutants} + g - C_3N_4 - h^+ \rightarrow g - C_3N_4 + CO_2 + H_2O + \text{intermediates} \quad (11.42)$$

$$\text{organic pollutants} + O_2 \rightarrow CO_2 + H_2O + \text{intermediates}. \quad (11.43)$$

The incorporation of the cation (transition or alkali metals) and anion (B, P, S, C, and N), heterojunction with another semiconducting material (Bi_2MoO_6, TiO_2, Fe_3O_4, Ag_3VO_4, ZnO Ag_2CO_3, Ag_2O, and AgX (X = Br, Cl, and I)) could improve the sorption ability of contaminants and decomposition of chemical pollutants from wastewater when exposed to visible light [39, 41–44]. Moreover, engineered defects in the g-C_3N_4 structure can change the band gap for outstanding utilization of visible light, while also acting as new active sites to initiate redox reactions, which improves the photocatalytic decomposition efficiency. The distinctive features of g-C_3N_4 (enormous surface area, surface functional groups, and rich active sites) could offer enhanced adsorption ability toward organic pollutant removal via various kinds of interaction mechanisms. Sample conditions (temperature, pH), morphology, and surface characteristics of g-C_3N_4 influence the adsorption efficiency. For example, $AlVO_4$/g-C_3N_4 nanocomposites were used for the specific cationic dye adsorption from an aqueous medium, and it was discovered that the sorption was influenced by pH. In another study, 3D-porous architecture g-C_3N_4-ZnO@graphene aerogel displayed enhanced sorption ability, light absorption capability, and efficient electron–hole separation (figure 11.6), thus, effectively enhancing the organic pollutant elimination in aqueous solutions [92]. In organic pollution management, such an advantage is commonly used in the sorption–photodecomposition of water contaminants.

11.4 Conclusion and future trends

The g-C_3N_4 nanomaterial has received a lot of attention in energy and environmental applications owing to its biocompatibility, stability, light-harvesting ability, low toxicity, and abundance. A detailed discussion of the fundamental concepts, mechanisms, and photocatalytic activity for surface modified (textural modification, cation/anion, hybridization, and sensitization) g-C_3N_4 was provided in this chapter. To produce defect-free g-C_3N_4 with superior crystallinity, the morphology may be considerably improved by choosing the right precursor, employing effective preparation techniques, and using selective heteroatom doping. To further enhance the photocatalytic efficiency of g-C_3N_4, however, there is still much research being done on how to modify its characteristics in the post-treatment method. Additionally,

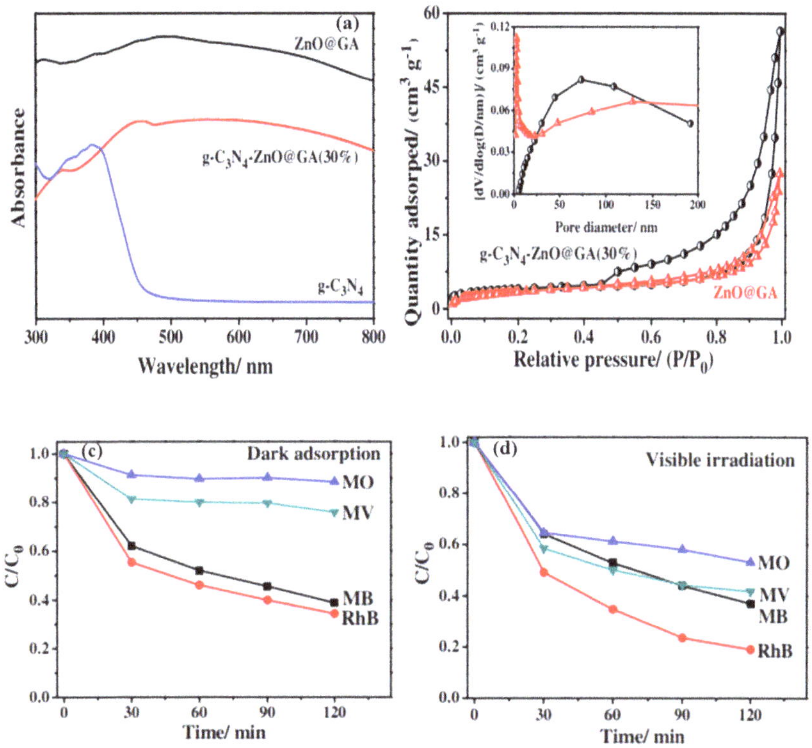

Figure 11.6. (a) Optical properties and (b) nitrogen adsorption–desorption isotherms and the pore size distribution. (c) Adsorption and (d) photodecomposition performance of 3D-porous structured nanocomposites. (Reproduced with permission from [92]. Copyright 2019 Elsevier.)

more work has to be done to create and adapt g-C_3N_4 catalysts in an economical and environmentally responsible manner. This is necessary for their scaling up.

References

[1] Wang X, Maeda K, Thomas A, Takanabe K, Xin G, Carlsson J M, Domen K and Antonietti M 2008 A metal-free polymeric photocatalyst for hydrogen production from water under visible light *Nat. Mater.* **8** 76–80

[2] Murugesan P, Narayanan S and Manickam M 2017 Experimental studies on photocatalytic reduction of CO_2 using AgBr decorated g-C_3N_4 composite in TEA mediated system *J. CO_2 Util.* **22** 250–61

[3] Chen D *et al* 2020 Photocatalytic degradation of organic pollutants using TiO_2-based photocatalysts: a review *J. Clean. Prod.* **268** 121725

[4] Murugesan P, Moses J A and Anandharamakrishnan C 2019 Photocatalytic disinfection efficiency of 2D structure graphitic carbon nitride-based nanocomposites : a review *J. Mater. Sci.* **54** 12206–35

[5] Fu J, Yu J, Jiang C and Cheng B 2018 g-C_3N_4-based heterostructured photocatalysts *Adv. Energy Mater.* **8** 1–31

[6] Naseri A, Samadi M, Pourjavadi A, Moshfegh A Z and Ramakrishna S 2017 Graphitic carbon nitride (g-C_3N_4)-based photocatalysts for solar hydrogen generation: recent advances and future development directions *J. Mater. Chem.* A **5** 23406–33
[7] Liebig J 1834 Uber einige Stickstoff—Verbindungen. Annalen der Pharmacie *Eur. J. Org. Chem.* **10** 1–47
[8] Vivek E, Arulraj A, Khalid M and Potheher I V 2022 Samarium hydroxide nanorolls anchored graphitic carbon nitride nanosheets: an active electrode material for supercapacitors *J. Alloys Compd.* **908** 164541
[9] Li S, Wang Z, Wang X, Sun F, Gao K, Hao N and Zhang Z 2017 Orientation controlled preparation of nanoporous carbon nitride fibers and related composite for gas sensing under ambient conditions *Nano Res.* **10** 1710–9
[10] Wen J, Xie J, Chen X and Li X 2017 A review on g-C_3N_4-based photocatalysts *Appl. Surf. Sci.* **391** 72–123
[11] Praus P, Svoboda L, Ritz M, Troppová I, Šihor M and Kočí K 2017 Graphitic carbon nitride: synthesis, characterization and photocatalytic decomposition of nitrous oxide *Mater. Chem. Phys.* **193** 438–46
[12] Vivek E, Arulraj A, Krishnan S G, Khalid M and Potheher I V 2021 Novel nanostructured $Nd(OH)_3$/g-C_3N_4 nanocomposites (nanorolls anchored on nanosheets) as reliable electrode material for supercapacitors *Energy Fuels* **35** 15205–12
[13] Zheng Y, Liu J, Liang J J, Jaroniec M and Qiao S Z 2012 Graphitic carbon nitride materials: controllable synthesis and applications in fuel cells and photocatalysis *Energy Environ. Sci.* **5** 6717–31
[14] Fanchini G, Tagliaferro A, Conway N M J and Godet C 2002 Role of lone-pair interactions and local disorder in determining the interdependency of optical constants of a-CN :H thin films *Phys. Rev.* B **66** 195415
[15] Cao S W, Liu X F, Yuan Y P, Zhang Z Y, Sen Liao Y, Fang J, Loo S C J, Sum T C and Xue C 2014 Solar-to-fuels conversion over In_2O_3/g-C_3N_4 hybrid photocatalysts *Appl. Catal.* B **147** 940–6
[16] Mao J, Peng T, Zhang X, Li K, Ye L and Zan L 2013 Effect of graphitic carbon nitride microstructures on the activity and selectivity of photocatalytic CO_2 reduction under visible light *Catal. Sci. Technol.* **3** 1253
[17] Dong F, Zhao Z, Xiong T, Ni Z, Zhang W, Sun Y and Ho W K 2013 *In situ* construction of g-C_3N_4/g-C_3N_4 metal-free heterojunction for enhanced visible-light photocatalysis *Appl. Mater. Interfaces.* **5** 11392–401
[18] Xia J, Di J, Yin S, Li H, Xu H, Xu L, Shu H and He M 2014 Solvothermal synthesis and enhanced visible-light photocatalytic decontamination of bisphenol A (BPA) by g-C_3N_4/BiOBr heterojunctions *Mater. Sci. Semicond. Process.* **24** 96–103
[19] Wang Y, Zhao S, Zhang Y, Fang J, Chen W, Yuan S and Zhou Y 2018 Facile synthesis of self-assembled g-C_3N_4 with abundant nitrogen defects for photocatalytic hydrogen evolution *ACS Sustain. Chem. Eng.* **6** 10200–10
[20] Hu C, Chu Y, Wang M and Wu X 2017 Rapid synthesis of g-C_3N_4 spheres using microwave-assisted solvothermal method for enhanced photocatalytic activity *J. Photochem. Photobiol.* A **348** 8–17
[21] Shogo K, Naveed A, Gabriela C and Yasuo I 2015 Photocatalytic conversion of carbon dioxide using Zn–Cu–Ga layered double hydroxides assembled with Cu phthalocyanine : Cu in contact with gaseous reactant is needed for methanol generation *Oil Gas Sci. Technol. J.* **70** 841–52

[22] Habisreutinger S N, Schmidt-Mende L and Stolarczyk J K 2013 Photocatalytic reduction of CO_2 on TiO_2 and other semiconductors *Angew. Chem. Int. Ed.* **52** 7372–408
[23] Dong G and Zhang L 2012 Porous structure dependent photoreactivity of graphitic carbon nitride under visible light *J. Mater. Chem.* **22** 1160–6
[24] Wang Y, Wang F, Zuo Y, Zhang X and Cui L F 2014 Simple synthesis of ordered cubic mesoporous graphitic carbon nitride by chemical vapor deposition method using melamine *Mater. Lett.* **136** 271–3
[25] Niu P, Yang Y, Yu J C, Liu G and Cheng H M 2014 Switching the selectivity of the photoreduction reaction of carbon dioxide by controlling the band structure of a g-C_3N_4 photocatalyst *Chem. Commun.* **50** 10837
[26] Su Q, Sun J, Wang J, Yang Z, Cheng W and Zhang S 2014 Urea-derived graphitic carbon nitride as an efficient heterogeneous catalyst for CO_2 conversion into cyclic carbonates *Catal. Sci. Technol.* **4** 1556
[27] Wang H, Sun Z, Li Q, Tang Q and Wu Z 2016 Surprisingly advanced CO_2 photocatalytic conversion over thiourea derived g-C_3N_4 with water vapor while introducing 200–420 nm UV light *J. CO_2 Util.* **14** 143–51
[28] Xia P, Zhu B, Yu J, Cao S and Jaroniec M 2017 Ultra-thin nanosheet assemblies of graphitic carbon nitride for enhanced photocatalytic CO_2 reduction *J. Mater. Chem.* A **5** 3230–8
[29] He Z, Wang D, Tang J, Song S, Chen J and Tao X 2017 A quasi-hexagonal prism-shaped carbon nitride for photoreduction of carbon dioxide under visible light *Environ. Sci. Pollut. Res.* **24** 8219–29
[30] Zhou J, Chen W, Sun C, Han L, Qin C, Chen M, Wang X, Wang E and Su Z 2017 Oxidative polyoxometalates modified graphitic carbon nitride for visible-light CO_2 reduction *ACS Appl. Mater. Interfaces* **9** 11689–95
[31] Kumar S, Karthikeyan S and Lee A 2018 g-C_3N_4-based nanomaterials for visible light-driven photocatalysis *Catalysts* **8** 74
[32] Yiming He M F, Lihong Z and Bo-Tao T 2015 A new application of Z-scheme Ag_3PO_4/g-C_3N_4 composite in converting CO_2 to fuel *Environ. Sci. Technol.* **49** 649–56
[33] Wang K, Li Q, Liu B, Cheng B, Ho W and Yu J 2015 Sulfur-doped g-C_3N_4 with enhanced photocatalytic CO_2-reduction performance *Appl. Catal.* B **177** 44–52
[34] Ye L, Wu D, Chu K H, Wang B, Xie H, Yip H Y and Wong P K 2016 Phosphorylation of g-C_3N_4 for enhanced photocatalytic CO_2 reduction *Chem. Eng. J.* **304** 376–83
[35] Yuan Y P, Cao S W, Sen Liao Y, Yin L S and Xue C 2013 Red phosphor/g-C_3N_4 heterojunction with enhanced photocatalytic activities for solar fuels production *Appl. Catal.* B **140–141** 164–8
[36] Liu B, Ye L, Wang R, Yang J, Zhang Y, Guan R, Tian L and Chen X 2017 Phosphorus-doped graphitic carbon nitride nanotubes with amino-rich surface for efficient CO_2 capture, enhanced photocatalytic activity and product selectivity *ACS Appl. Mater. Interfaces* **10** 4001–9
[37] Tahir B, Tahir M and Amin N A S 2017 Photo-induced CO_2 reduction by CH_4/H_2O to fuels over Cu-modified g-C_3N_4 nanorods under simulated solar energy *Appl. Surf. Sci.* **419** 875–85
[38] Yu J, Wang K, Xiao W and Cheng B 2014 Photocatalytic reduction of CO_2 into hydrocarbon solar fuels over g-C_3N_4–Pt nanocomposite photocatalysts *Phys. Chem. Chem. Phys.* **16** 11492
[39] Zhang G, Lan Z A and Wang X 2015 Merging surface organometallic chemistry with graphitic carbon nitride photocatalysis for CO_2 photofixation *Chem. Cat. Chem.* **7** 1422–3

[40] Maeda K, Sekizawa K and Ishitani O 2013 A polymeric-semiconductor–metal-complex hybrid photocatalyst for visible-light CO_2 reduction *Chem. Commun.* **49** 10127
[41] Kuriki R, Sekizawa K, Ishitani O and Maeda K 2015 Visible-light-driven CO_2 reduction with carbon nitride: enhancing the activity of ruthenium catalysts *Angew. Chem. Int. Ed* **54** 1–5
[42] Lin J, Pan Z and Wang X 2014 Photochemical reduction of CO_2 by graphitic carbon nitride polymers *ACS Sustain. Chem. Eng.* **2** 353–8
[43] Li K, Peng B, Jin J, Zan L and Peng T 2017 Carbon nitride nanodots decorated brookite TiO_2 quasi nanocubes for enhanced activity and selectivity of visible-light-driven CO_2 reduction *Appl. Catal.* B **203** 910–6
[44] Li M, Zhang L, Fan X, Wu M, Wang M, Cheng R, Zhang L, Yao H and Shi J 2017 Core–shell $LaPO_4$/g-C_3N_4 nanowires for highly active and selective CO_2 reduction *Appl. Catal.* B **201** 629–35
[45] Zhang X, Wang L, Du Q, Wang Z, Ma S and Yu M 2016 Photocatalytic CO_2 reduction over B_4C/C_3N_4 with internal electric field under visible light irradiation *J. Colloid Interface Sci.* **464** 89–95
[46] Shi H, Chen G, Zhang C and Zou Z 2014 Polymeric g-C_3N_4 coupled with $NaNbO_3$ nanowires toward enhanced photocatalytic reduction of CO_2 into renewable fuel *ACS Catal.* **4** 3637–43
[47] Shi H, Zhang C, Zhou C and Chen G 2015 Conversion of CO_2 into renewable fuel over Pt–g-C_3N_4/$KNbO_3$ composite photocatalyst *RSC Adv.* **5** 93615–22
[48] Li H, Gao Y, Wu X, Lee P H and Shih K 2017 Fabrication of heterostructured g-C_3N_4/Ag-TiO_2 hybrid photocatalyst with enhanced performance in photocatalytic conversion of CO_2 under simulated sunlight irradiation *Appl. Surf. Sci.* **402** 198–207
[49] Yang L, Huang J, Shi L, Cao L, Liu H, Liu Y, Li Y, Song H, Jie Y and Ye J 2018 Sb doped SnO_2-decorated porous g-C_3N_4 nanosheet heterostructures with enhanced photocatalytic activities under visible light irradiation *Appl. Catal.* B **221** 670–80
[50] Zhou S, Liu Y, Li J, Wang Y, Jiang G, Zhao Z, Wang D, Duan A, Liu J and Wei Y 2014 Facile *in situ* synthesis of graphitic carbon nitride (g-C_3N_4)-N-TiO_2 heterojunction as an efficient photocatalyst for the selective photoreduction of CO_2 to CO *Appl. Catal.* **B 159** 20–9
[51] Liu H, Zhang Z, Meng J and Zhang J 2017 Novel visible-light-driven $CdIn_2S_4$/mesoporous g-C_3N_4 hybrids for efficient photocatalytic reduction of CO_2 to methanol *Mol. Catal.* **430** 9–19
[52] Ong W J, Tan L L, Chai S P, Yong S T and Mohamed A R 2015 Surface charge modification via protonation of graphitic carbon nitride (g-C_3N_4) for electrostatic self-assembly construction of 2D/2D reduced graphene oxide (rGO)/g-C_3N_4 nanostructures toward enhanced photocatalytic reduction of carbon dioxide to methane *Nano Energy* **13** 757–70
[53] Yu K, Hu X, Yao K, Luo P, Wang X and Wang H 2017 Preparation of an ultrathin 2D/2D rGO/g-C_3N_4 nanocomposite with enhanced visible-light-driven photocatalytic performance *RSC Adv.* **7** 36793–9
[54] Li H, Gan S, Wang H, Han D and Niu L 2015 Intercorrelated superhybrid of AgBr supported on engineering photocatalytic activities for water purification and CO_2 reduction *Adv. Mater.* **27** 6906–13
[55] He Y, Zhang L, Fan M, Wang X, Walbridge M L, Nong Q, Wu Y and Zhao L 2015 Z-scheme SnO_{2-x}/g-C_3N_4 composite as an efficient photocatalyst for dye degradation and photocatalytic CO_2 reduction *Sol. Energy Mater. Sol. Cells* **137** 175–84

[56] Yu W, Xu D and Peng T 2015 Enhanced photocatalytic activity of g-C_3N_4 for selective CO_2 reduction to CH_3OH via facile coupling of ZnO: a direct Z-scheme mechanism *J. Mater. Chem.* A **3** 19936–47

[57] Li M, Zhang L, Fan X, Zhou Y, Wu M and Shi J 2015 Highly selective CO_2 photoreduction to CO over g-C_3N_4/Bi_2WO_6 composites under visible light *J. Mater. Chem.* A **3** 5189–96

[58] Jichao Wang Z L, Hong-Chang Y, Ze-Yu F, Lin Z, Jianshe W and Shuang-Quan Z 2016 Indirect Z-scheme BiOI/g-C_3N_4 photocatalysts with enhanced photoreduction CO_2 activity under visible light irradiation *ACS Appl. Mater. Interfaces* **8** 3765–75

[59] Bai Y, Ye L, Wang L, Shi X, Wang P, Bai W and Wong P K 2016 g-C_3N_4/$Bi_4O_5I_2$ heterojunction with I3-/I-redox mediator for enhanced photocatalytic CO_2 conversion *Appl. Catal.* B **194** 98–104

[60] Chen P, Wang H, Liu H, Ni Z, Li J, Zhou Y and Dong F 2018 Directional electron delivery and enhanced reactants activation enable efficient photocatalytic air purification on amorphous carbon nitride Co-functionalized with O/La *Appl. Catal.* B **242** 19–30

[61] Cui A W, Li J, Sun Y and Wang H 2018 Enhancing ROS generation and suppressing toxic intermediate production in photocatalytic NO oxidation on O/Ba Co-functionalized amorphous carbon nitride *Appl. Catal.* B **237** 938–46

[62] Cui W, Li J, Dong F, Sun Y, Cen W, Lee S C and Wu Z 2017 Highly efficient performance and conversion pathway of photocatalytic NO oxidation on SrO-clusters @ amorphous carbon nitride *Env. Sci. Technol.* **51** 10682–90

[63] Li A J, Dong X and Sun Y 2018 Tailoring the rate-determining step in photocatalysis via localized excess electrons for efficient and safe air cleaning *Appl. Catal.* B **239** 187–195

[64] Dong X, Li J, Xing Q, Zhou Y, Huang H and Dong F 2018 The activation of reactants and intermediates promotes the selective photocatalytic NO conversion on electron-localized Srintercalated g-C_3N_4 *Appl. Catal.* B **232** 69–76

[65] Xiong T, Wang H, Zhou Y, Sun Y, Cen W, Huang H, Zhang Y and Dong F 2018 KCl-mediated dual electronic channels in layered g-C_3N_4 for enhanced visible light photocatalytic NO removal *Nanoscale* **10** 8066–74

[66] Ehtisham Khan M, Hiep Han T, Mansoob Khan M, Rezaul Karim M and Hwan Cho M 2018 Environmentally sustainable fabrication of Ag@g-C_3N_4 nanostructures and their multifunctional efficacy as antibacterial agents and photocatalysts *ACS Appl. Nano Mater.* **1** 2912–22

[67] Tao Y, Ni Q, Wei M, Xia D, Li X and Xu A 2015 Metal-free activation of peroxymonosulfate by g-C_3N_4 under visible light irradiation for the degradation of organic dyes *RSC Adv.* **5** 44128–36

[68] Shanmugam V, Lakshmi A, Jayavel S and Sundar K 2018 Construction of high efficient g-C_3N_4 nanosheets combined with Bi_2MoO_6-Ag photocatalysts for visible-light-driven photocatalytic activity and inactivation of bacterias *Arab. J. Chem.* **13** 2439–55

[69] Thurston J H, Hunter N M, Wayment L J and Cornell K A 2017 Urea-derived graphitic carbon nitride (u-g-C_3N_4) films with highly enhanced antimicrobial and sporicidal activity *J. Colloid Interface Sci.* **505** 910–8

[70] Thurston J H, Hunter N M and Cornell K A 2016 Preparation and characterization of photoactive antimicrobial graphitic carbon nitride (g-C_3N_4) films *RSC Adv.* **6** 42240–8

[71] Xu J, Wang Z and Zhu Y 2017 Enhanced visible-light-driven photocatalytic disinfection performance and organic pollutant degradation activity of porous g-C_3N_4 nanosheets *ACS Appl. Mater. Interfaces* **9** 27727–35

[72] Ma S, Zhan S, Jia Y, Shi Q and Zhou Q 2016 Enhanced disinfection application of Ag-modified g-C_3N_4 composite under visible light *Appl. Catal.* B **186** 77–87
[73] Xu J, Gao Q, Bai X, Wang Z and Zhu Y 2018 Enhanced visible-light-induced photocatalytic degradation and disinfection activities of oxidized porous g-C_3N_4 by loading Ag nanoparticles *Catal. Today* **332** 227–35
[74] Liu C, Wang L, Xu H, Wang S, Gao S, Ji X, Xu Q and Lan W 2016 'One pot' green synthesis and the antibacterial activity of g-C_3N_4/Ag nanocomposites *Mater. Lett.* **164** 567–70
[75] Li G, Nie X, Chen J, Jiang Q, An P T, Wong P K, Zhang H, Zhao H and Yamashita H 2015 Enhanced visible-light-driven photocatalytic inactivation of *E. coli* using g-C_3N_4/TiO_2 hybrid photocatalyst synthesized using a hydrothermal-calcination approach *Water Res.* **86** 17–24
[76] Xu J, Li Y, Zhou X, Li Y, Da Gao Z, Song Y Y and Schmuki P 2016 Graphitic g-C_3N_4-sensitized TiO_2 nanotube layers: a visible-light activated efficient metal-free antimicrobial platform *Chem. Eur.* J. **22** 3947–51
[77] Wang W, Yu J C, Xia D, Wong P K and Li Y 2013 Graphene and g-C_3N_4 nanosheets cowrapped elemental α-sulfur as a novel metal-free heterojunction photocatalyst for bacterial inactivation under visible-light *Environ. Sci. Technol.* **47** 8724–32
[78] Zhang Q, Quan X, Wang H, Chen S, Su Y and Li Z 2017 Constructing a visible-light-driven photocatalytic membrane by g-C_3N_4 quantum dots and TiO_2 nanotube array for enhanced water treatment *Sci. Rep.* **7** 3128
[79] Li J, Yin Y, Liu E, Ma Y, Wan J, Fan J and Hu X 2017 *In situ* growing Bi $_2MoO_6$ on g-C_3N_4 nanosheets with enhanced photocatalytic hydrogen evolution and disinfection of bacteria under visible light irradiation *J. Hazard. Mater.* **321** 183–92
[80] Xia D, Wang W, Yin R, Jiang Z, An T, Li G, Zhao H and Wong P K 2017 Enhanced photocatalytic inactivation of *Escherichia coli* by a novel Z-scheme g-C_3N_4/m-Bi_2O_4 hybrid photocatalyst under visible light: the role of reactive oxygen species *Appl. Catal.* B **214** 23–33
[81] Deng J, Liang J, Li M and Tong M 2017 Enhanced visible-light-driven photocatalytic bacteria disinfection by g-C_3N_4-AgBr *Colloids Surf.* B **152** 49–57
[82] Li Y, Li Y, Ma S, Wang P, Hou Q, Han J and Zhan S 2017 Efficient water disinfection with Ag_2WO_4-doped mesoporous g-C_3N_4 under visible light *J. Hazard. Mater.* **338** 33–46
[83] Wang R *et al* 2018 Mechanism insight into rapid photocatalytic disinfection of *Salmonella* based on vanadate QDs-interspersed g-C_3N_4 heterostructures *Appl. Catal.* B **225** 228–37
[84] Devthade Vidyasagar R S, Sachin G G, Aditi K, Ashok G S and Suresh S U 2018 Microwave assisted *in situ* decoration of g-C_3N_4 surface with $CdCO_3$ nanoparticles for visible light driven photocatalysis *New J. Chem.* **42** 6322–31
[85] Sun L, Du T, Hu C, Chen J, Lu J, Lu Z and Han H 2017 Antibacterial activity of graphene oxide/g-C_3N_4 composite through photocatalytic disinfection under visible light *ACS Sustain. Chem. Eng.* **5** 8693–701
[86] Lin A T, Son Z, Wu Y, Chen L, Wang S, Fu F and Guo L 2018 Boron- and phenyl- codoped graphitic carbon nitride with greatly enhanced light responsive range for photocatalytic disinfection *J. Hazard Mater.* **358** 62–8
[87] Zhao H, Chen S, Quan X, Yu H and Zhao H 2016 Environmental integration of microfiltration and visible-light-driven photocatalysis on g-C_3N_4 nanosheet/reduced graphene oxide membrane for enhanced water treatment *Appl. Catal. B. Environ.* **194** 134–40

[88] Huang K, Li C, Yang J, Zheng R, Wang W and Wang L 2022 Platinum nanodots modified nitrogen-vacancies g-C_3N_4 Schottky junction for enhancing photocatalytic hydrogen evolution *Appl. Surf. Sci.* **581** 152298
[89] Zeng D, Wu P, Ong W, Tang B, Wu M, Zheng H, Chen Y and Peng D 2018 Construction of network-like and flower-like 2H-$MoSe_2$ nanostructures coupled with porous g-C_3N_4 for noble-metal-free photocatalytic H_2 evolution under visible light *App. Catal.* B **233** 26–34
[90] Zhang J, Mei J, Yi S and Guan X 2019 Constructing of Z-scheme 3D g-C_3N_4-ZnO @ graphene aerogel heterojunctions for high-efficient adsorption and photodegradation of organic pollutants *Appl. Surf. Sci.* **492** 808–17
[91] Iqbal W, Qiu B, Zhu Q, Xing M and Zhang J 2018 Self-modified breaking hydrogen bonds to highly crystalline graphitic carbon nitrides nanosheets for drastically enhanced hydrogen production *Appl. Catal.* B **232** 306–13
[92] Zhang J, Mei J, Yi S and Guan X 2019 Constructing of Z-scheme 3D g-C_3N-ZnO @ graphene aerogel heterojunctions for high-efficient adsorption and photodegradation of organic pollutants *Appl. Surf. Sci.* **492** 808–17

IOP Publishing

Materials Technology for the Energy and Environmental Nexus, Volume 1

R V Mangalaraja, Arunachalam Arulraj, Nalandhiran Pugazhenthiran and Chan Siew Hwa

Chapter 12

Integrating algae in personal care product and cosmetic industrial wastewater treatment—an approach to enhance the circular economy with value-added products

Anusha Gowri, Srinivasan Ravichandran, Athithyan Gunasekaran and Preethi Selvamuthu

The cosmetic industry has boomed in developing countries with a great impact on overall surface water pollution. Bioremediation can effectively remove microplastics (MPs), metals, and minerals, and can facilitate environmental sustainability. Sustainability has been an eminent problem in the past few decades, where the cosmetic industry is not an exception. Microbeads are solid primary microplastics with a diameter of less than 5 mm that are used in cosmetic products to exfoliate and/or cleanse the skin. After being used, they are discarded directly through wastewater treatment plants (WWTPs), where they have the potential to penetrate into the groundwater table and contribute to groundwater contamination. Once disposed of there is no effective way to recover the contaminants, since the environment does not permit complete biodegradation. Microalgae culture is an intriguing step in the treatment of wastewater since it enables a tertiary biotreatment together with the creation of potentially valuable biomass with variety of applications. It can be used effectively for the growth of inorganic nitrogen and phosphorus, and microalgae cultures provide an elegant answer to tertiary and quaternary treatments. In this chapter, we emphasise the MP biodegradation mechanism, the types of algae that have been used, and how MPs have an impact on them. We suggest that genetic engineering, the study of biodegrading enzymes, and uptake studies could be useful methods for enhancing MP breakdown.

doi:10.1088/978-0-7503-5724-1ch12

12.1 Introduction

During the past few decades the use of plastic products has increased drastically in the day-to-day life of every individual and owing to their extensive enduring properties they remain in the environment [1]. Plastic usage has been affecting both terrestrial and aquatic environments. Plastic items are breaking down slowly into smaller pieces through wear and tear and these minuscule plastic pieces, which are less than 5 mm in length or diameter (smaller than a grain of rice) and have been named microplastics (MPs), are found significantly in oceans and in other aquatic ecosystems [2]. Industrial effluents are also a major pollutant of water and other resources and raise concerns for human health regarding fresh water consumption and, further, the marine species consumed by humans have in recent years been highly contaminated with MPs. This increases the chance of the transfer of plastic additives and adsorbed pollutants to the human food chain and in trace amounts to blood in the human body. India has endorsed a United Nations (UN) resolution to end plastic pollution and forge an international legally binding agreement by 2024.

The wastewater of the cosmetics industry contains many microplastics and thereby possesses many complex safety issues. The safety assessment of personal care products (PCPs) and ingredients that are representative and pose complex safety issues needs to be carried out [3, 4]. PCPs are generally applied to human skin and mainly produce local exposure, but skin penetration or use in the oral cavity, face, lips, eyes, and mucosa may also produce systemic human exposure [5]. The plastic ingredients used for the scrubbing or exfoliating function in PCPs are most commonly referred to as plastic microbeads. These tiny plastics persist in the environment and have a damaging effect on marine life, the environment, and human health. In the past, plastic microbeads were also used in a few speciality toothpastes. Microbeads made of polyethylene are by far the most frequent used, but in addition, polyurethane, ethylene-vinyl acetate copolymer (EVA), etc, have also been used. Microbeads are found in the riverbeds, Arctic ice, natural fertilisers, soils, and even in drinking water with a noticeable amount of microplastics. Microbeads are usually $> 0.1\ \mu$m and $\leqslant 1$ mm in size and can be manufactured with different shapes from various types of polymers. These microbeads and microplastics are tremendously difficult to degrade in the environment, in addition they will keep accruing in living organisms and/or adsorb toxic matter causing toxicity or conservational risks [6–8]. The toxicity assessment of MPs in aquatic ecosystems is difficult for both established and progressive methods for controlling MPs pollution. Thus, improving the ecological status of water sources is a growing focus for many developed and developing nations, in particular with regard to reducing nitrogen and phosphorus in the wastewater effluent [9].

In recent years, mixotrophic microalgae have received increased interest in implementations for wastewater treatment. Microalgae are primary producers which account for about 50% of production and play a wide role in recycling materials and also maintaining the energy in aquatic ecosystems. Accordingly, the growth and healthiness or microalgae regulate the constancy of aquatic food networks [10]. Microalgae are simple in their growth compared to other larger

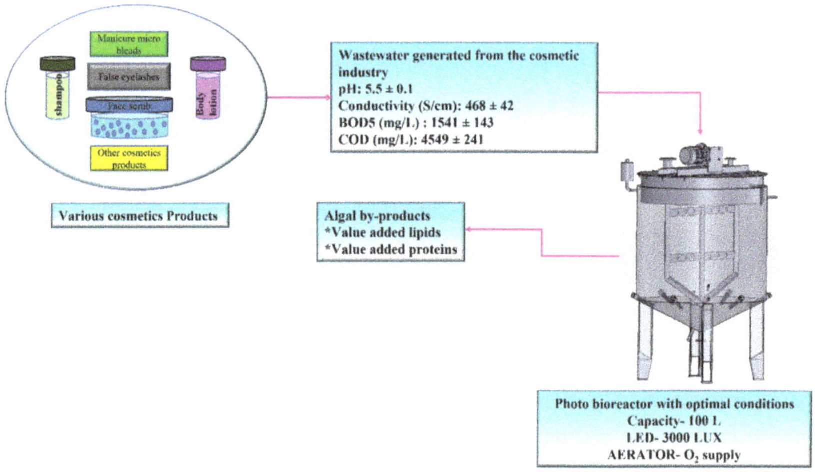

Figure 12.1. Schematic of cosmetic wastewater treatment enhanced in a photobioreactor by microalgae.

plants, which makes them sensitive to outside pollution. Microalgae have been considered to be an eco-friendly, efficient, and conservative treatment method which can be used widely for MP degradation [11, 12]. Additionally, the microalgae, a kind of biomass, are renewable sources and their use in the treatment of cosmetic wastewater is economical. Also, they are viable for CO_2 bio-fixation and are capable of utilising inorganic and organic carbon as well as inorganic P and N for their health and surplus production, which reduces these chemicals' concentrations in the water. The major benefit of integrating microalgae into cosmetic wastewater treatment is O_2 generation by photosynthesis, which is essential for heterotrophic microorganisms for biodegrading carboniferous materials (figure 12.1) [13, 14].

12.2 Wastewater composition in the cosmetics industry

Over the past few years, the cosmetics industry has grown steadily in many developing nations, but knowledge on the toxicity of the wastewater produced from the industry is limited. The eco-friendly cosmetic industries are predicted to reach US $5.25 billion by 2030 based on a growth prediction with a compound annual rate of 5.18% during the forecast period of 2019–30. The worldwide cosmetics market is in roughly third place in terms of its rapid growth [15]. Pharmaceuticals and personal care products (PPCPs) are the two components of a broader chemical puzzle employed in the cosmetics sector. PPCPs are used to maintain the most hazardous ingredients found in cosmetic products, such as contrast agents, hormones, preservatives, beta blockers, sunscreen UV filters, anti-inflammatory drugs, cleansers, disinfectants, and detergents [16]. The cosmetics industrial wastewater consists of high total organic carbon levels, chemical oxygen demand (COD) (>100 000 mg l^{-1}), BOD5, high concentrations of petroleum ether extract, organic nitrogen and phosphorus, suspended solids, oils and fats, and detergents [17]. Most of the compounds present in cosmetic wastewater are characterised as non-polar and thus cause a huge threat to the environment since it takes a huge amount of time for biodegradation [18]. Biodegradability is the major factor that determines whether the material or a product is eco-friendly or not. Waste materials with poor biodegradability could lead to serious damage to the environment if not discarded properly. Based on the data from 1950 to 2015, about ~8300 megatons of synthetic polymers were produced and of these only 4900 megatons were disposed of in landfills. The life cycle analysis (LCA) of a material can show its impact on the environment from obtaining its raw materials, through production, dispersal, and usage, to disposal, although it does not yield an ethical account, covering the social and economic aspects. The LCA gives details for use in decision-making associated with the development of ecological designs and products, enhancements to the generation system, and choice of product at the user level. Additionally, ~50% of plastics are related to hostile chemical monomers and secondary products which might cause irreparable environmental consequences. Some plastics become MPs in the degradation process, which shows the further need for the proper plastic waste management [19–21]. Table 12.1 shows the composition of cosmetic wastewater estimated by Friha *et al* in 2014 [22].

Microplastics react as an adsorbent material and also act as carriers for several harmful chemicals, for instance polycyclic aromatic hydrocarbons (PAH), heavy metals, pesticides and persistent organic pollutants (POP), and polychlorinated biphenyls (PCBs). Hence these materials in combination with MPs can cause biomagnification and bioaccumulation in the food chain, with several harmful effects on aquatic ecosystems, which include reproductive and metabolic changes, subcellular toxicity, and oxidative stress. Consequently, microplastics generation disturbs ecosystems in unrestrained and asymmetrical ways which lead to an unmanageable cycle. Based on these issues, there is a clear need for a change in actions related to the optimisation of products and production methods by the cosmetics industry. Microalgae are aerobic or anaerobic microorganisms which convert the essential nutrients present in wastewater along with solar energy to

Table 12.1. Composition of cosmetic wastewater. (Reproduced with permission from [22]. Copyright 2014 Elsevier.)

Characteristics	Maximum	Minimum	Mean	Standard Deviation (SD)*
pH	6.08	4.93	5.51	0.52
EC (mS cm^{-1})	2.89	2.43	2.61	0.17
BOD_5 (g l^{-1})	1.7	1	1.62	0.23
Soluble COD (g l^{-1})	16.66	9.34	13.57	3.15
Ca	81.95	45.96	63.95	15.33
K	165.99	9.34	13.57	3.15
Na	501	151	263.5	124.16
Mg	353	83	178	95.22
Cr	1.48	1.24	1.35	0.09
Zn	2.11	0.268	1.15	0.63

create biomass [19]. Additionally, cosmetics need to be prioritised more than drugs since they are used excessively by larger populations for longer periods of time [23].

12.3 Microplastic biodegradation by microalgal species

One of the emerging pollutants that is detrimental to the environment is MPs. The primary MPs are widely used in industries such as pharmaceuticals, facial cleansers, and cosmetics resulting from the excessive use of petroleum derivatives. When UV light breaks down macro- and mesoplastics, secondary MPs are created [24]. As mentioned, MPs are tiny particles with a diameter of 5 mm or less. MPs can be created specifically in the micrometre range (i.e. primary MPs) or derived from the breakdown of MPs (i.e. secondary MPs) [25]. They can be found in plastic pre-production pellets (nurdles), 'microbeads' in personal care products, and sand-blasting abrasives [26]. When the wastewater is treated, microplastics are frequently removed with the sludge, which is commonly used as an organic fertiliser. Microalgae can utilise plastic as a source of carbon, and when the plastic additives are released, they help microorganisms flourish by providing a food source. The microalgae and MP interaction can modify the polymer properties significantly with a huge positive effect on aquatic environments. This process can be categorised in two ways, i.e. the biodegradability of the plastic polymer, and the change in polymer density leading to enhanced sinking properties. Table 12.2 explains the various types of microalgae used for yielding biomass in the respective wastewater treatment and its bioreactor type (Figure 12.1).

In this regard, bio-fouling can be formed by bacteria and microalgae on microplastics, since their particle sizes are tiny and range from 1 to 5 mm [27]. Various studies have described that the bio-fouling methods of MPs may modify their properties significantly, to be precise, the adsorption ability appears to be improved. For example, In 2020 Wang *et al* described that the chemical and physical properties of MPs were altered with the expansion of biofilm, which resulted in modified adsorption properties of MPs for tetracycline and copper. Likewise, it has been found that the ions from water adsorbed onto the surface of biofilms were not bound strongly compared to the ion-exchange polymers. This shows that the ions are more effortlessly leached from the biofilms. Hence, the microalgae adhesion on MP surfaces is of critical importance in the adsorption and desorption of MP pollutants.

12.4 Microalgal species and their value-added products

Microalgae produce biochemical substances such as proteins, glucose, lipids, carotenoids, and pigments. The algal cell metabolic rate is inferred indirectly from the biochemical make-up of microalgae. In the cosmetics industry, microalgal biomass is utilised as thickeners, water-binding agents, and antioxidants. For example, *Arthrospira* and *Chlorella* are frequently found in skin care products. The extracts of various microalgal species, including *Chondrus crispus*, *Mastocarpus stellatus*, *Ascophyllum nodosum*, *Alaria esculenta*, *Spirulina platensis*, *Nannochloropsis oculata*, *Chlorella vulgaris*, and *Dunaliella salina*, are frequently

Table 12.2.

Author	Year	Material	Species	Biomass (DW $g/m^2/day$)	Suspended Medium	Reactor	Ref
Derakhshan *et al.*,	2018	Kenaf Fiber	Filamentous green algae	>5g/L	effluent	hybrid microalgal membrane photobioreactors	[54]
Qi Zhang *et al.*,	2017	Pine Sawdust	Mix culture (S. obliquus FACHB-416, C. vulgaris FACHB-32, O. tenuisFACHB-1052)	2.53	BG-11 Medium	Flat plate bioreactor	[55]
Qi Zhang *et al.*,	2017	Rice Husk	Mix culture (S. obliquus FACHB-416, C. vulgaris FACHB-32, O. tenuis FACHB-1052)	1.47	BG-11 Medium	Flat plate bioreactor	[55]
Qi Zhang *et al.*,	2017	Oak Sawdust	Mix culture (S. obliquus FACHB-416, C. vulgaris FACHB-32, O. tenuisFACHB-1052)	1.78	BG-11 Medium	Flat plate bioreactor	[55]
Qi Zhang *et al.*,	2017	Sugar Cane Baggase	Mix culture (S. obliquus FACHB-416, C. vulgaris FACHB-32, O. tenuisFACHB-1052)	2.16	BG-11 Medium	Flat plate bioreactor	[55]
Tourlouki *et al.*,	2020	Glass Reinforced Fiber Polymer	Scenedesmus rubescens	0.77	BG-11 Medium	Glass Reinforced Fiber Polymer reactor	[56]
Luo *et al.*,	2018	Polyester-mesh	N/a	2.41-2.82 (SBP)	Artificial wastewater		[57]

						Integrated photo-bioelectro-chemical system	
Garbowski *et al.*,	2020	Pine Bark	Mix culture (Cyanobacteria, Chlorophyta- coccoid and filamentous, Bacillariophyceae, Xantophyceae)	n/a	sewage water	Acrylic photobioreactor	[58]
Lee *et al.*,	2015	Polyethylene screen	Mix culture (Scenedesmus, Chlorella, Pediastrum, Nitzschia, Cosmarium, filamentous microalgae)	5.35-5.55	effluent wastewater	raceway pond system	[59]
Lee *et al.*,	2015	Nylon mesh	Mix culture (*Scenedesmus*, *Chlorella*, Nitzschia *Pediastrum*, *Cosmarium*, filamentous microalgae)	9.1-13.5	effluent wastewater	raceway pond system	[59]
Irving *et al.*,	2011	Polyurethane	*Scenedesmus obliquus, Chlorella vulgaris*		Bold Basal Medium		[60]
Choudhary *et al.*,	2017	Nonwoven Spunbond fabric	*Chlorella, Phormidium* mix culture	4	TAP/wastewater	Biofilm reactor	[61]
Johnson *et al.*,	2009	Polystyrene foam	*Chlorella sp.*	2.57	soil extract medium/ dairy wastewater	lab scale shaker system	[62]
Johnson *et al.*,	2009	Cardboard	*Chlorella sp.*	1.47	soil extract medium/ dairy wastewater	lab scale shaker system	[62]
Johnson *et al.*,	2009	Polyethylene fabric	*Chlorella sp.*	0.58	soil extract medium/ dairy wastewater	lab scale shaker system	[62]
Johnson *et al.*,	2009	Loofah sponge	*Chlorella sp.*	1.28	soil extract medium/ dairy wastewater	lab scale shaker system	[62]

used to make the value-added products (VAPs), including face and skin care products, anti-aging creams, and emollients.

To shield themselves from the UV rays, microalgae produce a number of organic metabolites, including sporopollenin, scytonemin, and amino acids that resemble the mycosporine. These high VAPs are used extensively in a variety of industries, including pharmaceuticals, therapeutics, human nutrition, food technology, functional foods, antibiotics, and green plastics, among others, demonstrating the myriad ways that microalgal biomass is put to use after being harvested on a large scale in various categories [28]. Because they do not compete with each other for the resources to create various value-added chemicals, microalgae are a suitable feedstock. The current position of these microalgal-based VAPs has gone from the research level to the industrial and commercial levels. The market potential of the products in question is shown by the current status (commercial and laboratory scale) of microalgal-derived products in relation to their cost [29]. The biggest disadvantage of using microalgae products is that they do not produce as much of the in-demand isotope compounds such as phycobiliproteins and beta-carotene. The most recent advancements and newsworthy research trends focus on the scientific and technological applications of microalgae, and microalgal applications for VAPs are one of these [15]. Therefore, by concentrating the efforts in research and development to optimise the product output at lower input costs, it is possible to build economically viable VAPs (table 12.3).

These components from algal biomass can be isolated one at a time and used to make various products with added value. The recovered lipids, for instance, might be trans-esterified to create bio-diesel. In general, naive biomass (NB) can produce 65% of the residual biomass after lipid extraction, which is rich in carbohydrates (starch and cellulose), lipids, and some proteins [42]. After being saccharified, the residual biomass after lipid extraction (RBLE) can be used as a substrate for the fermentation of ethanol. Through anaerobic digestion, the residual biomass after extraction of lipids and sugars (RBLSE) can generate biogas. In order to produce microalgal biomass sustainably, several products need be extracted successively from the microalgal biomass [43].

12.5 Methods to enhance MP breakdown by microalgae

In general, microalgae can adhere to MP surfaces and generate enzymes which deteriorate the MP polymer bonds, and change them into water, CO_2, and other biomasses. To date, researches have shown that microalgae can be predominantly efficient in biodegrading bisphenol A (BPA) and both low- and high-density polyethylene (PE) materials in plastic bags, bottles, and fabrics. Furthermore, it is recorded that about a 58.9% reduction of carbon composition of low-density PE has been successfully achieved using blue-green microalgae in 2020. In this regard, there are a few wastewater treatment methods in the field to enhance the biodegradation of MPs [44].

Table 12.3. Algae species and the VAPs obtained to achieve a circular bioeconomy.

Algae	By-product	Yield	Reference
Botryococcus braunii	Bio-diesel	17.85%	[30]
Chlorella protothecoides	Bio-diesel	29.4%–1.5%	[31]
Cladophora fracta	Bio-diesel	14.2%–0.8%	[31]
Chlorella kessleri	Methane	62%	[32]
Euglena gracilis	Methane	93%	[32]
Chlamydomonas reinhardtii cc1036	Bio-hydrogen	0.48 mmol l^{-1} h^{-1}	[33]
Chlamydomonas reinhardtii	Bio-hydrogen	0.13 mmol l^{-1} h^{-1}	[34]
Chlamydomonas reinhardtii	Bio-hydrogen, biogas, methane	0.094 mmol l^{-1} h^{-1}, 587%DW, 111%	[35]
Gloeocapsa alpicola	Bio-hydrogen	1.6 mmol l^{-1} h^{-1}	[36]
Botryococcus braunii	Bio-oil	29%–75%DW	
Chlorella sp.	Bio-oil	28%–32%DW	[38]
Chlamydomonas reinhardtii	Biogas	587 ± 9 ml g^{-1} VS	[32]
Chlorella kessleri	Biogas	335 ± 8 ml g^{-1} VS	[32]
Ulva lactuca	Bio-ethanol	55%–60%	[39]
Tetraselmis sp. CS-362	Bio-ethanol	26%	[40]
Hantzschia DI-160	Bio-oil	66%DW	[32]
Phaeodactylum tricornutum	Bio-oil	20%–30%DW	[38]
Schizochytrium sp.	Bio-oil	50%–77%DW	[38]
Nitzschia sp.	Bio-oil	45%–47%DW	[38]
Botryococcus braunii	Bio-oil	25%–75%DW	[38]
Chlamydomonas reinhardtii	Biogas	587 ± 9 ml g^{-1} VS	[32]
Chlorella kessleri	Biogas	335 ± 8 ml g^{-1} VS	[32]
Spirogyra neglecta	Biogas	0.23 ml g^{-1} VS	[41]

12.5.1 Sludge treatment method

Sludge treatment is one of the most cost-effective methods compared to other complex systems owing to the low concentration PCP levels, which are not sufficiently high to withstand the microorganism's growth and in addition the MP breakdown plays a key role in the biodegradation [45]. The treatment efficiency may be disturbed by operative aspects, i.e. sludge retention times (SRTs) and hydraulic retention times (HRTs). The eradication rates of most PCPs are enhanced by extended HRTs, although the downside of this method is the inability to extract the PCPs [46, 47]. Luis Malvar *et al* evaluated seven types of PCPs, along with their key metabolites, observing them using several stabilisation methods, dehydration, composting, and lagooning. The analysed samples were evaluated for 16 compounds, and it was found that they have similar distributions as in primary sludge, despite having samples from different wastewater treatment locations, with metabolic ratios of widely held compounds. The various stability behaviours were noticed in each compound and some persevered in all of the reviewed technologies, while others were extremely degradable [48].

12.5.2 Anaerobic and aerobic methods

The anaerobic method occurs if the material has an oxygen deficit and this method involves microalgae for degrading cosmetic waste into gases such as nitrogen and methane and other by-products. This method has the unique characteristic of involving only a little energy for the extraction of the helpful nutrients. Furthermore, the anaerobic treatment requires observing some of the working conditions which need quite a long time to complete the conversion process [49–51]. Likewise, the aerobic method involves organic contaminants and microalgae but the process uses oxygen from water for degradation under aerobic conditions. It is noticeable that oxygen is crucial for the conversion which will be purged into the water constantly during the entire process. Additionally, some of the other factors such as temperature, time, pH, and microalgal characteristics should also be noted throughout the process. This technique is beneficial for removing dissolved, suspended, or volatile substances, along with chemical and biological oxygen, phosphates, nitrates, and others. The aerobic method permits less organic waste of about ~90%, although the additional biowaste should be further treated using other methods, which is relatively costly [49, 51].

12.5.3 Bioreactors

The biodegradation of cosmetic wastewater using membrane bioreactors is a promising treatment procedure with better results compared to the other treatment processes (figure 12.2). It involves membrane filtration and biodegradation in a synchronised way and gives high flexibility towards possible alteration which is essential to modify biological processes [52]. The process converts the organic components to biomass, which decreases membrane fouling and tends to increase the retrieval rate. The membrane in the bioreactor generates an effective barrier which retains the microalgae and solids in the tank. This method has numerous advantages, such as less sludge formation and higher quality rate of treated water, in addition to being stable and effortlessly expandable. In this regard, the membrane

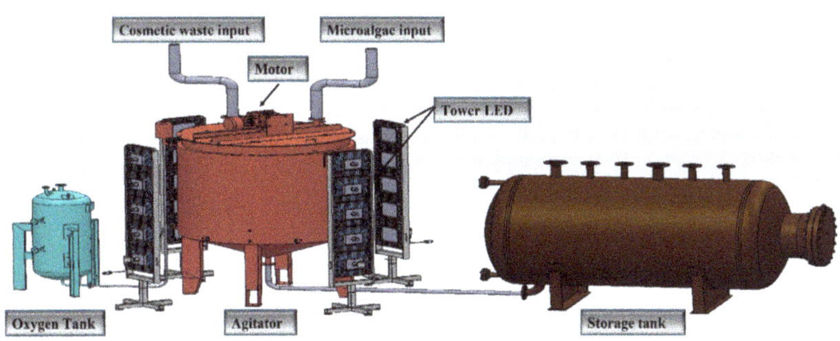

Figure 12.2. Pilot scale photobioreactor to treat cosmetic wastewater.

bioreactor is appropriate for recusant wastewater for which long duration SRTs are required, allowing the efficient eradication of organic contaminants that biodegrade slowly. However, fouling is a substantial drawback of this method as it requires high energy [22].

According to recent research, the membrane bioreactor method was found to be a major removal system for the eradication of contaminants. In addition, biodegradation and wetlands are other common methods. Weiss and Reemtsma *et al.* estimated that the wastewater treated using a laboratory-scale membrane bioreactor showed 50% extraction of evaluated materials, whereas one-step treatment with a higher than traditional stimulated sludge treatment with aerobic and anaerobic stages gave an extraction rate of 61%–94%. In addition, the treatment method shows low waste concentrations in the range of 22%–56% and 7 h of HRT appears to be adequate for the elimination of other trace elements [53].

12.6 Conclusion

The sustainable method of cosmetic wastewater treatment needs to be eco-friendly, and technologically and economically viable. In this regard involving microalgae in biological cosmetic wastewater treatment options shows promising output in the biodegradation of organic contaminants. Accordingly, the use of microalgae in the treatment reduces carbonaceous, phosphorous, and nitrogenous materials and also allows the process to function at a minimal footprint with respect to greenhouse gas generation and energy consumption compared to regular biological cosmetics wastewater treatment methods. The foremost encounter is the microalgae application into the wastewater treatment to be done in a non-sterile environment. Accordingly, researchers have focused on culturing microalgae with other microbes such as bacteria, fungi, and yeasts, for the development of stable groups. Hence, the forthcoming research must evaluate the organic carbon augmented wastewater treatment method with microalgae and bacterial combinations, and enhance the efficiency of the treatment using cost-effective approaches.

References

[1] Gu Y, Julia Z and Jeremiah A J 2020 Polymer networks: from plastics and gels to porous frameworks *Angew. Chem. Int. Ed.* **59** 5022–49

[2] Thompson R C, Ylva O, Richard P M, Anthony D, Steven J R, Anthony W G J, Daniel M G and Andrea E R 2004 Lost at sea: where is all the plastic? *Science* **304** 838–8

[3] Scopetani C, Cincinelli A, Martellini T, Lombardini E, Ciofini A, Fortunati A, Pasquali V, Ciattini S and Ugolini A 2018 Ingested microplastic as a two-way transporter for PBDEs in *Talitrus saltator Environ. Res.* **167** 411–7

[4] Isobe A 2016 Percentage of microbeads in pelagic microplastics within Japanese coastal waters *Mar. Pollut. Bull.* **110** 432–7

[5] Issac M N and Balasubramanian K 2021 Effect of microplastics in water and aquatic systems *Environ. Sci. Pollut. Res.* **28** 19544–62

[6] Gu Y, Julia Z and Jeremiah A J 2020 Polymer networks: from plastics and gels to porous frameworks *Angew. Chem. Int. Ed.* **59** 5022–49

[7] Samanta P, Dey S, Kundu D, Dutta D, Jambulkar R, Mishra R, Ghosh A R and Kumar S 2022 An insight on sampling, identification, quantification and characteristics of microplastics in solid wastes *Trends Environ. Anal. Chem.* **36** e00181

[8] Nam H C and Park W H 2019 Eco-friendly poly(lactic acid) microbeads for cosmetics via melt electrospraying *Int. J. Biol. Macromol.* **157** 734–42

[9] Eriksen M, Laurent C M L, Carson H S, Thiel M, Moore C J, Borerro J C, Galgani F, Ryan P G and Reisser J 2014 Plastic pollution in the world's oceans: more than 5 trillion plastic pieces weighing over 250,000 tons afloat at sea *PLoS One* **9** e111913

[10] Ramadass K, Megharaj M, Venkateswarlu K and Naidu R 2017 Toxicity of diesel water accommodated fraction toward microalgae, *Pseudokirchneriella subcapitata* and *Chlorella* sp. MM3 *Ecotoxicol. Environ. Saf.* **142** 538–43

[11] Prata J C, da Costa J P, Lopes I, Duarte A C and Rocha-Santos T 2019 Effects of microplastics on microalgae populations: a critical review *Sci. Total Environ.* **665** 400–5

[12] Bergstrom A K and Karlsson J 2019 Light and nutrient control phytoplankton biomass responses to global change in northern lakes *Glob. Change Biol.* **25** 2021–9

[13] Bhattacharya P, Lin S, Turner J P and Ke P C 2010 Physical adsorption of charged plastic nanoparticles affects algal photosynthesis *J. Phys. Chem.* C **114** 16556–61

[14] Lu T, Zhang Q, Zhang Z, Hu B, Chen J, Chen J and Qian H 2021 Pollutant toxicology with respect to microalgae and cyanobacteria *J. Environ. Sci.* **99** 175–86

[15] Shen Y, Xu X, Zhao Y and Lin X 2014 Influence of algae species, substrata and culture conditions on attached microalgal culture *Bioprocess. Biosyst. Eng.* **37** 441–50

[16] Hwang J H *et al* 2014 Photoautotrophic hydrogen production by eukaryotic microalgae under aerobic conditions *Nat. Commun.* **5** 1–6

[17] Lukman Abidemi B *et al* 2018 Treatment technologies for wastewater from cosmetic industry: a review *Int. J. Chem. Biomol. Sci.* **4** 69–80

[18] Cubas A L V, Bianchet R T, Reis dos I M A S and Gouveia I C 2022 Plastics and microplastic in the cosmetic industry: aggregating sustainable actions aimed at alignment and interaction with UN sustainable development goals *Polymers (Basel)* **14** 4576

[19] Kabir E, Kaur R, Lee J, Kim K H and Kwon E E 2020 Prospects of biopolymer technology as an alternative option for non-degradable plastics and sustainable management of plastic wastes *J. Clean. Prod.* **258** 120536

[20] Ness B, Urbel-Piirsalu E, Anderberg S and Olsson L 2007 Categorising tools for sustainability assessment *Ecol. Econ.* **60** 498–508

[21] Okoye C O, Addey C I, Oderinde O, Okoro J O, Uwamungu J Y, Ikechukwu C K, Okeke E S, Ejeromedoghene O and Odii E C 2022 Toxic chemicals and persistent organic pollutants associated with micro- and nanoplastics pollution *Chem. Eng. J. Adv.* **11** 100310

[22] Friha I, Karray F, Feki F, Jlaiel L and Sayadi S 2014 Treatment of cosmetic industry wastewater by submerged membrane bioreactor with consideration of microbial community dynamics *Int. Biodeterior. Biodegrad.* **88** 125–33

[23] Juliano C and Magrini G A 2017 Cosmetic ingredients as emerging pollutants of environmental and health concern. A mini-review *Cosmetics* **4** 11

[24] Calero M, Godoy V, Quesada L and Martín-Lara M Á 2021 Green strategies for microplastics reduction *Curr. Opin. Green Sustain. Chem.* **28** 100442

[25] Ungureanu N and Vladut V M I P 2021 Nutrient removal from wastewater by microalgae chlorella vulgaris *Acta Tech. Corviniensis-Bull. Eng.* **9** 65–72

[26] Park J B K, Craggs R J and Shilton A N 2011 Wastewater treatment high rate algal ponds for biofuel production *Bioresour. Technol.* **102** 35–42
[27] Banayan Esfahani E, Asadi Zeidabadi F, Bazargan A and McKay G 2019 The modified bardenpho process *Handbook of Environmental Materials Management* (Berlin: Springer)
[28] Kaur S, Khattar J I S, Singh Y, Singh D P and Ahluwalia A S 2019 Extraction, purification and characterisation of Phycocyanin from *Anabaena fertilissima* PUPCCC 410.5: as a natural and food grade stable pigment *J. Appl. Phycol.* **31** 1685–96
[29] Shen Y, Fan Z, Chen C and Xu X 2015 An auto-flocculation strategy for *Chlorella vulgaris Biotechnol. Lett.* **37** 75–80
[30] Drexler I L C, Bekaan S, Eskandari Y and Yeh D 2013 Feasibility assessment of cultivating native and non-native algae species at different stages of a municipal treatment plant *86th Annu. Water Environ. Fed. Tech. Exhib. Conf.* vol 9 pp 5402–13
[31] Demirbaş A 2009 Production of biodiesel from algae oils *Energy Sources* A **31** 163–8
[32] Mussgnug J H, Klassen V, Schlüter A and Kruse O 2010 Microalgae as substrates for fermentative biogas production in a combined biorefinery concept *J. Biotechnol.* **150** 51–6
[33] Laurinavichene T V, Fedorov A S, Ghirardi M L, Seibert M and Tsygankov A A 2006 Demonstration of sustained hydrogen photoproduction by immobilized, sulfur-deprived *Chlamydomonas reinhardtii* cells *Int. J. Hydrogen Energy* **31** 659–67
[34] Gibbs M, Gfeller R P and Chen C 1986 Fermentative metabolism of *Chlamydomonas reinhardii Plant Physiol.* **82** 160–6
[35] Tano K, Kamuanga M, Faminow M D and Swallow B 2007 Lipid productivity of algae grown on dairy wastewater as a possible feedstock for biodiesel *Ecol. Econ.* **45** 393–407
[36] Garnett T *et al* 2013 Sustainable intensification in agriculture: premises and policies *Science* **341** 33–4
[37] Nautiyal P, Subramanian K A and Dastidar M G 2014 Recent advancements in the production of biodiesel from algae: a review *Reference Module in Earth Systems and Environmental Sciences* (Amsterdam: Elsevier)
[38] Chisti Y 2007 Biodiesel from microalgae *Biotechnol. Adv.* **25** 294–306
[39] Prajapati S K, Kumar P, Malik A and Vijay V K 2014 Bioconversion of algae to methane and subsequent utilization of digestate for algae cultivation: a closed loop bioenergy generation process *Bioresour. Technol.* **158** 174–80
[40] Charrier B *et al* 2008 Development and physiology of the brown alga *Ectocarpus siliculosus*: two centuries of research *New Phytol.* **177** 319–32
[41] Baltrenas P and Misevičius A 2015 Biogas production experimental research using algae *J. Environ. Heal. Sci. Eng.* **13** 1–7
[42] Chavan R and Mutnuri S 2021 Domestic wastewater treatment by constructed wetland and microalgal treatment system for the production of value-added products *Environ. Technol.* **42** 3304–17
[43] Powers S E and Baliga R 2010 Sustainable algae biodiesel production in cold climates *Int. J. Chem. Eng.* **2010** 102179
[44] Wang Y, Wang X, Li Y, Li J, Wang F, Xia S and Zhao J 2020 Biofilm alters tetracycline and copper adsorption behaviors onto polyethylene microplastics *Chem. Eng. J.* **392** 123808
[45] Onesios K M, Yu J T and Bouwer E J 2009 Biodegradation and removal of pharmaceuticals and personal care products in treatment systems: a review *Biodegradation* **20** 441–66
[46] Vergili I 2013 Application of nanofiltration for the removal of carbamazepine, diclofenac and ibuprofen from drinking water sources *J. Environ. Manage.* **127** 177–87

[47] Oulton R L, Kohn T and Cwiertny D M 2010 Pharmaceuticals and personal care products in effluent matrices: a survey of transformation and removal during wastewater treatment and implications for wastewater management *J. Environ. Monit.* **12** 1956

[48] Luis Malvar J, Luis Santos J, Martín J, Aparicio I and Alonso E 2020 Occurrence of the main metabolites of pharmaceuticals and personal care products in sludge stabilization treatments *Waste Manag.* **116** 22–30

[49] Bolisetty S, Peydayesh M and Mezzenga R 2019 Sustainable technologies for water purification from heavy metals: review and analysis *Chem. Soc. Rev.* **48** 463–87

[50] Chan Y J, Chong M F, Law C L and Hassell D G 2009 A review on anaerobic–aerobic treatment of industrial and municipal wastewater *Chem. Eng. J.* **155** 1–18

[51] Gupta V K, Ali I, Saleh T A, Nayak A and Agarwal S 2012 Chemical treatment technologies for waste-water recycling—an overview *RSC Adv.* **2** 6380

[52] Tay J H, Yang P, Zhuang W Q, Tay S T L and Pan Z H 2007 Reactor performance and membrane fltration in aerobic granular sludge membrane bioreactor *J. Membr. Sci.* **304** 24–32

[53] Weiss S and Reemtsma T 2008 Membrane bioreactors for municipal wastewater treatment—a viable option to reduce the amount of polar pollutants discharged into surface waters? *Water Res.* **42** 3837–47

[54] Derakhshan Z, Mahvi A H, Ehrampoush M H, Ghaneian M T, Yousefinejad S, Faramarzian M, Mazloomi S M, Dehghani M and Fallahzadeh H 2018 Evaluation of kenaf fibers as moving bed biofilm carriers in algal membrane photobioreactor *Ecotoxicol. Environ. Saf.* **152** 1–7

[55] Zhang Q, Wang L, Yu Z, Zhou T, Gu Z, Huang Q, Xiao B, Zhou W, Ruan R and Liu Y 2020 Pine sawdust as algal biofilm biocarrier for wastewater treatment and algae-based byproducts production *J. Clean. Prod.* **256** 120449

[56] Tourlouki K, Tsavatopoulou V, Alexandropoulos D, Manariotis I D and Mazzucato S 2020 A novel microalgae harvesting method using laser micromachined glass fiber reinforced polymers *In Photonics MDPI* **7** 42

[57] Luo S, Sampara P S S and He Z 2018 Effective algal harvesting by using mesh membrane for enhanced energy recovery in an innovative integrated photobioelectrochemical system *Bioresour. Technol.* **253** 33–40

[58] Garbowski T, Charazińska S, Pulikowski K and Wiercik P 2020 Application of microalgae cultivated on pine bark for the treatment of municipal wastewater in cylindrical photobioreactors *Water Environ. J.* **34** 949–959

[59] Seung-Hoon Lee, Hee-Mock Oh, Beom-Ho Jo, Sang-A Lee, Sang-Yoon Shin, Hee-Sik Kim, Sang-Ah Lee and Chi-Yong Ahn 2014 Higher Biomass Productivity of Microalgae in an Attached Growth System using Wastewater *J. Microbiol. Biotechnol.* **24** 1566–1573

[60] Irving T E and Allen D G 2011 Species and material considerations in the formation and development of microalgal biofilms *Appl. Microbiol. Biotechnol.* **92** 283–294

[61] Choudhary P, Malik A and Pant K K 2017 Mass-scale algal biomass production using algal biofilm reactor and conversion to energy and chemical precursors by hydropyrolysis *ACS Sustain. Chem. Eng.* **5** 4234–4242

[62] Johnson M B 2009 *Microalgal biodiesel production through a novel attached culture system and conversion parameters* (Doctoral dissertation: Virginia Tech)

www.ingramcontent.com/pod-product-compliance
Ingram Content Group UK Ltd.
Pitfield, Milton Keynes, MK11 3LW, UK
UKHW051900100926
8556IPUK00005B/711

9 780750 357227